This book by one of the world's foremost philosophers in the fields of epistemology and logic offers an account of suppositional reasoning relevant to practical deliberation, explanation, prediction, and hypothesis testing.

Suppositions made "for the sake of the argument" sometimes conflict with our beliefs, and when they do, some beliefs are rejected and others retained. Thanks to such hypothetical belief contravention, adding content to a supposition can undermine conclusions reached without it. Subversion can also arise because suppositional reasoning is ampliative. These two types of nonmonotonicity are the focus of this book. A detailed comparison of nonmonotonicity appropriate to both belief-contravening and ampliative suppositional reasoning reveals important differences that have been overlooked.

In arguing that the distinction between belief contravening and inductive nonmonotonicity plays a far greater role in deliberation and decision than it is given credit for, this major study will be required reading for all philosophers and logicians concerned with conditionals, decision theory, and inductive inference. It will also interest those in artificial intelligence who work on expert systems, default reasoning, and nonmonotonic reasoning.

For the Sake of the Argument

For the Sake of the Argument

Ramsey Test Conditionals, Inductive Inference, and Nonmonotonic Reasoning

ISAAC LEVI
Columbia University

CAMBRIDGE
UNIVERSITY PRESS

Published by the Press Syndicate of the University of Cambridge
The Pitt Building, Trumpington Street, Cambridge CB2 1RP
40 West 20th Street, New York, NY 10011-4211, USA
10 Stamford Road, Oakleigh, Melbourne 3166, Australia

First published 1996

Printed in the United States of America

Library of Congress Cataloging-in-Publication Data
Levi, Isaac, 1930–
For the sake of the argument : Ramsey test conditionals, inductive
inference, and nonmonotonic reasoning / Isaac Levi.
p.   cm.
Includes bibliographical references (p.   ).
ISBN   0-521-49713-2 (hbk.)
1. Hypothesis.   2. Commonsense reasoning.   3. Conditionals (Logic)
4. Induction (Logic)   5. Inference.   I. Title.
BC183.L48   1996
160–dc20   95-2436

A catalog record for this book is available from the British Library.

ISBN 0-521-49713-2 hardback

For
Jonathan and David

Contents

Preface

This book has its origins in conversations I had with André Fuhrmann at the meetings of the International Congress of Logic, Methodology, and Philosphy of Science in Uppsala in 1991. These exchanges led to a joint essay (Fuhrmann and Levi, 1994) that discussed some peculiarities of conditional reasoning when induction is taken into consideration. One sin begets another and I began contemplating the writing of a long paper combining the version of Ramsey test conditionals I had already advocated with ideas on inductive inference I had discussed ever since I had published *Gambling with Truth* (Levi, 1967).

As a preliminary, however, it seemed desirable to spell out somewhat more elaborately than I had done before the view of Ramsey test conditionals I favor. At around the same time, John Collins and I started a reading group at Columbia that included as regular participants Markko Ahtisaari, Horacio Arló Costa, John Danaher, Scott Shapiro, and, for a brief period when he was visiting Columbia, André Fuhrmann. John Collins had presented to the group his account of the structural differences between revision of belief as understood by Alchourrón, Gärdenfors, and Makinson (1985) and revision by imaging in a sense parasitic on the possible-worlds semantics for conditionals pioneered by D. Lewis (1973). In the course of his presentations, Collins defended the view that imaging was best suited to characterizing suppositional reasoning whereas AGM revision is suited to changing beliefs. I had already argued (Levi, 1991) that the AGM formalism was inadequate as an account of rational belief change. And I was quite convinced that imaging was ill suited to capture suppositional reasoning – especially in applications to practical deliberations. Thanks to Collins's thought-provoking presentation, it occurred to me that many of the complaints I had lodged against the AGM theory as a theory of rational belief change (Levi, 1991) do not apply to that theory understood as a basis for suppositional reasoning. It seemed to me that AGM theory should be reconsidered from this latter perspective. Combining this consideration with the idea that conditional judgments of modality are expressions of suppositional reasoning "for the sake of the argument" provided the basis for developing my approach to conditionals from what I hope is a fresher point of view.

Chapters 1 through 4 develop this theme. The discussion includes a more elaborate treatment of iterated conditionals than I had thought feasible. I owe a debt to S. O. Hansson, whose own work on this topic provoked me to rethink the matter. But most important, the characterization of validity for the logic of conditionals proposed by P. Gärdenfors in his 1988 book led Arló Costa and myself to work out ideas concerning the distinction between positive and negative validity for epistemic Ramsey test conditionals (with kibitzing from Fuhrmann, 1993). It is reported in a manuscript by Arló Costa and Levi (1995) and in Chapter 4.

Makinson and Gärdenfors (1991) explore ways of interpreting the AGM logic of revision as an account of nonmonotonic reasoning. Arló Costa and Shapiro (1991) examine connections between nonmonotonic logic as understood by students of artificial intelligence and the logic of conditionals. Inductive reasoning is also widely recognized to be nonmonotonic in some sense or other, and my collaboration with Fuhrmann points to some peculiar aspects of inductive nonmonotonicity. In Chapters 5 and 6, I defend the thesis that inductive nonmonotonicity ought not to be confused with the belief-contravening monotonicity based on AGM revision or the improvement I call Ramsey revision. These chapters explain inductively extended suppositional reasoning, inductively extended Ramsey test conditionals, and inductively extended nonmonotonic implication utilizing ideas about induction I have been elaborating and defending since the 1960s but applying them in a new setting that yields some new results.

Because contemporary discussion of nonmonotonic reasoning has developed largely in the artificial intelligence and computer science communities, it seemed desirable to make at least a partial critical survey of work done there from the perspective developed in the first six chapters. Attention is focused on work on default reasoning by R. Reiter (1980) and D. Poole (1988) in Chapter 7 by utilizing the insightful commentaries found in Makinson (1994) and Makinson and Gärdenfors (1991). Chapter 8 discusses the relevance to nonmonotonic reasoning of assessments of probability with special attention to the views of H. E. Kyburg and J. Pearl and nonprobabilistic assessments of degree of belief, disbelief or surprise, and possibility that were first elaborated by G. L. S. Shackle (1949) and subsequently reinvented by L. Zadeh, L. J. Cohen, D. Dubois, and H. Prade and W. Spohn, among many others.

Chapter 9 addresses the systematic attempts to formulate and classify logics of nonmonotonic inference pioneered by D. Gabbay (1985) and D. Makinson (1989), with special attention paid to the work of S. Kraus,

D. Lehmann and M. Magidor (1990), Lehmann and Magidor (1992), and P. Gärdenfors and D. Makinson (1993).

I claim that the intended application of default reasoning and kindred sorts of nonmonotonic reasoning seems to be to contexts calling for inductive reasoning. Yet there is a widespread tendency to try to assimilate this inductive nonmonotonicity to nonmonotonicity of the belief-contravening variety. I am convinced by the efforts of Gärdenfors and Makinson (1993) and by the somewhat related work of Lehmann and Magidor (1992) on "rational" nonmonotonic inference relations that, if the assimilation were legitimate, inductive nonmonotonicity could be explicated in terms of AGM revision (or the improved Ramsey version).

I am not, however, convinced by the efforts at assimilation. Within the framework for induction elaborated in Chapters 5 and 6, conditions required to achieve the integration successfully are identified. I contend that to require satisfaction of these conditions is to require inductive reasoners to be bolder than they ought to be. If this is right, the scope of applicability of nonmonotonic reasoning of the inductive variety and, hence, of default reasoning is reduced almost to the vanishing point. In developing this argument, attention is also directed to the advantages of relating probability and Shackle measures to inductive reasoning along the lines I favor over the alternative views in the literature.

I am neither a logician nor a computer scientist. I appeal to logicians or computer scientists who read this book to exercise both patience and tolerance while I meander through topics only some of which will seem relevant at the outset to their concerns and which, when relevant, are presented without the rigor to which many of them are accustomed. My preoccupation is in the final analysis with identifying standards of rational health in reasoning. Logicians, mathematicians, and computer scientists make an important contribution to identifying what those standards might be and to the design of technologies that can contribute to enhancing our limited abilities to realize these standards. I am interested in defending a view not of what these standards might be but what they should be.

Philosophers interested in the topics discussed in this book may also find my exhortations distasteful on the grounds that the task of philosophy is to explicate and analyze the standards of rationality as we find them in our "ways of life" and our linguistic and conceptual "intuitions." The standard reflecting a given way of life trivially entails recommendations concerning how we ought to think (in that way of life), but such prescriptions are a by-product of the conceptual analysis. Intuitions tend, however, to be controversial and, hence, insufficient to furnish a viable standard of rational health. Conceptual analysis cannot fill the gap unless the standards of rationality are already in place. I agree that prescription should respect those judgments that are for the time being noncontroversially settled. I seek, when I can, to identify presystematic judgments that seem to be shared widely if not universally. But I refuse to pretend that I can justify via noncontroversial judgments and conceptual necessities the correctness of my views to those who are resolutely skeptical of the perspective I favor without begging the issue. Given my concern to elaborate the standards for suppositional, conditional, or nonmonotonic reasoning I favor, I have not engaged in protracted discussions of alternative philosophical perspectives on these topics. I have rested content with indicating where I stand with respect to these alternatives. This benign neglect of the views of others should not be understood as a mark of contempt. It stems from a conviction that the best I can do in the way of an argument is to elaborate the fragment of a prescriptive standard for rational health discussed in this book and to invite those who think that such standards are the product of some sort of conceptual or biological necessity to stretch their conceptual imaginations. The alternatives they endorse yield prescriptive standards rival to the standards I favor. Appeal to conceptual analysis alone cannot resolve the conflict between these standards one way or the other.

In any case, the long paper I set out to write has turned into a long book. This book is a response to interactions with the ideas of the several friends, students, and colleagues I have explicitly mentioned. I owe a special debt of gratitude to the participants in the reading group for the pleasures of intellectual exchange and the instruction that accompanies it. I take responsibility for the way what I have learned from others is assembled into a systematic view.

I am also grateful to two referees from Cambridge University Press, both of whom offered encouragement accompanied by serious criticism that I have tried to take to heart in finishing the final draft. Henry Kyburg read and commented on the first draft. He, as always, has been the best of colleagues and friends even though on many (though not all) of the issues addressed here he and I are seriously at odds with one another.

The first draft of this book was completed while I was on sabbatical leave at Cambridge in the spring of 1993. I am grateful to the Fellows of Darwin College, the Faculty of Philosophy of Cambridge University, and my friend and colleague Hugh Mellor for making it possible for me to work in such pleasant and intellectually stimulating surroundings.

The second draft of the book was written while I was visiting the Institute for Advanced Studies of the Hebrew University for four months in 1994 as part of a seminar on rationality. The comfortable facilities, helpful cooperation of the staff at the institute (especially D. Shulman, E. Sharon, S. Danziger, D. Aviely, and Y. Yaari), and the extremely lively and stimulating topics related to "interactive rationality" presented by distinguished students of game theory, cognitive psychology, and philosophy afforded me the opportunity to work on finishing the book while appreciating how much of relevance to the subject of this book (especially pertaining to the notion of "common knowledge" and suppositional reasoning about the views of others) I have left unsaid. I am very grateful.

My stay at the institute is the first time I have visited Jerusalem for more than a few days. Its beautiful long views of rocky arid land have altered my usual distaste for desert landscapes. And in spite of my resolutely secular attitude and abiding opposition to both Jewish and Palestinian nationalism, I have acquired a deep affection for that religion-ridden, contentious, and contended-for city.

I have dedicated this book to my two sons, Jonathan and David, as an expression of my admiration for the contributions they have already made in their chosen careers, my pride in their integrity and character, and my love. I owe my good fortune in having such sons as in so much else to my wife, Judith.

Jerusalem
July 1994

For the Sake of the Argument

1 Introduction

1.1 Supposing and Believing

Keeping a firm grip on the difference between fact and fiction entails much more than distinguishing between what is judged true, what is judged false, and what hangs in suspense. Any agent who has a clear "sense of reality" distinguishes between what he or she *fully believes* to be true and what he or she *supposes* to be true for the sake of the argument. But both what is fully believed and what is supposed furnish the basis for a tripartite distinction between what is judged true, what is judged false, or what hangs in suspense.

Given an agent's state of full belief at a given time, some propositions are judged possibly true because they are not ruled out by the state of full belief and others are judged impossible. In this epistemic sense of serious possibility, h is judged true relative to the agent's state of full belief **K** if and only if ~h is not a serious possibility, h is judged false if and only if ~h is judged true, and the question of the truth of h hangs in suspense if and only if both h and ~h count as serious possibilities.

If a would-be investor is uncertain as to whether the government will propose an investment tax credit for long-term investments in U.S. firms, the uncertainty will have some relevance to the investor's conclusion as to how to make investments. Whether his judgments of uncertainty are to be construed as judgments of (determinate or indeterminate) credal probability or as judgments of uncertainty of some nonprobabilistic sort, such uncertainty presupposes a judgment on the part of the decision maker that propositions concerning which he or she is uncertain are possibly true and possibly false in the sense of serious possibility. When someone rules out the logical possibility that George Bush won the 1992 presidential election over Bill Clinton as a serious possibility, there is no further question to answer concerning the agent's confidence in this judgment. The agent is maximally certain that George Bush did not win the election. If the agent is in suspense as to whether Bush won or lost so that neither prospect is ruled out as a serious possibility, the alternatives are believed with varying degrees of confidence or are judged probable to varying degrees.

1

When it comes to judgments of value, the same point holds. If h is judged better than h' in the context of a deliberation concerned with the issue of seeking to realize h or seeking to realize h', h and h' are both judged to be seriously possible.

In general, positive probability and comparative desirability both presuppose possibility. And when probability and desirability are understood to be subjective or personal probability and desirability, the possibility is subjective or personal as well. Serious possibility is personal possibility in that sense.[1]

The inquirer's judging it epistemically impossible that h is false does not preclude his judging it epistemically possible that he will change his mind in the future and, indeed, do so for good reason. The inquirer might, as far as the inquirer knows, cease being sure that h is true at some future moment and might do so justifiably – just as the inquirer might in the future add new items to his state of full belief and become sure that g.

Thus, agents can change their beliefs and one way they can do so is through intelligently conducted inquiry. When they change their beliefs (whether through intelligently conducted inquiry or not), they alter the distinction to which they are committed between serious possibility and impossibility or between what is judged true, judged false, and left in suspense. They thereby alter their vision of reality (without, in general, changing any relevant aspect of reality other than their beliefs about it).

However, in deciding whether and how to change their minds via intelligently conducted inquiry, inquirers rely on the distinction between serious possibility and impossibility they endorsed prior to making the change. If a change in the standard for serious possibility is justified, the justification is based on the standard for serious possibility endorsed before the change is instituted.

Thus, not only practical deliberation but intelligently conducted inquiry concerned with changing states of full belief relies on having a sense of reality embodied in a distinction between what is possibly true and what is not.

Both practical deliberation and intelligently conducted inquiry also depend on the imaginative use of suppositional reasoning. In suppositional reasoning, one transforms the initial state of full belief by adding to it a proposition supposed to be true for the sake of the argument. Like a change in the state of full belief, the result is an altered way of distinguishing between the possible and impossible or between what is judged true, what is judged false, and what is held in suspense. But the transformed state is not the inquirer's new state of full belief. The inquirer need not have changed his full beliefs at all. A new state of full belief is simulated in the sense that the ramifications of moving to that state are explored without actually making the change. The transformation of the initial state is the product of adding a proposition *supposed to be true purely for the sake of the argument*.

Suppositional reasoning in this sense resembles the writing of fiction that posits a certain kind of situation and invites the reader to "suspend disbelief" and proceed as if the "premises" of the story were true. Readers with fervid imaginations may confuse fiction with reality. And because transformations of belief states by suppositional reasoning yield pseudo–belief states that can be used to perform functions structurally similar to those performed by states of full belief, even those who engage in more careful reflection on the character of belief change may have difficulty with the distinction.

Practical deliberation, like the writing of fiction, calls for fantasizing. One of the tasks confronting the deliberating decision maker is the evaluation with respect to his or her goals and the values of options he or she recognizes to be available. To reason about a given option, the decision maker may add to his or her state of full belief the *supposition*, adopted purely for the sake of the argument, that the option is chosen and implemented and consider what the consequences *would be*.

During the process of deliberation, the agent should be in suspense concerning which of the available options he or she will choose. If the agent is convinced that he or she would not choose a given policy, choosing and implementing are not serious possibilities according to that agent's state of full belief and the policy is not optional for that agent (see Levi, 1992).

The deliberating agent needs to be able to assess the implications of choosing the policy even though he or she is in doubt as to whether it will be chosen and implemented. Supposition for the sake of the argument is designed to allow the agent to do just that. Prior to deciding between rival job offers, Jones might reflect on what the consequences of interest to him might be of his joining a corporate law firm rather than working for the Legal Aid Society. That is to say, Jones can suppose that he has joined the corporate law firm and consider what might be the case under that supposition without in any sense believing that he will join the law firm.

Moreover, the addition of the supposition to the agent's state of full belief does not require jettisoning any convictions already fully believed. The result of this modification of the state of full belief by supposition is a new potential state of full belief containing the consequences of the supposition added and the initial state of full belief. However, the shift from the initial state of full belief to this new one is purely for the sake of the argument. It is an exercise in fantasy and does not represent a genuine change in belief state. Yet, relative to the potential state of full belief derived by supposition, a distinction between serious possibility and impossibility can be made and propositions judged true, judged false, or held in suspense.

When the decision maker decides to implement a given option, from the decision maker's perspective the issue is settled. Even if the option will be implemented at some later time, the decision maker comes to fully believe that it will be implemented. Thus, Jones can decide today that he will accept the offer of a position in the corporate law firm. Once the decision is reached, the issue is settled and Jones is no longer in doubt as to what he will do. The decision maker comes to full belief not by inference from what he or she believes but in virtue of his or her choice (Levi, 1991, sec. 3.8;1992).

Prior to making a decision, the agent can and often does *suppose* that an option is implemented without coming to full belief that it is implemented. On deciding to implement an option, the agent comes to full belief that the option is implemented. Supposition is no longer in order. However, the state of full belief reached by adding the information that the option is chosen and is going to be implemented is the same potential state of full belief as the suppositional state reached by supposing for the sake of the argument that the option is implemented. That is why suppositional reasoning is so useful in practical deliberation.

Suppositional reasoning is also relevant to inquiry where the aim includes a concern to modify the state of full belief. Such inquiry involves identifying potential answers to a given question at the "abductive" phase of inquiry. Such potential answers are conjectures. Both they and their negations are serious possibilities held in suspense. The concern is to change the state of full belief by incorporating one of these potential answers (or its negation) into the state of full belief.[2] To do this, however, it will often be necessary to explore experimental consequences of the truth of the several competing answers.

As in the case of identifying the consequences of options in a practical decision problem, the inquirer is in suspense as to which of the recognized potential answers is true. In elaborating on the experimental implications of the conjecture, the inquirer may, once more for the sake of the argument, suppose that conjecture h is true. That is to say, the inquirer may add h to his current belief state **K** or the corpus (theory or deductively closed set of sentences in regimented language $\underline{L}$) $\underline{K}$ representing **K** and examine the logical consequences of doing so. That is to say, $\underline{K}$ is *expanded* by adding h and forming the logical closure of $\underline{K}$ and h. If the inquirer subsequently adopts the potential answer represented by expanding $\underline{K}$ through adding h as his or her state of full belief, the formal structure of the shift from $\underline{K}$ is the same as when $\underline{K}$ is expanded by supposing h to be true for the sake of the argument. But, once more, it is madness to confuse expansion by supposing h to be true with expansion by coming to believe that h is true. The latter kind of change takes place when the agent terminates inquiry with adopting an answer to the question under investigation. The former kind occurs as part of the process of exploring the merits of alternative potential solutions to the problem.

Suppositional reasoning looms large in special ways in two more particular contexts of inquiry. Making predictions or retrodictions is often conditional on the truth of propositions that are not believed to be true. This conditionality is suppositional conditionality. Supposing for the sake of the argument that interest rates will remain low in the next quarter, GNP will rise by 3 percent – so someone might judge.

Explanation makes even more substantial use of suppositional argument. The inquirer fully believes that the litmus paper was immersed in acid and turned red. When asked why it turned red, the inquirer cites the fact that it was immersed in acid and some principles covering chemical interactions. One of the functions this explanation serves is to support a piece of suppositional reasoning. Here the supposition is information the inquirer already fully believes – to wit that the litmus paper was immersed in acid. Even so, the claim is suppositional and, in a sense, doubly so. For the inquirer is required first to pretend to doubt where the inquirer is in no doubt. And then, from this fantasized context of doubt, the agent is to further suppose for the sake of the argument that the litmus paper was immersed in acid and to determine from this supposition whether the litmus paper turned red or not.

Thus, suppositional reasoning is important to many different types of reflection. In none of these contexts is supposing that h equivalent to believing that h. Sometimes the supposition "contravenes" the current state of belief. Sometimes it agrees with the current state of full belief. And sometimes the current state reflects suspense concerning the status of the supposition. Supposing for the sake of the argument is clearly different from believing.

Changing full beliefs calls for some sort of accounting or justification. Supposition does not. If an agent had genuinely come to believe fully the truth of h, he or she could be asked reasonably for a justification of the change showing why someone in his or her initial belief state and facing the problem under consideration should change his or her belief state in the given manner. If the change is merely suppositional, however, no such demand for justification is required. The only respect in which supposition may be critically assessed concerns its relevance to the problem at hand and the legitimacy of the reasoning from the supposition to further conclusions.

Earlier (Levi, 1991, sec. 3.6), I complained that the account of rational belief change found in Alchourrón, Gärdenfors, and Makinson (1985) and Gärdenfors (1988) failed to provide an adequate account of the conditions under which a change in a state of full belief by expansion is legitimate. Gärdenfors (1988) does allow that when new information is acquired as an input from an external source such as sensory input or the testimony of witnesses, expansion by adding the new information may be legitimate; but he provides no account of expansion by (inductive) inference. And there is no discussion of the conditions under which sensory input or testimony from other oracles may be used or forbidden in legitimate expansion. Essentially expansion is characterized as adding new information to $\underline{K}$ and closing under deduction with little in way of guidance as to the conditions under which such expansion is justified.

The absence of an account of the conditions under which expansion is justified is a serious lacuna in a theory of rational belief change. Unlike genuine expansion, suppositional expansion needs no justification. If the theory of rational belief change of Gärdenfors and of Alchourrón, Gärdenfors, and Makinson were an account of change by suppositional addition of information to states of full belief, the complaint I lodged against their account of expansion in (Levi, 1991) would be misdirected. In point of fact, they have applied their ideas to both types of change: genuine change in belief and suppositional change.

The importance of distinguishing between suppositional belief change and justified belief change to the assessment of the ideas of Alchourrón, Gärdenfors, and Makinson is even greater in those cases where some new information h is added to $\underline{K}$ that is inconsistent with $\underline{K}$.

Gärdenfors insists that whenever consistent new information h is added inconsistent with $\underline{K}$ to $\underline{K}$, the transformation of $\underline{K}$ should be a *minimal revision* – that is, a transformation of $\underline{K}$ to a new corpus or deductively closed theory that is consistent, contains h, and removes from $\underline{K}$ only the minimum (in a sense to be specified) amount sufficient to secure this result.[3]

I contended that a justified inductive expansion could never warrant expansion adding information inconsistent with the initial belief state (Levi, 1980, 1991). Routine expansion exemplified by expansion through observation or expert testimony could. If the only legitimate response to routine expansion that injected conflicting information into the inquirer's belief set were the incorporation of that new conflicting information and the jettisoning of old information, minimal revision would be the basic kind of belief change Gärdenfors has taken it to be. (I conceded as much in Levi, 1991.) Sometimes, however, the input information h is called into question and the background information retained. Sometimes new input and background information are both questioned. In studying rational genuine belief change, we need an account of the conditions under which agents are justified in choosing one of these three kinds of options over the others. (See also Hansson, 1991, p. 40, and sections 8.3–8.6 of this book.) But the need for justification arises only when the inquirer has already expanded by adding new information inconsistent with the information already available. So the problem to be faced concerns how to extricate oneself from such conflict when it arises. I have suggested representing it as a problem of choosing between rival *contractions* of the inconsistent belief state.[4] In effect, Alchourrón, Gärdenfors, and Makinson are committed to one solution: Retain the new information and modify the background information as little as possible. If this were acceptable, there would be no special problem of how to extricate oneself from inconsistency when it arises distinct from the question of identifying the conditions under which it is legitimate to expand into inconsistency. The two problems collapse into the single question of determining when minimal revision is justified. But if there are three kinds of reactions to inadvertent expansion into inconsistency and each of them may be warranted in some kinds of situation, the claim that minimal revision is a basic kind of belief change for the purposes of justification becomes untenable.

Once more, all of this seems irrelevant when the new information h inconsistent with $\underline{K}$ is added to it by supposition for the sake of the argument. There is no longer any need to justify adding h to the initial corpus. Questioning h is no longer optional. Supposing h forbids it (at least for the sake of the argument). The inquirer is required, for the sake of the argument, to remove elements in the background precluding consistent admission of h.

There are many contexts where supposition for the sake of the argument calls for belief contravention of some kind – for example, for questioning and removing some aspect of the background information for the sake of the argument. Economists, psychologists, and decision theorists have often been interested in "eliciting" an agent's judgments of subject or credal probability and utility judgments from information concerning the agent's choices among appropriately constructed "lotteries." Flesh and blood agents rarely face decision problems where the options available to them consist of all options that need to be taken into account in successful elicitation. Elicitations might need, for example, to consider the agent's judgments concerning how he or she would choose under the belief-contravening supposition that the agent had "mixtures" or "roulette lotteries" over any finite set of options available to him or her.

More generally, a full understanding of the content of laws and theories in the sciences often calls for thinking through consequences of belief-contravening suppositions in *Gedankenexperimenten*.

Thus, belief-contravening suppositional transformations of belief states like suppositional transformations that are not belief contravening are important to deliberation and inquiry just as genuine changes in belief state are. And such changes remain changes from an initial state of full belief to a potential state of full belief that has not been implemented. Confusion between suppositional changes and genuine changes in belief state can be as serious here as in cases where no belief contravention is involved.

These considerations suggest that the account of rational belief change proposed in Gärdenfors (1988) and Alchourrón, Gärdenfors, and Makinson (1985) is untouched by the criticisms I offered in 1991 when their proposals are understood as an account of *suppositional* as opposed to sincere belief change – even if the difficulties with their view I raised call for substantial modification and supplementation of their ideas when their approach is presented as an account of rational change in genuine belief.

Because the proposals advanced by Gärdenfors and Alchourrón, Gärdenfors, and Makinson have been applied to both sincere and suppositional belief change, the critique of these proposals understood as applied to sincere belief change should be supplemented by a consideration of supposition.

One way to explore this topic is by examining the role of suppositional reasoning in explicating conditional judgment. This is the way, I believe, Gärdenfors intended to proceed when offering an account of conditionals (Gärdenfors, 1978) and this is the way I understood the matter (Levi, 1984a, ch. 12).

1.2 Supposition and Conditionals

Frank Ramsey once wrote:

If two people are arguing "if p will q?" and are both in doubt as to p, they are adding p hypothetically to their stock of knowledge and arguing on that basis about q. (Ramsey, 1990, 155 n. 1)

Many authors have used this quotation as the basis for attributing to Ramsey a certain criterion for the acceptability of conditionals in terms of some transformation of the agent's state of full belief. Thus, if the "stock of knowledge" or state of full belief is **K** (representable in some regimented language $\underline{L}$ by a deductively closed set of sentences $\underline{K}$ in $\underline{L}$ called a *corpus* or *theory*), the transformation is a belief state **K*** representable in $\underline{L}$ by a corpus $\underline{K}^*$ that is constrained to contain p and to be related to $\underline{K}$ in some definite manner.

Ramsey explicitly characterized the transformation he had in mind for a special case – to wit, when the initial corpus $\underline{K}$ contains neither p nor ~p and is in this sense "open" with respect to p. The inquirer in belief state **K** is then in doubt or suspense as to the truth value of p. Ramsey's proposal was to add p (which by hypothesis is consistent with $\underline{K}$) to $\underline{K}$ and form the deductive closure. If q is in the resulting expansion $\underline{K}^+_p$, the conditional "if p then q" is assertible according to $\underline{K}$. Clearly, in this case, Ramsey's account of conditionals qualifies as a suppositional account of their acceptability in the epistemic sense according to which supposition is a transformation of the agent's belief state by adding new information to it – for the sake of the argument.

In the footnote from which the citation from Ramsey was taken, Ramsey goes out of his way to say that in belief-contravening cases the question "if p will q?" "ceases to mean anything" to the parties to the dispute unless the question is reconstrued as a question about "what follows from certain laws or hypotheses."

Thus, Ramsey acknowledged that in *belief-contravening* cases where ~p is in $\underline{K}$ (and, perhaps, in *belief-conforming* cases where p is in $\underline{K}$), his suppositional account of "if" sentences was incomplete. Instead of attempting to extend the suppositional interpretation of "if" sentences to those cases, Ramsey introduced a different view of such sentences as rendering verdicts concerning "what follows from certain laws or hypotheses."

Ramsey did not always emphasize how different his approach to open and to belief-contravening cases was. In order to support the conditional judgment "if p is the case, q will be the case" for open conditionals, the current belief state should contain the assumption that either p is false or q true (Ramsey, 1990, p. 154). No lawlike generalization is deployed or, in logic, need be accepted in the belief state. Yet, Ramsey seems to think that some such generalization is required. His reasons seem to be (1) that open and belief-contravening conditionals should be considered from a single perspective and (2) that belief-contravening conditionals probe what follows from generalizations (p. 155).

That open and belief contravening conditionals should be treated in a unitary fashion if possible seems sensible advice. One possibility is that conditionals should be treated as the product of reasoning from suppositions. If the suppositional interpretation of conditionals that Ramsey seems to have endorsed in the citation for the open case is extended to belief-contravening cases, derivability from laws does not appear to be a necessary feature for the acceptability of conditionals as Ramsey claimed. In supposing that p when ~p is fully believed, the inquirer needs to modify his or her belief state so that the claim "~p or q" is retained when ~p is removed. This *may* involve the retention of laws from which the conditional may be "derived" but it need not do so.

The derivability-from-laws view that Ramsey endorsed for belief-contravening conditionals is not a suppositional interpretation. In belief-contravening cases, suppositional reasoning focuses attention on the criteria for what must be removed from the belief state in order to allow the supposition to be consistently added. The net effect is a transformation of the inquirer's belief state (for the sake of the argument) by first removing information and then adding the supposition.

Derivability-from-laws interpretations insist that the only thing that matters is whether the conditional is derivable from true laws or reliable habits and a suitable selection of "facts" specifying initial and boundary conditions for application of the laws. Of course, the inquirer who makes belief-contravening conditional judgments must invoke laws that he or she believes to be reliable and singular propositions he or she judges to be facts and, hence, true. Derivability-from-laws interpretations insist that "if p then q will be true" is acceptable for the inquirer just in case q is derivable from p and the singular propositions judged true via inferences licensed by the judged-to-be-reliable laws.

Thus, in the open case where the inquirer is in suspense concerning p but is certain that either p is false or q is true, Ramsey's criterion for open conditionals licenses an affirmative answer to "if p, will q?" This is so whether or not q can be derived from p and generalizations the inquirer judges to be laws. The derivability-from-laws approach requires such a derivation. Ramsey avoided inconsistency by restricting the derivability from laws approach to the belief-contravening case.

Thus, when Ramsey's sketchy remarks about conditionals are trimmed to make them consistent, they seem to advocate two rather distinct approaches: (1) an approach according to which open conditionals express results of suppositional reasoning, and (2) an approach according to which belief-contravening conditionals express what follows from given laws.[5] Ramsey himself expressly introduces the derivability-from-laws interpretation in order to cover the belief-contravening cases that he did not see how to address from a suppositional point of view. This book focuses on suppositional reasoning. It argues that suppositions can, indeed, be belief contravening and, as a consequence, it is possible to extend Ramsey's interpretation of open conditionals as expressions of suppositional reasoning to the belief-contravening conditionals as well. I take it that so-called Ramsey test criteria for the acceptability of conditionals are designed to do just that.

Extending Ramsey's suppositional interpretation of open conditionals to the belief-contravening and belief-conforming cases can be carried out in many different ways yielding criteria for the acceptability of conditionals that are quite different from one another. It would be foolish to claim that one of these extensions of Ramsey's proposal has a unique claim to be *the* Ramsey test. But at least we can say this much: Any such criterion for acceptability that does not embrace Ramsey's explicitly stated proposal for the open case is not a Ramsey test in the sense in which a conditional is an expression of the result of hypothetical or suppositional reasoning. This point has often been and continues to be ignored.[6]

Anyone wishing to bask in the glory of Ramsey by calling tests for the acceptability of conditionals "Ramsey tests" even when they violate the minimal condition of adequacy just cited are free to do so – provided it is clear that Ramsey's practice is not in line with theirs. The crucial question after all is not what Ramsey thought but whether suppositional interpretations of conditionals capture important applications of conditional reasoning.

In section 1.1, I have mentioned several settings in which *fantasizing* as V. H. Dudman calls it or reasoning from suppositions in contexts where the truth of the supposition is held in suspense seems an important facet of practical deliberation and scientific inquiry.

In practical deliberation, it is commonplace that reasoning from the supposition that a policy is implemented to judgments concerning what the consequences will or might be can be couched in terms of conditionals. Jones might say: "If I join the corporate law firm, I might become rich; but if I work for Legal Aid, I will be poor."

Similarly suppositional reasoning from conjectures to experimental consequences can be expressed in terms of conditionals.

There are, of course, many contexts where fantasizing for the sake of the argument calls for belief contravention of some kind – that is, for questioning some aspect of the background information for the sake of the argument. We have reviewed some of these contexts in section 1.1. Once more it is not far-fetched to think of such suppositional reasoning as being expressed in terms of conditional judgments judged to be "counterfactual."

Thus, understanding conditionals as expressions of suppositional reasoning as Ramsey proposed doing in the case where the contents of the "if" clause are "open" and, hence, held in suspense may be a useful way of understanding the force of conditional judgments, on the one hand, and probing the character of suppositional reasoning on the other.

This suggests a second condition of adequacy for accounts of criteria for the acceptability of conditionals according to Ramsey tests – to wit, that conditionals be understood as expressions of reasoning from suppositions adopted for the sake of the argument to judgments of serious possibility or impossibility relative to the "hypothetically" adopted state of full belief resulting from adopting the supposition.

To many readers, my decision to emphasize Ramsey's treatment of open conditionals rather than his suggestions concerning belief-contravening conditionals may seem a willful suppression of familiar facts. It has often been pointed out that causal explanations appeal to generalizations that support the acceptability of conditionals whose "if" clauses conflict with what the inquirer judges to be so. Generalizations that do not support such conditionals are often alleged to yield defective explanations. In this setting also, there is some need to give an account of belief-contravening conditionals. It may be suggested, however, that the conditionals supported by causal generalizations cannot be expressions of suppositional reasoning. As J. Mackie pointed out (1962), the "condensed arguments" expressing suppositional reasoning cannot be assertions. Hence, they cannot be entailed by causal generalizations. Advocates of the causal reasoning approach to conditionals may ask how causal laws can support conditionals unless their approach is endorsed.

Causal laws support conditionals. They do not do so, however, because such laws entail them. According to suppositional accounts of belief-contravening conditionals, suppositional reasoning involves removing (for the sake of the argument) assumptions from the current belief state so that transformed belief state becomes open with respect to the content of the "if" clause of the conditional. Having suitably "contracted" the belief state, the content of the "if" clause is added suppositionally and the appropriate judgment made.

In "contracting" the belief state, however, the inquirer has several choices as to which items to remove in order to simulate suppositionally the "open" case. Some propositions believed true are judged to be better entrenched, less corrigible, or informationally more valuable than others. Belief-contravening conditionals are supported by causal generalizations believed by the inquirer and judged to be lawlike because the inquirer will be reluctant to give up a generalization he or she judges to be a law when called on to question items in the background in order to allow room for the supposition. Lawlike claims are informationally valuable and, hence, relatively immune to being given up when there is pressure to question background information. Such relative invulnerability to being given up (leading to their being highly "incorrigible" or "well entrenched") combined with an appeal to appropriate extensions of Ramsey test criteria to belief-contravening cases can be deployed to provide an account of the respects in which laws support conditionals. But the acceptability of conditionals depends not only on the relatively high entrenchment of lawlike generalizations. If the inquirer believes that the surface of the table is currently dry and oxygen is currently present but the match is not struck on the surface of the table and does not light and then supposes that at that time the match is struck, the inquirer will, of course, need to give up the claim that the match was not struck. Will he or she need to give up the claim that the match did not light? Given background assumptions about the laws of nature that are not to be given up because of their informational value, the claim that the match did not light may still fail to be given up provided the claim that oxygen is present or that the surface was dry is given up. The critical question here concerns how to evaluate loss of informational value or degree of entrenchment or incorrigibility. Appeal to reasoning from lawlike generalizations simply does not help. The suppositional view insists that the retention of lawlike generalizations and the retention of assumptions about collateral assumptions both reflect the demands for information of the inquiring agent. Ramsey was right to take note of the role of lawlike generalizations in much belief-contravening suppositional reasoning. But his focus was too restrictive.

In cases where fantasizing for the sake of the argument rests on suppositions as yet unsettled as far as the agent is concerned, the transformations of belief state demanded are of the kind recommended by Ramsey in his test for the acceptability of conditionals. Here there is no need to give any beliefs up for the sake of the argument. One does not need to weigh losses of informational value. One does not need always to appeal to causal generalizations. In such open cases, Ramsey was proposing an account of conditional judgment where the acceptability of conditionals is determined by the cogency of suppositional reasoning. No matter what we wish to say about the belief-contravening fantasies, our account ought to cohere with Ramsey's requirement for the "open" case where both the supposition and its negation are serious possibilities from the inquirer's point of view.

In Chapter 2, I shall describe three versions of the Ramsey test. Two of them satisfy the requirement Ramsey laid down for the acceptability of open conditionals. The third satisfies the requirement provided the open character of the "if" clause is relative to a point of view that represents a consensus between several different perspectives. The first two of these versions seem to aim at capturing the same kind of conditional appraisal of serious possibility. For reasons that shall emerge subsequently, I shall call them "informational-value-based Ramsey tests." The third version (or family of versions) I shall call the "consensus Ramsey test." I shall argue that verdicts of informational-value-based Ramsey tests are expressed in English by what V. H. Dudman (1983, 1984, 1985) calls "conditionals," whereas the verdicts of consensus Ramsey tests are expressed by "hypotheticals." According to this account, all "if" sentences customarily classified by contemporary philosophers as indicatives and as subjunctives are explicated by various versions of the Ramsey test satisfying the condition of adequacy mentioned earlier.[7]

This view stands at odds with the ideas of authors like Stalnaker and Lewis who seek to make room for imaging tests for the acceptability of many important "if" sentences. I shall briefly contrast the three versions of the Ramsey test I identify with the imaging test that is derivative from D. Lewis's semantics for conditionals.

One of the issues under dispute concerns whether conditionals or hypotheticals can figure as ingredients in states of full belief – that is, whether the acceptability of a conditional is reducible to believability. P. Gärdenfors (1978, 1988), who endorses a version of a Ramsey test of the sort I call "informational value based," agrees with those who take imaging seriously that conditionals are "objective" in this sense. By appealing to the condition of adequacy for suppositional reasoning and the acceptability of conditionals favored by Ramsey for open cases, I shall argue that the "trivialization" results Gärdenfors has identified for Ramsey test conditionals are best understood as due to combining a suppositional reasoning approach to conditional judgment with objectification of conditionals by allowing them to be objects of belief. This is no problem for advocates of imaging who seem happy to embrace such objectification. Advocates of the Ramsey approach need to show how one can live without the objectification.

There is a second point of dispute concerning the status of imaging. I contend that accounts of conditionals that violate the condition of adequacy proposed by Ramsey for the "open" case sever conditional judgment from the kind of suppositional reasoning relevant to practical deliberation. At present, accounts of conditionals and conditional logic based on the "possible worlds" approach of Lewis or Stalnacker remain formalisms without application.[8] The most interesting candidate for an application is furnished by Katsuno and Mendelzon (1992). I argue, however, that this application of the formalism is an account of some abstract features of change of states of systems over time. It is not a rival to suppositional interpretations of conditionals.

Suppositional reasoning for the sake of the argument by adding supposition h involves transforming one's current state of full belief in one of three kinds of situation: (1) when neither h nor ~h is a consequence of $\underline{K}$, (2) when h is a consequence but ~h is not, and (3) when h is not a consequence but ~h is. The revision transformation cannot, of course, be an imaging transformation or, indeed, any kind of transformation that fails to meet the condition of adequacy specified for case (1). This, however, leaves considerable latitude concerning the properties a revision operator should have. One kind of revision operator is the consensus revision transformation relevant to the appraisal of hypotheticals. Our main focus, however, is on two other kinds: the informational-value-based varieties. One of these revision transformations satisfies the requirements of the theory proposed by Alchourrón, Gärdenfors, and Makinson (1985) and Gärdenfors (1988). I call it *AGM* revision. The other kind of revision transformation violates some of the postulates for AGM revision. I shall call it *revision according to Ramsey*. AGM and Ramsey revision agree that in case (1), the transformation is an expansion – the formation of the set of logical consequences (the logical closure) of $\underline{K}$ and h. In case (3) the transformation is equivalent to removing ~h in an admissible manner from $\underline{K}$ to form a corpus (the admissible contraction of $\underline{K}$ removing ~h and expanding this contraction by adding h). In case (2), the AGM revision of $\underline{K}$ is $\underline{K}$ itself. The Ramsey revision of $\underline{K}$ is the expansion by adding h of the admissible contraction removing h.

The properties characterizing these two types of revision provide us with two somewhat different understandings of supposition for the sake of the argument. Chapter 2 explains the difference between these two types of revision and the types of suppositional reasoning characterized by them. It also argues that revision according to Ramsey is preferable when attention is focused on suppositional reasoning in practical deliberation and scientific inquiry.

Chapter 3 offers an account of the acceptability of iterated conditionals that avoids regarding conditionals as objects of belief. If the acceptability of conditional judgments depends on the legitimacy of the suppositional reasoning expressed by such judgments, a logic of conditionals should reflect the constraints imposed on suppositional reasoning.

Chapter 4 discusses the properties of conditional logics for Ramsey test conditionals based on AGM and on Ramsey revision. I argue that efforts by Gärdenfors to mimic the formal properties of Lewis's favorite version of conditional logic VC (Lewis, 1973, sec. 6.1 and especially p. 130) using variants on AGM revision–based Ramsey tests fail whether the logic is designed for formal languages with or without iterated conditionals. The failure runs deeper when Ramsey revision is deployed; but it is important even when AGM revision is used. The discussion is based on a characterization of logical validity for conditionals that extends Ramsey's insistence in a distinction between the logic of consistency and the logic of truth for probability judgment to the case of conditional modal judgment.

1.3   Belief-Contravening and Inductive Nonmonotonicity

Suppositional reasoning is *nonmonotonic*. Given a set of background beliefs $\underline{K}$, supposing that p is true for the sake of the argument may warrant judging q true relative to the transformed corpus; but supposing p ∧ r true for the sake of the argument may require suspending judgment regarding q or, in some cases, judging ~q true relative to the same $\underline{K}$.

The term *nonmonotonic* has become a part of the vocabulary of theoretically reflective computer scientists. Logicians and computer scientists (including Makinson and Gärdenfors) interested in this topic have tended to focus on the sorts of nonmonotonicity arising in suppositional reasoning due to belief contravention.

However, a different form of nonmonotonicity has been familiar to philosophers for many centuries though not under that name. Inductive inferences are nonmonotonic. Relative to background corpus $\underline{K}$ and additional data e (forming a corpus that is the expansion $\underline{K}^+_e$ of $\underline{K}$ by adding e), an inquirer might expand further inductively by adding h. But if the additional data had contained f as well as e, the warrant for h may have been undermined so that suspense is recommended or undercut so that ~h is recommended. A. Fuhrmann and I (1993) have pointed out that both computer scientists and philosophers have neglected "inductively extended conditionals" and the modes of suppositional nonmonotonic reasoning associated with them. Consider the claim that if Lyndon Johnson had run against Nixon in 1968, he would have been reelected president. One cannot ground acceptance of this conditional on causal generalizations. The fact that polls taken during the 1968 campaign seem to show that Johnson outpaced Nixon might serve as the basis for nondeductive or inductive reasoning supporting such a judgment. Without such reasoning, the judgment could not be acceptable according to advocates of a pure suppositional account of conditionals.

Chapters 5 and 6 compare the two kinds of nonmonotonicity. On the understanding that belief-contravening reasoning cannot be inference, it is argued that the kind of nonmonotonicity in suppositional reasoning cannot be inference; that the only kind of inference that can be nonmonotonic must be ampliative (i.e., reasoning that adds information to the belief state not entailed by the belief state and does so without belief contravention); and that inductive inference understood broadly as including ampliative inference from data to hypotheses, from statistical hypotheses to outcomes of experiments, and from sample to predictions as well as other forms of theory choice constitute the typical forms of ampliative inference.

Inductive inferences can, of course, be based on an appeal to sincerely held full beliefs as well as to such beliefs supplemented by suppositions adopted for the sake of the argument. They can, therefore, be expressed by "inductively extended" conditionals. Chapters 5 and 6 explore the properties of inductively extended suppositional reasoning and inductively extended conditionals and contrast the standards of nonmonotonic reasoning generated when induction is taken into account as compared with situations where attention is restricted to belief-contravening nonmonotonicity.

Nonmonotonicity is of interest in computer science to some degree because of the interest among computer scientists in default reasoning. The intended applications of default reasoning suggest that such reasoning is supposed to be ampliative or inductive. Yet, much (though not all) of the recent discussion has tended to think of default reasoning as belief contravening. Chapters 7 and 9 represent my own efforts to figure out what default reasoning and its relation to nonmonotonic reasoning might be. In Chapter 7, I review some ideas of R. Reiter and D. Poole and propose a reconstruction of default reasoning as inductive inference when the inquirer is maximally bold and when the results of inductive inference are fed into the background evidence and the process of inductive inference is reiterated until a fixed point is reached where no more information can be obtained. In Chapter 9, this same conclusion is reached concerning the recent proposal of Gärdenfors and Makinson for reasoning from defaults using expectations. If this reconstruction is sound, an account of default reasoning that links it with probabilistic and decision theoretic thinking will have been obtained.

Chapter 8 considers approaches to inference that insist on assigning finer-grained evaluations to conclusions of suppositional and inductive reasoning than possible and not possible. Probabilistic and nonprobabilistic measures will be reviewed. The results will be connected to the discussions of inductive inference in Chapters 7 and 9.

2   Unextended Ramsey Tests

2.1   Contractions

Belief states are not linguistic entities. Like mechanical states of physical systems, however, belief states of agents may be represented symbolically.

Agents need not speak the languages used to represent their belief states any more than physical systems speak the languages we use to represent their states. To be sure, agents sometimes represent their own belief states linguistically. But whether they can or cannot do so, they do use language to express or reveal their beliefs and other psychological attitudes. The use of language to represent states of full belief is different from its use to manifest or display such states (Levi, 1991, ch. 2). The use of highly regimented and technical language sometimes facilitates representation of belief states in systematic discussion even though such language may not be used effectively in manifesting belief states.

I have argued elsewhere (Levi, 1991, ch. 2) that a conceptual framework consisting of potential states of full belief available to an inquirer or to a group of inquirers should have a structure where the potential states of full belief are partially ordered according to a consequence relation yielding, at the very least, a Boolean algebra.

Given a conceptual framework K and a regimented language $\underline{L}$, it may be possible to map sentences in $\underline{L}$ into the potential belief states in K so that, if sentence h logically implies sentence g in $\underline{L}$, the belief state represented by g is a consequence of the belief state represented by h in the partial ordering of elements of K. The belief state represented by h is precisely that belief state having as consequences according to the Boolean algebra of potential belief states just those potential belief states represented by logical consequences of h. Hence, we can take the set of logical consequences of h in $\underline{L}$ to be a representation in $\underline{L}$ alternative to h itself of the belief state represented by h.

Every finitely axiomatizable theory or deductively closed set in $\underline{L}$ corresponds to a representation by a single sentence of a potential belief state. However, some deductively closed sets in $\underline{L}$ may not be finitely axiomatizable. Such sets may represent potential states of full belief even if these states are not represented by sentences in $\underline{L}$. For this reason, if a potential state of full belief **K** is representable in $\underline{L}$ at all, it will be represented by a "theory" or deductively closed set of sentences $\underline{K}$ in $\underline{L}$. I shall call $\underline{K}$ a potential *corpus*. An inquirer's belief state at a given time is representable by that inquirer's corpus in $\underline{L}$.

Strictly speaking, a potential corpus $\underline{K}$ does not represent a unique belief state. Two or more belief states are representable by the same $\underline{K}$ in $\underline{L}$ because $\underline{L}$ lacks the resources to provide a discrimination between the belief states. We may, in principle, always do better by enriching $\underline{L}$. But there is no need to suppose that we have at our disposal a language rich enough to represent all potential states of full belief by distinct potential corpora. For our purposes, it will be enough to focus attention on changes in states of full belief that we are capable of representing as changes in potential corpora expressible in some fixed $\underline{L}$.

The set of potential corpora should form a Boolean algebra. The strongest or most informative element is the set of all sentences in the given language $\underline{L}$. $\underline{L}$ represents the inconsistent belief state **0**. The weakest element is the set $\underline{T}$ of logical truths in $\underline{L}$. It represents the state of maximal ignorance **1**. The set of potential belief states is also a Boolean algebra. Unlike the set of potential corpora, it is not assumed to be "atomic." There are maximally consistent potential corpora but there need be no strongest potential belief states weaker than **0** (Levi, 1991, ch. 2).

Suppositional reasoning and Ramsey test conditionals involve transformations of potential states of full belief. It would be possible to discuss such transformations directly; but, for the most part, I shall, instead, speak of transformations of the potential corpora that represent such belief states. Conditionals are, after all, linguistic expressions used to manifest certain kinds of attitudes or judgments. Ramsey tests are supposed to offer some clarification of the kinds of attitudes that conditional sentences express. And Ramsey tests are typically formulated as involving transformations of sets of sentences where such sets are designed to represent potential belief states. There may be many sets that can represent the same potential state of full belief. One kind of set can always be used. That is a set closed under logical consequence of the sort I am calling a potential corpus. This is the kind of representation I shall be using.

Before formulating various versions of the Ramsey test, certain preliminary definitions may prove helpful.

Corpus $\underline{K}'$ is an *expansion* of corpus $\underline{K}$ if and only if $\underline{K} \subseteq \underline{K}'$. Under these conditions, $\underline{K}'$ is a *contraction* of $\underline{K}$. Given a corpus $\underline{K}$ and a sentence h, the *expansion* $\underline{K}^+_h$ of $\underline{K}$ *by adding* h is the smallest potential corpus containing the logical consequences $Cn\,(\underline{K} \cup \{h\})$ of $\underline{K}$ and h.

Given that the set of potential corpora consist of all deductively closed theories in $\underline{L}$ (or all such theories that are contained in some weakest theory), $\underline{K}_h^+ = Cn(\underline{K} \cup \{h\})$. In Chapters 3 and 4, we shall use expansion transformations in contexts where some deductively closed theories in $\underline{L}$ do not count as potential corpora. In those contexts, this simple definition of expansion of $\underline{K}$ by adding h cannot be used. $\underline{K}_h^+$ is, for this reason, defined as the smallest deductively closed potential corpus in $\underline{L}$ containing $Cn(\underline{K} \cup \{h\})$. Such an expansion may be guaranteed to exist for every potential corpus $\underline{K}$ and sentence h in $\underline{L}$ by postulating that the set of potential corpora is closed under expansion. Expansion so understood covers both the cases where expansion adding h to $\underline{K}$ is identical $Cn(\underline{K} \cup \{h\})$ and those cases where it sometimes is not.

Let "h > g" abbreviate sentences of the form: If h is (were, had been) true, g will be (would be, would have been) true. (See section 2.7 for a more elaborate discussion of conditional and hypothetical "if" sentences.)

When neither h nor ~h is in $\underline{K}$, Ramsey's proposal for the acceptability of such conditionals discussed in section 1.2 states that the conditional h > g is acceptable[1] according to $\underline{K}$ if and only if g is in $\underline{K}_h^+$. In order to extend Ramsey's suppositional interpretation of open conditionals to belief-contravening and belief-conforming conditionals, however, we need to introduce the notion of a contraction transformation to supplement that of an expansion.

$\underline{K}$ is a *contraction* of $\underline{K}'$ if and only if $\underline{K}'$ is an expansion of $\underline{K}$.
A *contraction* of $\underline{K}$ *removing h* is any contraction of $\underline{K}$ of which h is not a member if h is not a logical truth. Otherwise it is $\underline{K}$ itself.

$C(\underline{K}, h)$ is the set of contractions of $\underline{K}$ removing h.

A contraction removing h from $\underline{K}$ is *saturatable* if and only if the result of adding ~h to the contraction is maximally consistent in $\underline{L}$.

$S(\underline{K}, h)$ is the set of saturatable contractions removing h from $\underline{K}$.

An element of $S(\underline{K}, h)$ is *maxichoice* if and only if for every g in $\underline{K}$ but not in the contraction ~g ∨ h is in the contraction.

A maxichoice contraction removing h from $\underline{K}$ is a saturatable contraction that is contained in no other saturatable contraction removing h from $\underline{K}$.

Every potential contraction removing h from $\underline{K}$ (every member of $C(\underline{K}, h)$) is the intersection of a subset of the set of saturatable contractions removing h from $\underline{K}$ (members of $S(\underline{K}, h)$).

If an inquiring agent were to modify genuinely his or her belief state by contraction, he or she would cease being convinced of information initially considered as settled and certain. In this sense, information would be lost. I suggest that loss of information is a cost to the inquirer who changes belief state via contraction. In belief-contravening supposition, the belief state is not genuinely contracted. Nonetheless, the inquirer needs to consider which of the potential contractions in $C(\underline{K}, h)$ to deploy in suppositional reasoning; and the choice should seek to minimize the loss of valuable information. For this reason, we need to explore ways of representing informational value.

In doing so, the question of which kinds of information are important or valuable will be left entirely open. That will depend to a considerable degree on the problem under consideration. We can, to be sure, make some general comments. Explanatorily useful information is often prized over explanatorily useless information. Simplicity (in some sense or other) is favored over complexity. Predictive power and testability may figure in as well. At the level of generality deployed here, however, we need not pay too close attention to these matters. It is desirable, however, that we consider some minimal requirements that assessments of informational value should satisfy.

Let M be a probability measure defined over all potential corpora in K. The *informational value* of a corpus $\underline{K}$ relative to M is equal to $1 - M(\underline{K})$.

If an agent were constrained to contract his or her current corpus $\underline{K}$ by removing h, the agent ought, so I contend, to choose a contraction in $C(\underline{K}, h)$ that minimizes informational loss in a sense yet to be explained. One possible proposal for the index to be minimized is the loss in informational value in the sense just defined. This proposal satisfies the following condition of adequacy for assessments of informational value:

Condition of adequacy for assessments of informational value:
If $\underline{K}'$ is an expansion of $\underline{K}$, $\underline{K}'$ is required to carry at least as much informational value as $\underline{K}$.

$1 - M(\underline{K})$ is not the only function of M that meets this condition of adequacy. $1/M(\underline{K})$ and $-\log M(\underline{K})$ do so as well. The discussion of inductive expansion in Chapter 6 supports the usefulness of $1 - M(K)$ as a quantitative index of informational value (see also Levi, 1984a, ch. 5). In the current discussion of contraction, only the ordinal properties of assessments of informational value are relevant. An ordering with respect to informational value orders potential corpora in the reverse order to the order induced by some probability measure M.

Of course, different probability measures will induce different orderings with respect to informational value. As indicated previously, I shall avoid protracted discussion of considerations favoring one ordering rather than another. However, it is worth emphasizing that probability measures used to determine informational value need not be the same as probability measures used to represent degrees of belief or credal probabilities. And they may differ also from probabilities used to represent physical probabilities or chances. M-functions are *informational-value*-determining probabilities.

Minimizing loss of informational value would require the agent to choose an element in $C(\underline{K}, \text{h})$ for which the M-value is a minimum. At least one maxichoice contraction has this property.[2] There can be more than one maxichoice contraction carrying minimal M-value and, moreover, there can be saturatable contractions that are not maxichoice that carry minimal M-value. However, no nonsaturatable contraction can carry minimal M-value. Such a contraction must be the intersection of saturatable contractions neither of which is a contraction of the other. Its M-value must be greater than the M-values of the saturatable contractions in the intersection. Hence, its informational value must be less than the informational value of these saturatable contractions. Adopting a nonsaturatable contraction cannot minimize loss of informational value (Levi, 1991, secs. 4.3–4.4).

Thus, minimizing loss of informational value requires choosing a saturatable contraction. Doing so, however, violates presystematic judgment. If the agent who contracts by removing h were then to expand by adding ~h, the resulting belief state would be maximally opinionated. To *require* that this be so as a result of first removing h and then adding ~h is absurd whether belief change is undertaken in earnest or as part of a fantasy for the sake of the argument (Alchourrón and Makinson, 1982; Levi, 1991, secs. 4.3–4.4).

To avoid this unpleasant result, it would be desirable to be in a position to recommend choosing the contraction strategy in $C(\underline{K}, \text{h})$ that is the intersection of all saturatable contractions that minimize loss of informational value. As we have seen, however, the recommended contraction strategy fails to minimize loss of informational value. Hence, the index of informational value should be replaced by another index that allows nonsaturatable contractions to be optimal contraction strategies.

On the other hand, it remains desirable that the new index should remain an index of the value of information in the sense that it continues to satisfy the condition of adequacy laid down earlier. This means that the nonsaturatable contraction $\underline{K}_{\text{h}}^{\wedge}$ that is the intersection of all saturatable contractions that bear maximum M-value (among saturatable contractions in $C(\underline{K}, \text{h})$) cannot carry more informational value than the saturatable contractions from which it is constituted.

If this nonsaturatable contraction $\underline{K}_{\text{h}}^{\wedge}$ is to carry optimal value according to the new index of informational value, it must carry the same informational value as its constituents.

Moreover, it must carry more informational value than the nonsaturatable contraction that is the intersection of $\underline{K}_h^{\wedge}$ with another saturatable contraction carrying less informational value.

In general, we shall need to introduce a measure of damped informational value that guarantees that the damped informational value of the intersection of a set of saturatable contractions is the minimum damped informational value of an element of that set. Since every potential contraction in $C(\underline{K}, h)$ is the intersection of a set of saturatable contractions, the damped informational value of a potential contraction is determined by the damped informational values of the saturatable contractions that constitute it. If we set the damped informational values of saturatable contractions in $C(\underline{K}, h)$ equal to their undamped informational values, we obtain the following index of damped informational value (Levi, 1991, sec. 4.4):

> The *damped informational value* of a potential contraction $\underline{I}$ is $1 - m$ where m is the largest value of M assigned a saturatable contraction in the set of saturatable contractions whose intersection is $\underline{I}$.

It is then possible for several contractions to minimize loss of damped informational value including not only each saturatable contraction that does so but the intersection of all such contractions as well. If we break ties among optimal contraction strategies by endorsing the weakest of the contractions minimizing loss of damped informational value (there must be exactly one such weakest contraction), the following contraction is recommended.

> The *admissible contraction* $\underline{K}_h^-$ *removing h from consistent $\underline{K}$* will be the intersection of all saturatable contractions that minimize loss of damped informational value.[3] If h is a logical truth, $\underline{K}_h^- = \underline{K}$. Furthermore, if h is not in $\underline{K}$, $\underline{K}_h^- = \underline{K}$.

Suppose ~h is in $\underline{K}$. $\underline{K}_{\sim h}^-$ removes ~h from $\underline{K}$ with minimum loss of informational value and the result is one where neither ~h nor h is in $\underline{K}$. In effect, the admissible contraction converts $\underline{K}$ into a corpus open with respect to h so that the Ramsey test for the acceptability in the form originally conceived by Ramsey can be applied.

However, when h is in $\underline{K}$, $\underline{K}_{\sim h}^- = \underline{K}$. $\underline{K}$ is not open with respect to $\underline{K}$. It contains h. If we want to obtain an open corpus, we need to remove h. This suggests that we introduce a new kind of contraction – one that is always guaranteed to yield a corpus containing neither h nor ~h from consistent $\underline{K}$.

> The *Ramsey contraction of consistent $\underline{K}$ with respect to* $h = (\underline{K}_{\sim h}^-)_h^- = \underline{K}_h^{-r}$.

Notice that Ramsey contractions like admissible contractions are based on assessments of damped informational value that are, in turn, derived from the informational value determining probability measure M. However, the Ramsey contraction of consistent $\underline{K}$ with respect to h is also the Ramsey contraction of consistent $\underline{K}$ with respect to ~h. Both contractions require that $\underline{K}$ be rid of h via optimal contraction if it is present in $\underline{K}$ and that it be rid of ~h if that is present. This kind of contraction of consistent $\underline{K}$ is guaranteed to be open with respect to h and ~h with a minimum loss of damped informational value.[4]

One important qualification should be added to this account of informational-value-based contraction. The set $C(\underline{K}, h)$ of potential contractions removing h was represented as the set of meets of all subsets of $S(\underline{K}, h)$ = the set of saturatable contractions in $\underline{L}$. Expanding a member of $S(\underline{K}, h)$ by adding ~h results in a maximally consistent corpus in $\underline{L}$.

Let g_r assert that object a is red, g_x assert that object a has color x (different from red), and g_d assert that object a has no color. Consider the following two scenarios:

Scenario 1: $\underline{K}$ contains ~h. It also contains g_r. Consequently, it contains ~g_d.

Scenario 2: $\underline{K}'$ contains ~h but, unlike $\underline{K}$, does not contain g_r. It does, however, contain ~g_d asserting that object a does have a color without specifying what it is.

Consider the saturatable contractions removing ~h from $\underline{K}$ or K'. These contractions will retain exactly one sentence of the form ~h $\vee$ z where z is a maximally specific description of the color of the object a in $\underline{L}$ or a declaration that the object a is without color. These include cases where ~h $\vee$ g_r is retained, ~h $\vee$ g_d is retained, and where ~h $\vee$ g_x is retained for every color x recognized in $\underline{L}$.

The recommended contraction removing ~h from $\underline{K}$ is the one that retains every such disjunction carrying minimum M-value among the disjunctions in the given set.

It is quite plausible to suppose that many agents rank ~h $\vee$ g_r equal with ~h $\vee$ g_x (for each color x) in M-value and, hence, in informational value. ~h $\vee$ g_d might be ranked anywhere in comparison with the other disjunctions. Contracton by removing ~h would then lead to removing ~h $\vee$ g_r and g_r from $\underline{K}$. Expanding by adding h would not restore g_r so that the revision of $\underline{K}$ removing ~h and adding h would not contain the claim that object a is red. This analysis also holds for contraction removing ~h from $\underline{K}'$ and then expanding by adding h.

Many would object to this result for scenario 1. If h is a claim about the population of St. Louis, Missouri, it is wrong to remove g_r asserting that object a is red from $\underline{K}$. ~h $\vee$ g$_r$ should be retained and g$_r$ should be in the revision adding h.

This objection does not apply to the removal of ~h from $\underline{K}'$ in scenario 2 where the analysis looks more plausible.

Whatever the merits of the intuitions driving the objection, the responses of the objectors seem coherent enough to warrant our seeking a way to allow the response in scenario 2 and the objector's response in scenario 1 to be legitimate. The problem is that given the informational values specified, h $\vee$ g$_r$ should be removed in both scenarios and not just in the second one.

The remedy for this is not to suggest different assessments of M-value for the two scenarios. These assessments should quite plausibly be the same. But as long as they are the same, the same responses should be favored in both scenarios.

Instead of tinkering with the M-function, another alternative is available. I have supposed that the set of contraction strategies available in the two scenarios is the same – namely, the set of intersections of saturatable contraction strategies removing ~h from $\underline{K}$ (from $\underline{K}'$).

Keep in mind, however, that in the first scenario prior to contraction there is no issue – not even a minor one – as to what the color of object a is. The inquirer has settled that it is red. In contemplating a contraction of $\underline{K}$ removing ~h, the inquirer may be able to think about finer-grained partitions of logical space than that into g$_r$ and ~g$_r$. But if addressed with the issue of removing ~h and being compelled to remove some nonempty subset of the sentences ~h $\vee$ g$_r$, ~h $\vee$ g$_x$ and ~h $\vee$ g$_d$ whose joint truth is logically equivalent to the truth of ~h, the inquirer may very well consider as options for contraction only whether or not in removing ~h, doubt should be cast on the truth of g$_r$. If the inquirer refused to cast doubt, he or she would remove ~h $\vee$ ~g$_r$. If the inquirer considering doubting g$_r$ for the sake of the argument, he or she would remove ~h $\vee$ g$_r$ and perhaps, ~h $\vee$ ~g$_r$ as well. But if ~h $\vee$ g$_r$ were removed and ~h $\vee$ ~g$_r$ retained, this would count as a saturatable contraction even if every disjunction ~h $\vee$ ~g$_x$ or ~h $\vee$ ~g$_d$ was retained. In evaluating damped informational value, the inquirer would begin by appealing to the M-values and informational values of the three contraction strategies I am suggesting are available. But now it is clear that the informational value of retaining ~h $\vee$ g$_r$ must be greater than the informational value of retaining ~h $\vee$ ~g$_r$. According to the suggestion under consideration, the informational value of the latter option is its undamped informational value and that is much greater than the damped = undamped informational value of the former according to the specifications of scenario 1. If the inquirer were recognizing all potential contraction strategies in $C(\underline{K}, h)$ as options, the informational value of retaining ~h $\vee$ ~g$_r$ without commitment to the sort of nonred thing a$_r$ is supposed to be would be smallest informational value assigned a hypothesis g_x or g_d. This value, by the stipulations of the model is no less than that of g_r.

In the second scenario, the agent whose corpus is $\underline{K}'$ is in suspense as to what the color of a is. In such a case, the agent might regard ascertaining the color of the object a, if it has one, as an issue and so in contraction removing ~h want to regard all the saturatable contractions in $S(\underline{K}', h)$ as available options. For this reason, ~h $\vee$ ~g_r will be given up as will ~h $\vee$ ~g_x for every color x in the range recognized.

Thus, the difference between scenario 1 and 2 is a difference not in the assessment of informational losses but in the set of contraction strategies the agent regards as available for choice. The determination of the set of contraction strategies is a reflection of a different dimension of what I am inclined to call the agent's demands for information from the one captured by the assessment of informational or M-value.

According to the modification of the account of optimal contraction I am suggesting, the set of contraction strategies need not be $C(\underline{K}, h)$ but $C(\underline{K}, h, V)$. Here V is a partition of the set of maximally consistent corpora into exclusive and exhaustive alternatives. It is the *basic partition* (Levi, 1991, pp. 123–4). A contraction removing h from $\underline{K}$ is saturatable relative to V if and only if the expansion adding ~h of the contraction removing h entails the truth of an element of V. Damped informational value is relativized to V as well as $\underline{K}$, h, and M the notion of an admissible contraction modified accordingly.

When, in scenario 1, ~h is removed in the context of genuine belief change (e.g., in the context of having expanded into inconsistency by adding h via observation as in Levi, 1991, sec. 4.8), the concern is to identify a potential contraction from inconsistency that retains h. In that setting, such a contraction is, roughly speaking, the result of removing ~h from $\underline{K}$ and then expanding the result by adding h. The inquirer may not want to lose the undamped informational value resulting from losing g_r unnecessarily. He or she might do so, if it turned out that the undamped informational value of ~g_r is as great as that of ~g_r. But otherwise not. This feature of the inquirer's demands for information can be represented by taking him or her as focusing on no alternatives to g_r except ~g_r and adopting a coarse-grained basic partition V. The M-value of ~h $\vee$ g_r is much lower than the M-value of ~h $\vee$ ~g_r so that the latter will be treated as carrying not only less undamped informational value than the former but less damped informational value as well. On the other hand, when expansion into inconsistency from $\underline{K}$ occurs, the inquirer may well be interested in the unknown color of a at the time so that neither h $\vee$ g_r nor h $\vee$ ~g_r is retained.

I do not propose this account as suggesting a mandatory way to select a basic partition but mention it merely as an acceptable way.

In the context of suppositional reasoning that is of interest here, the issue of the choice of V has a bearing on the treatment of iterated conditionals discussed in sections 3.7 and 4.4. One might for example endorse corpus $\underline{K}'$, suppose that g_r, and then suppose that h. In such a case, the demands for information including the choice of V are those embraced relative to $\underline{K}'$ whatever they may be and remain fixed throughout the suppositional exercise.

For the most part, reference to the relativity to a basic partition V will be suppressed. In the context of suppositional reasoning, it arises only in the context of iterated conditional reasoning and there it shall be taken for granted that it is held fixed. But it is important to keep in mind here that it may be subject to greater variability than I acknowledged in (Levi, 1991) in the context of genuine belief change (although less so in the context of iterated suppositional reasoning) and that the variation can take place while the M-function remains fixed.

Sometimes the aim in contraction might be not to minimize loss of informational value but to identify the shared agreements between the agent and some other agent or group of agents. In such a case, contraction does not stipulate a sentence to be removed from $\underline{K}$ with a minimum loss of damped informational value but rather demands contraction to common ground. This leads to the following notion of contraction:

The *consensus contraction* of consistent $\underline{K}$ with respect to consistent $\underline{K}'$ is the intersection $\underline{K} \cap \underline{K}'$ of $\underline{K}$ and $\underline{K}' = K_{\underline{K}'}^{-c}$.

Consensus contraction is not based on assessments of informational value but on the contents of a reference corpus $\underline{K}'$.

2.2 Revisions for Ramsey Tests

We often change our minds by adding new information via expansion. We change our minds in the other direction when we lose confidence in settled assumptions. A serious and rational inquirer might lose convictions because the inquirer has stumbled into inconsistency and seeks to escape or wants to give a deserving theory a hearing without begging questions for or against it (Levi, 1991, ch. 4).

An inquirer can stumble into inconsistency because he or she made observations or consulted with experts or witnesses and, as a result, expanded the set of beliefs to include information incompatible with his or her initial beliefs.

In response to such inadvertent expansion into inconsistency, the inquirer might remove some of his or her initial convictions while retaining the new information. The net effect is what Gärdenfors calls a "revision" of the initial belief state.

A second possible response is to bring both the new information and background beliefs into question. In that case, the net effect is a contraction of the initial state.

A third possible response is, of course, to question the new information rather than the old and, hence, to remain essentially with the status quo ex ante.

In the first case, the revision can be represented formally as the expansion by adding the new information of a corpus that is an admissible contraction of the initial corpus removing the background information that is challenged.

In the second case, the net effect of retreating from expansion into inconsistency is an admissible contraction of the old information brought into question.

In both cases, the notion of an admissible contraction plays a central role.

Contraction may also be reasonable when the inquirer is presented with a theory inconsistent with his or her initial beliefs that, nonetheless, is explanatorily powerful or in some other way is sufficiently attractive to merit a hearing. The initial corpus is contracted admissibly by removing its implication that the theory is false.

We are not currently preoccupied with justified belief change but with belief change for the sake of the argument. If p is inconsistent with the corpus $\underline{K}$ and we want to *suppose* it true for the sake of the argument in the sense of supposition cohering with Ramsey's idea that p is added "hypothetically" to $\underline{K}$, we need to prepare $\underline{K}$ so that the supposition that h is true may be consistently introduced. This is done by removing the offending ~h from $\underline{K}$. We must contract by removing ~h.

Observe that ~h is not being removed because the inquirer has good reason to do so. Indeed, the contraction is not in earnest so that no such reason need be given. It is stipulated for the sake of the argument that ~h is to be given up.

Nonetheless, given the stipulation that ~h is to be removed from $\underline{K}$ "for the sake of the argument" rather than for good reason, the problem of deciding how to contract subject to the constraint that ~h is to be removed from $\underline{K}$ is substantially the same in serious inquiry and in fantasizing. In both cases, the contraction should be the admissible contraction of $\underline{K}$ removing ~h.

Once ~h is removed via admissible contraction, the initial belief state is transformed to one where h is held in suspense (for the sake of the argument). Contexts where h is held in suspense are precisely those where Ramsey did provide an explicit account of the acceptability of conditionals and with it of suppositional reasoning. If neither h nor ~h is in $\underline{K}$, supposing that h and concluding from this that g is deducing g from $\underline{K}$ and h. If ~h is in consistent $\underline{K}$, supposing that h and concluding from this that g is deducing g from $\underline{K}_h^-$ relative to which h is held in suspense.

The question arises: Can one coherently suppose that h when one already fully believes that h?

One response might be that there is little point in talking of supposing that h in such a case except as a convenient way of permitting supposition to be allowable regardless of whether h is believed, disbelieved, or held in suspense. If the current corpus $\underline{K}$ already contains h, supposing that h amounts to identifying the consequences of the current belief state. Supposing that h is degenerate in the same sense that expanding $\underline{K}$ by adding h is degenerate because the result is identical with $\underline{K}$.

There is a second more interesting understanding of supposition as it applies to cases where the inquirer fully believes that h. When the agent is invited to suppose that h is true in such circumstances, he or she is urged to feign doubt as to the truth of h by removing h (not ~h) from $\underline{K}$ according to an admissible contraction $\underline{K}_h^-$ ($= \underline{K}_h^{-\tau}$). In such a state of doubt, supposing that h is true makes perfectly good sense.

Thus, in a situation where the inquirer is prepared to explain why g (describing the occurrence of some event) is true by appealing to the truth of h while fully believing that both h and g are true, a necessary condition for supporting the claim that h explains g relative to his or her belief state $\underline{K}$ is showing that g is in $(\underline{K}_h^{-\tau})_h^+$. That is to say, the inquirer supposes for the sake of the argument that h is true by performing a Ramsey contraction of $\underline{K}$ giving up h and then determining whether g is in the expansion of that contraction. Of course, this bit of suppositional reasoning may not be sufficient to secure an explanation of g by h. But there is some plausibility to the view that it is necessary.

Thus, we have two notions of supposition. The more restrictive one regards supposing that h is true to be degenerate or incoherent when h is fully believed. The less restrictive one denies that supposing that h is degenerate or incoherent in this case but, as in the case where ~h is fully believed, takes supposing as requiring a feigned shift to suspense between h and ~h through an admissible contraction minimizing loss of damped informational value. Both conceptions of supposition are, therefore, *informational value* based. The first is the *restrictive* informational-value-based conception. The second is the *nonrestrictive* informational-value-based conception. The second nonrestrictive version seems more in the spirit of Ramsey's proposal because it understands suppositional reasoning as first transforming the belief state to one where the supposition is left in doubt whether it is believed true or believed false and then proceeding as Ramsey suggested for this case.

There is yet a third conception of supposition that might be adopted. One might be motivated to suppose that h is true for the sake of the argument in a context where a rival view to the inquirer's is advanced. The rival view might agree with the inquirer's view in holding h in suspense. It might differ in judging h true where the inquirer judges ~h true or the disagreement might go the other way. In all three cases, the inquirer might be asked to suppose that h relative to the consensus corpus of shared agreements. The aim is then to investigate from common ground with the rival view the merits of the conjecture that h is true. In this setting the original Ramsey idea should be applied to $\underline{K}_{K'}^{-c}$.

The three approaches to supposing for the sake of the argument just sketched correspond to three different kinds of transformation of $\underline{K}$ that shall be identified as three different kinds of *revision*:

For consistent $\underline{K}$, *the Ramsey revision* of $\underline{K}$ with respect to h = $\underline{K}^{*r}_h = (K^{-r}_h)^+_h$.

I name this type of revision after Ramsey because the nonrestrictive informational-value-based conception of suppositional reasoning it expresses seems to generalize the kind of reasoning Ramsey thought relevant to assessing conditionals in the "open" case in a straightforward manner.

For consistent $\underline{K}$, *the AGM revision* $\underline{K}^*_h$ of $\underline{K}$ with respect to h = $(K^-_{\neg h})^+_h$.

The classic paper by Alchourrón, Gärdenfors, and Makinson (1985) explores the properties of *. Hence, the epithet "AGM revision."

The reasons for the name of the next type of revision are obvious.

For consistent $\underline{K}$, *the consensus revision* of $\underline{K}$ with respect to $\underline{K}'$ is $K^{*cK'}_h = (K^{-c}_{K'})^+_h$.

According to Ramsey, a conditional expresses hypothetical or suppositional reasoning at least in the case where the supposition is held in suspense. The three different notions of supposition identified previously have been associated with three different kinds of revision transformations of belief state. Corresponding to these three conceptions of supposition, there are three versions of the Ramsey test that generalize Ramsey's understanding of conditionals.

The Positive Ramsey Test according to Ramsey (PRTR):

h > g is acceptable relative to consistent $\underline{K}$ iff g is a member of K^{*r}_h .

The Positive Ramsey Test according to Gärdenfors and Levi (PRTGL):

h > g is acceptable relative to consistent K iff g is a member of K^*_h.

The Positive Consensus Ramsey Test (PRTC):

h > g is acceptable relative to consistent $\underline{K}$ iff g is a member of $\underline{K}^{*cK'}_h$.

The original Ramsey proposal for the open conditional is covered by the first two tests and very nearly so by the third. In this sense, they are all generalizations of Ramsey's idea. PRTR generalizes in the most straightforward way. It stipulates that when the supposition is not left in doubt, the belief state should be transformed into one where it is before Ramsey's original proposal is to be applied. PRTGL captures the idea introduced by Levi (1984a, ch. 12) and Gärdenfors (1978).

Neither the original proposal nor these three generalizations cover conditionals of the form "if p is (were, had been) true, then q may be (might be, might have been) true." They apply to "if p is (were, had been) true, then ~q will be (would be, would have been) true." Keeping in mind that suppositional reasoning seeks to base judgments of serious possibility and impossibility on transformations of the current corpus of full belief through the adoption of suppositions for the sake of the argument, it is clear that the Ramsey test should be extended to the case where q is a serious possibility relative to the suppositional corpus. But that will be so as long as the negation of q is not a consequence of the suppositional corpus. Since, moreover, "might" is the dual of "must," we should expect "if p, maybe q" to be the dual of "p > q." Hence, "if p, maybe q" should be equated with "~(p > q)."

Supplying acceptance conditions for "may" or "might" conditionals is tantamount to supplying acceptance conditions for the negations of "will" or "would" conditionals. To obtain acceptance conditions for the negations of conditionals, we introduce three negative Ramsey tests corresponding to each of the three positive Ramsey tests:

The Negative Ramsey Test according to Ramsey, Gärdenfors and Levi, and consensus respectively (NRTR, NRTGL, NRTC):

~(h > g) is acceptable relative to consistent K if and only if g is not a member of $\underline{K}^{*r}_h$, $\underline{K}^*_h$, $K^{*cK'}_h$ respectively.

The *Ramsey test according to Ramsey* (RTR) is the conjunction of the positive and negative Ramsey tests according to Ramsey. Similarly, we have the *Ramsey test according to Gärdenfors and Levi* (RTGL) and the *consensus Ramsey test* (RTC).

In case $\underline{K}$ is inconsistent, both h > g and ~(h > g) are acceptable relative to $\underline{K}$ according to RTR and RTGL. The situation with RTC is discussed in section 2.5 in connection with (K*c).

Ramsey originally formulated his test for situations where the truth value of h is unsettled relative to $\underline{K}$. I have argued that the criterion he formulated for that case ought to be satisfied by any criterion for the acceptability of conditionals expressing suppositional reasoning relevant to practical deliberation and scientific inquiry.

Condition of Adequacy for Ramsey Tests:

If h $\notin \underline{K}$ and ~h $\notin \underline{K}$, h > g (~(h > g)) is acceptable relative to $\underline{K}$ according to the Ramsey test if and only if g is (is not) in $\underline{K}_h^+$.

For the Ramsey test to work as intended in this core application where neither h nor ~h is in $\underline{K}$, $\underline{K}_h^+$ should satisfy two conditions:

Closure under Expansion: For every $\underline{K}$ in the set K of potential corpora and every h such that h $\notin \underline{K}$ and ~h $\notin K$, $\underline{K}_h^+$ is a potential corpus in K.

Consistent Expandability: For every potential corpus $\underline{K}$ in K and every h such that h $\notin \underline{K}$ and ~h $\notin K$, $\underline{K}_h^+$ is consistent.

When the set of potential corpora consists of all deductively closed theories for the given language, closure under expansion and consistent expandability are satisfied. As pointed out in section 2.1, under this condition, $K_h^+ = Cn(\underline{K} \cup \{h\})$.

Failure of closure of K under expansion implies that the Ramsey test will fail to apply in a situation where the truth of h is unsettled. This is precisely the kind of situation for which Ramsey designed his criterion. Our condition of adequacy for Ramsey tests ought to preclude such failure.

Failure of consistent expandability means that even though K is consistent and neither h nor ~h is in $\underline{K}$, both h > g and h > ~g are acceptable relative to K according to the Ramsey test. Any version of a Ramsey test that leads to this result allows for incoherent satisfaction of the condition of adequacy. It seems reasonable to require coherent satisfaction of the condition.

The consensus Ramsey test RTC does not strictly speaking satisfy the condition of adequacy. When the truth value of h is unsettled relative to $\underline{K}$, ~(h > g) will be acceptable according to RTC if (but not only if) g is not in the expansion of $\underline{K}$ by adding h. However, h > g need not be acceptable even though g is in this expansion. For h > g to be acceptable, g must be in the expansion of the consensus corpus K ∩ K' by adding h. Only in the special case where $\underline{K} = \underline{K}'$ will RTC satisfy the condition of adequacy.

Nonetheless, I am inclined to group RTC together with the other Ramsey tests. Consensus Ramsey tests are designed for situations where inquirers wish to reason for the sake of the argument not from their own perspectives but from the perspective of a consensus corpus. When neither the supposition h nor its negation belongs to the consensus, from the point of view of the consensus (though not, in general, from the points of view the participants), the issue is open as to the truth of h and, relative to the consensus corpus $\underline{K} \cap K'$, RTC does satisfy the condition of adequacy.

All three versions of the Ramsey test deploy revision criteria for the acceptability of conditionals. In order to secure coherent satisfaction of the condition of adequacy, however, they presuppose that revisions are expansions by adding h of contractions of $\underline{K}$. The assumption that they are nontrivially applicable also presupposes that such revisions represent belief states (are potential corpora) and, in the case of revisions according to Gärdenfors and according to Ramsey that when $\underline{K}$ and h are consistent, the revision is consistent.

Once more RTC is deviant. If h is inconsistent with the consensus corpus, the expansion of the consensus corpus by adding h is inconsistent, even though the consensus corpus, $\underline{K}$, and h are all individually consistent.

However, if we keep in mind the intended applications of RTC, this deviation is of minor import. In reasoning for the sake of the argument from consensus, the normal practice is to restrict consideration to suppositions that are open with respect to the consensus.

To be sure, sometimes one does consider suppositions that contravene the consensus. But the aim of such exercises is to point out that the suppositions are consensus contravening by showing that, from the consensual perspective, they lead to absurdity.

Consider, for example, the judgment "If Her Majesty is present at this party, I am a monkey's uncle." Given that the consensus contains the negation of the claim that the agent making the judgment is a monkey's uncle, the sentence registers the judgment that Her Majesty's presence at the party is not a serious possibility according to the consensus.

Suppositional reasoning from consensus allows for reductio arguments of a sort precluded when suppositional reasoning is from admissible or Ramsey contractions. In spite of this special feature of consensual revision, it always yields a consistent corpus provided the supposition h is consistent with the consensus even if it contravenes elements of $\underline{K}$.

All three versions of the Ramsey test express judgments of subjective, epistemic, or, as I call it, *serious* possibility relative to a transformation of $\underline{K}$. Given any consistent corpus $\underline{K}$ closed under logical consequence, h is seriously possible relative to $\underline{K}$ if and only if ~h is not in $\underline{K}$. h is impossible relative to $\underline{K}$ if and only if ~h is in $\underline{K}$.

Judgments of serious possibility and impossibility relative to the current corpus $\underline{K}$ are representable via definitional fiat by $\Diamond h = {\sim}(T > {\sim}h)$ and ${\sim}\Diamond h = T > {\sim}h$ where T is any logical truth expressible in $\underline{L}$ and RTR or RTGL is used as the standard of acceptability.

~(T > ~h) and T > ~h express judgments of serious possibility relative to the corpus that is the intersection of K and K' (the consensus corpus relative to K and K') when RTC is used. When K' = K, such judgments are judgments of serious possibility and impossibility relative to the current corpus and are equivalent to ◇ h and ~ ◇ h respectively.

I shall argue later that to judge a conditional acceptable is neither to believe it fully nor to judge it probable to some degree or other. For the present, however, this question may be left unsettled. What should, however, be kept in mind is that accepting a conditional h > g is to judge ~g impossible conditional on h and accepting ~(h > g) is judging ~g possible conditional on h. Such conditional judgments of serious possibility are psychological attitudes like full belief, judgments of credal probability, value judgments, and judgments of desirability. They should not be confused with linguistic devices used to represent such judgments or with the linguistic performances that manifest such attitudes. Accepting a conditional is not engaging in some kind of linguistic performance like uttering or inscribing a conditional sentence token or, indeed, being disposed to do so.

2.3 Informational-Value-Based Ramsey Tests

Judgments of subjective, epistemic, or serious possibility play an important role in scientific inquiry and practical deliberation. They define the space of possibilities over which judgments of subjective, epistemic, or credal probability are made. If h is judged impossible relative to K, h ought to be assigned 0 credal probability. (The converse does not, however, obtain. See section 8.3.)

Conditional judgments of serious possibility define spaces of possibility relative to transformations $T(\underline{K})$ of the current corpus $\underline{K}$ over which credal probabilities can be defined relative to $T(\underline{K})$. Thus, when a decision maker faces a choice between options $\underline{O}_1, \ldots, \underline{O}_n$, the deliberating agent does not take for granted which of the $\underline{O}_i$'s he or she will implement. Implementing any one of them is a serious possibility. In order to evaluate these options, he or she needs to identify the possible consequences of each option – that is, what might happen if he or she implements $\underline{O}_i$. That is to say, the agent is called on to judge serious possibility relative to $\underline{K}^+_{Oi}$ for each i. Relative to each such corpus, the agent may assign credal probabilities to the seriously possible hypotheses concerning the outcome of $\underline{O}_i$. This was the application of conditional judgments that was paramount in Ramsey's mind when he proposed his version of the Ramsey test for situations where the supposition is left in suspense.

Conditional judgments of serious possibility are frequently taken to be coarse-grained characterizations of judgments of conditional credal or subjective probability. This can be misleading. Consider the case where the agent is in suspense as to the truth of h. Judgments of serious possibility conditional on h are judgments of serious possibility relative to the expansion of the agent's corpus $\underline{K}$ by adding h. Not only are these the judgments of serious possibility that the agent is currently prepared to endorse in case he or she expands by adding h, but if he or she is rational, he or she is obliged to do so in case the expansion is implemented. Let $Q(./h)$ be the agent's conditional credal probability relative to $\underline{K}$. The conditional probability function represents the agent's judgment as to what kinds of bets are fair and what sorts of risks are reasonable to take on gambles that are called off in case h is false. Such conditional credal probability judgments are well defined when h is consistent with $\underline{K}$ but carry credal probability 0. Conditional credal probability is not belief contravening. To obtain belief contravention, h must be inconsistent with $\underline{K}$. But conditional probability is not well defined in case h is inconsistent with $\underline{K}$. Relative to $\underline{K}$ all bets conditional on h (where h is inconsistent with $\underline{K}$) are sure to be called off. For this reason alone such conditional probability judgments are not belief contravening as conditional judgments of serious possibility can be.[5]

Conditional judgments of serious possibility are useful in practical deliberation not only when the condition is open but when the condition is belief contravening. In attempting to calibrate utilities and probabilities in the context of a given decision problem, it is often useful to elicit information concerning how the agent would choose were his or her options different than they are.

The tendency of authors seeking to extend the applicability of Ramsey's original proposal for open conditionals to belief-contravening conditionals has been to contract $\underline{K}$ containing ~h by removing ~h and then adding h to the contraction. If the addition of h were regarded as an expansion, the result would be the inconsistent corpus $\underline{L}$. The usual response is to say that $\underline{K}$ undergoes a *minimal revision* to accommodate h without contradiction.

Modifying an idea of Mackie's (1962), I suggested that the acceptability of conditionals appeals to expansions of admissible contractions (Levi, 1984a, ch. 12). Gärdenfors (1978) proposed minimal-revision-based standards of acceptability where revisions are according to AGM. The test for the acceptability that emerges is the Ramsey test according to Gärdenfors and Levi (or RTGL). It very nearly satisfies the requirements of decision theorists interested in hypothetical choice.

There is one deficiency. Suppose agent X is offered a gamble on the outcome of the toss of a fair coin where he wins \$1,000 if the coin lands heads and loses \$1,000 if the coin lands tails. Let utility be linear in dollars. The expected value is $500. X has to choose between this gamble and receiving $700 for sure. X has foolishly (given his beliefs and values) accepted the gamble and won $1,000. Y points out to him that his choice was foolish. X denies this. He says: "If I had accepted the gamble, I would have won $1,000. If I had refused, I would have won $700." This line of thinking is an impeccable application of RTGL. Because h (the gamble is accepted and the coin is tossed) is in X's corpus $\underline{K}$, optimal contraction removing ~h has no impact on $\underline{K}$ and the subsequent expansion also makes no alteration in $\underline{K}$. The claim "X wins $1,000" is in $\underline{K}$. Hence, "If I had accepted the gamble, I would have won $1,000" is acceptable by X according to RTGL. The second conditional is also acceptable according to RTGL. X's restrospective argument for the rationality of his choice is vindicated by the use of RTGL.

But surely the argument is fallacious. The acceptability of "If X had accepted the gamble, X would have won $1,000" according to RTGL is undeniable. But even though X won the gamble, it was foolish of X to have taken it. RTGL conditionals are, in this context, irrelevant to the appraisal of the rationality of X's decision.

Using RTR avoids this difficulty. The idea behind this test is that the initial contraction step is a transformation of the current corpus $\underline{K}$ to a corpus that is open with respect to the content of the "if" clause. h is removed from $\underline{K}$. Hence, so is the claim g that $1,000 is won on the gamble. Once the corpus is "prepared" in this way, Ramsey's original suggestion can be applied at the expansion step.

The question then arises. Should restoring h lead to restoring all lost elements from $\underline{K}$? That is to say, should admissible contraction removing h satisfy the recovery postulate?

Recovery:

   If h is in $\underline{K}$, $(\underline{K}_h^-)_h^+ = \underline{K}$.

If recovery fails, X no longer has the rationalization of his choice available when RTGL is used. RTR is sensitive to the status of recovery. RTGL is not.

RTGL stipulates that before expansion by adding h, $\underline{K}$ has to be "prepared" by removing ~h when it is in $\underline{K}$. RTR requires, in addition, that if h is in $\underline{K}$ it needs to be removed as well. In both cases, contraction is prompted by the need to remove a specific proposition from $\underline{K}$ seen as an obstacle to deploying the expansion suggested by Ramsey originally. When contraction is generated this way, there is a need to choose between alternative contraction strategies that meet the requirement of removing the given proposition. This means that contraction strategies need to be evaluated according to some standard. Gärdenfors and I have agreed that such evaluation takes into account the incorrigibility, immunity to being given up, or entrenchment of statements in the corpus being subjected to contraction (Levi, 1980, 1991; Gärdenfors, 1988). I have, however, interpreted both incorrigibility and entrenchment as reflecting the amount of informational value that is lost in the contraction (Levi, 1991). For this reason, revisions according to AGM and according to Ramsey may be called *informational-value-based* revisions.

It is useful to take note of some structural properties that contractions and revisions exhibit. Gärdenfors (1988, sec. 3.4) lists eight postulates for contraction originating in Alchourrón, Gärdenfors, and Makinson (1985). I have introduced three distinct kinds of contraction operators. Two of them, admissible contraction and Ramsey contraction, are defined in terms of an informational-value-determining probability measure M defined over potential corpora in K. My intention is not to take Gärdenfors's postulates as legislative for contraction and then to seek interpretations that satisfy them. Instead, I consider potentially useful conceptions of contraction (such as admissible and Ramsey contraction) and examine which of these postulates are satisfied.

AGM Postulates for Contraction

(K-1) For any h and $\underline{K}$, $\underline{K}_h^-$ is a corpus.

(K-2) $\underline{K}_h^- \subseteq \underline{K}$.

(K-3) If h $\notin$ K, then $\underline{K}_h^- = \underline{K}$.

(K-4) If not $\vdash$ h, then h $\notin \underline{K}_h^-$

(K-5) If h $\in \underline{K}$, $\underline{K} \subseteq (\underline{K}_h^-)_h^+$.

(K-6) If $\vdash$ h $\equiv$ h', $\underline{K}_h^- = \underline{K}_{h'}^-$,

(K-7) $\underline{K}_h^- \cap \underline{K}_g^- \subseteq \underline{K}_{h\,\&g}^-$.

(K-8) If h $\notin \underline{K}_{h\,\&g}^-$, then $\underline{K}_{h\,\&g}^- \subseteq \underline{K}_h^-$.

An important consequence of (K-2) is the following:

(K-2w) if h $\in \underline{K}$, $(K_h^-)_h^+ \subseteq \underline{K}_h^+ (= \underline{K})$.

This condition is a trivial and noncontroversial consequence of the intended meaning of contraction.

(K-5) asserts the converse of (K-2w) and together they yield the recovery postulate mentioned earlier.

The example of the coin known to have been tossed and to have landed heads is a clear counterinstance to recovery. Let the corpus be contracted by giving up the claim that the coin was tossed. The claim that it landed heads will be given up and will not be restored if the claim that the coin was tossed is restored. There are other kinds of counterinstances as well (see

Levi, 1991). This suggests strongly that (K-5) should be given up.[6] This conclusion is also supported by a consideration of the way admissible contraction and Ramsey contraction are defined.

Consider all the saturatable contractions removing h from $\underline{K}$. Given any such saturatable contraction, it will contain for every f in $\underline{L}$ either h ∨ f or h ∨ ~f but not both.

For any such saturatable contraction, either it is maxichoice or there is a unique maxichoice contraction removing h from $\underline{K}$ that is stronger than it is.

The fact that the maxichoice contraction is stronger than the saturatable but not maxichoice contraction does not mean that the maxichoice contraction carries greater (damped or undamped) informational value. That depends on whether the M-function assigns greater value to the maxichoice contraction or equal value. Both assignments are coherent.

If greater value is assigned invariably to maxichoice contractions, then (K-5) will be satisfied and recovery will be obeyed. In that case, an admissible contraction will be the intersection of all optimal maxichoice contractions. That is precisely the characterization of contractions provided by Alchourrón, Gärdenfors, and Makinson (1985).

However, if the M-function is allowed to assign equal value to a non-maxichoice but saturatable contraction and its maxichoice counterpart, the intersection of the saturatable contraction and its maxichoice counterpart will, of course, be the nonmaxichoice but saturatable contraction. It is then easy to see that recovery can be violated.

Thus, the informational-value-based account of contraction just sketched is designed to furnish a decision theoretic rationale for favoring one contraction strategy over another. It insists that an admissible contraction strategy removing h from $\underline{K}$ should be an optimal option (if one exists) among all the available options with respect to the goals of contraction. It identifies the set of available options as consisting of *all* contractions removing h from $\underline{K}$ rather than imposing restrictions in advance as Alchourrón, Gärdenfors, and Makinson (1985) do by considering only contractions that are intersections of subsets of the set of maxichoice contractions, and it explicates the goal of the exercise by specifying the features of a utility or value function for evaluating the available contractions.[7]

From the decision theoretic perspective, it is not enough to identify a set of postulates that admissible contractions should satisfy as is currently customary. Where there are disputes concerning the merits of some postulate or other (such as recovery), the current practice can resort to presystematic practice and nothing else. Providing a decision theoretic rationale allows for another constraint on our assessment of characterizations of admissible contractions. This constraint is mandated by the pragmatist perspective infusing this essay according to which changes in belief state are to be justified by showing how well they "adapt means to ends."

The decision theoretic perspective pays for itself in the discussion of recovery. Presystematic judgment does call recovery into question as we have seen. The decision theoretic perspective I have proposed offers a rationale for failures of recovery when they occur and helps identify some of the missteps advocates of recovery have taken.

Both admissible contraction and Ramsey contraction violate (K-5). Admissible contraction satisfies all other postulates. Ramsey contraction violates (K-3) as well as (K-5) but satisfies the other postulates. (See Hansson, 1993.) (K-3) may be replaced by the following:

(K-3w) If $h \notin K$ and $\sim h \notin \underline{K}$, then $K_h^{-r} = \underline{K}$.

Although both RTGL and RTR are informational value based, they are, nonetheless, different. The difference can be traced to the recovery postulate.

To see this, it is useful to introduce postulates for AGM revision (Gärdenfors, 1988, sec. 3.3). It is important to keep in mind, however, that revisions as understood here are expansions of contractions where the kind of revision being constructed depends on the kind of contraction being used in the definition. Our concern then is to determine which of the postulates for AGM revision are satisfied by the various definitions we have introduced. In formulating the postulates, the operator * for AGM revision will be used. However, we shall also consider applications of the postulates with the Ramsey revision operator *r substituting for *.

$\underline{K}_h^*$ is a function from the set of potential corpora and the set of sentences in $\underline{L}$ to the set of theories. Revisions of the inconsistent corpus are not well defined because neither admissible contractions nor Ramsey contractions of inconsistent corpora are well defined. This does not mean that contractions of inconsistent corpora are unintelligible but only that they are not admissible or Ramsey contractions. (Contractions of inconsistent corpora are discussed in Levi, 1991, ch. 3.) In any case, interest in revision of inconsistent corpora is not salient in determining acceptability conditions for conditionals. For this reason, the inconsistent corpus will be left out of the domain of the function. The following postulates are constraints on such revision functions proposed by Gärdenfors (1988):

Postulates for AGM Revision

(K*1) For any sentence h and consistent $\underline{K}$, $\underline{K}_h^*$ is a corpus.

(K*2) $h \in \underline{K}_h^*$. (success)

(K*3) $K*h \subseteq K_h^+$. (inclusion)

(K*4) If $\sim$h $\notin$ $\underline{K}$, $\underline{K}^+_h \subseteq \underline{K}^*_h$. (preservation)

(K*5) $\underline{K}^*_h$ is inconsistent iff $\vdash \sim$h. (consistency)

(K*6) If $\vdash$ h $\equiv$ h′, then $\underline{K}^*_h = K^*_{h'}$.

(K*7) $\underline{K}^*_{h\&g} \subseteq (\underline{K}^*_h)^+_g$.

(K*8) If $\sim$g $\notin$ $\underline{K}^*_h$, then $(\underline{K}^*_h)^+_g \subseteq \underline{K}^*_{h\&g}$. (restricted weakening)

Gärdenfors shows that revision according to AGM satisfies these eight postulates whether or not contraction satisfies (K-5) as long as it satisfies the other seven postulates for optimal contraction (1988, p. 69).

Postulates for Revision according to Ramsey

Postulates for revision according to Ramsey include (K*1) – (K*3) and (K*5) – (K*7) together with the following modifications of (K*4) and (K*8):

(K*r4) If neither h nor $\sim$h is a member of $\underline{K}$, $\underline{K}^+_h \subseteq K^{*r}_h$.

(K*r8) If neither g nor $\sim$g is a member of $\underline{K}^{*r}_h$, then $(K^{*r}_h)^+_g \subseteq \underline{K}^*_{h\&g}$.

The failures of (K*4) and (K*8) are attributable to the failure of (K-5). The fact that (K-3) needs to be replaced by (K-3w) has no impact on the revision postulates. Consequently, if recovery or (K-5) were added to the requirements on contraction, the postulates for Ramsey revision would coincide with those for AGM revision.

To obtain (K*4) from (K*r4) one needs to add the following postulate:

(K*4w) If h is a member of $\underline{K}$, $\underline{K}$ (= $\underline{K}^+_h$) $\subseteq K^{*r}_h$.

This is the condition Gärdenfors calls "weak preservation." If (K-5) and, hence, recovery holds, (K*4w) obtains. Given (K*r4) and (K*r8) we get both (K*4) and (K*8). Under these conditions, RTGL and RTR yield identical results. However, the two versions of the Ramsey test are not, in general, equivalent.

Although there is no logical obstacle to recognizing two distinct kinds of conditional modal appraisal corresponding to the two tests, I am inclined to think of both of them as rival accounts of the same kind of conditional modal appraisal. Consequently, we can reasonably ask whether one test serves our purposes better than the other. RTR is better than RTGL in the case of the foolish gamble. Other considerations argue in favor of it as well.[8]

2.4 Consistent Expandability and Preservation

Once more recall that the condition of adequacy for Ramsey tests states that when neither h nor $\sim$h is in $\underline{K}$, h > g is acceptable according to $\underline{K}$ if and only if g is in $\underline{K}^+_h$. We expect the condition of adequacy to be coherently applicable – that is, when neither h nor ~h is in $\underline{K}$, either h > g or ~(h > g) is not acceptable. I pointed out in section 2.2 that such coherent applicability presupposes that the consistent expandability condition should be satisfied.

In order to extend the Ramsey test to cases where either h or ~h is in $\underline{K}$, a revision criterion stating that h > g is acceptable according to $\underline{K}$ if and only if g is in the (minimal or admissible) revision of $\underline{K}$ by adding h is introduced. When neither h nor ~h is in $\underline{K}$, a revision criterion satisfying the condition of adequacy for Ramsey tests implies that expansion of $\underline{K}$ by adding h and revision of $\underline{K}$ by adding h are identical when neither h nor ~h is in $\underline{K}$. If the condition of adequacy is coherently satisfiable (so that consistent expandability is satisfied), (K*r4) is satisfied as well.

Notice that (K*4w) or "weak preservation" is not implied by the condition of adequacy combined with the revision criterion. The combination of (K*4w) and (K*r4) is equivalent to (K*4) and (K*4) is itself equivalent to the following:

Preservation:

If g is in $\underline{K}$ and h is consistent with $\underline{K}$, g is in $\underline{K}^*_h$.

If we wish to endorse a Ramsey test utilizing a revision criterion meeting the condition of adequacy, the revision operator has to satisfy (K*r4) but, as the example of the Ramsey revision and RTR shows, does not have to meet (K*4w).

Gärdenfors (1986) claims to have demonstrated a conflict between the preservation condition and RTGL if the revision operator is not to be trivialized in a sense that will be explained later. More specifically, the conflict is between (K*r4) and RTGL. But according to the account I have just given, (K*r4) is presupposed by the assumption that RTGL coherently satisfies the condition of adequacy. Hence, if Gärdenfors's argument is correct, RTGL is incoherent. That is to say, RTGL implies that AGM revision has two incompatible properties:

(i) Satisfaction of (K*r4).

(ii) Satisfaction of the *inclusion monotonicity condition*:

if $\underline{K} \subseteq \underline{J}, \underline{K}^*_h \subseteq \underline{J}^*_h$.

Gärdenfors thinks that RTGL presupposes the inclusion monotonicity condition but not (K*r4). As I have argued, however, satisfaction of (K*r4) is a condition of adequacy for Ramsey tests. Gärdenfors sees the issue as one of choosing between Ramsey test criteria for conditionals and the part of the preservation condition characterized by (K*r4). That is, however, misleading. If his argument is correct, either inclusion monotonicity or (K*r4) must go. Either way, the Ramsey test is abandoned.

There are, however, side assumptions made by Gärdenfors in his proof that are open to question. Rather than conclude that RTGL is incoherent, it is preferable to conclude that one of these side constraints is objectionable. We shall examine this matter more closely later on.

2.5   Consensus-Based Ramsey Tests

Sometimes a proposition is supposed to be true for the sake of the argument relative to a background of shared agreements rather than relative to a contraction of the current corpus minimizing loss of damped informational value. We may seek to make modal judgments relative to common ground between us where all controversial items are expunged from the corpus relative to which modal judgments are to be made. Or we may seek to make modal judgments relying on some mathematical or scientific theory we embrace endorsed as part of the current doctrine together with "conditions" specified in "if" clauses.

In all such cases, we may be represented as contracting from K to the intersection of K and some other corpus K' and then expanding by adding the "conditions."

Smith who has corpus $\underline{K}$ and is convinced that Oswald killed Kennedy confronts Jones who has corpus $\underline{K}'$ and is convinced that Oswald did not kill Kennedy. If Smith and Jones move to common ground – that is, to the claims they share in common (the intersection of $\underline{K}$ and $\underline{K}'$) – they will both be in suspense as to whether Oswald killed Kennedy or not. But they both will agree that somebody killed him. Hence, If they expand the consensus corpus by adding the claim "Oswald did not kill Kennedy," they will obtain a corpus containing the claim that somebody different from Oswald did. Here consensus is between the opinions of Smith and another person, Jones. Sometimes, however, Smith is concerned with consensus with some authoritative doctrine (e.g., Euclidian geometry, quantum mechanics). The revisions generated in this way are *consensus-based revisions* and the associated Ramsey tests are *consensus-based Ramsey tests*.

Like Ramsey revision but unlike AGM revision, consensus-based revision violates (K*4) and (K*8). In contrast to Ramsey revision, however, it fails to satisfy (K*r4) and (K*r8). And unlike both AGM revision and Ramsey revision, it violates (K*5). However, it satisfies a weaker version of this postulate that asserts that the consensus revision of $\underline{K}$ by adding h is inconsistent if and only if h is inconsistent with the consensus.

(K*5c) $K*_h^{cK'}$ is inconsistent if and only if $\underline{K} \cap \underline{K}' \vdash \sim h$.

Checking these claims will be left to the reader to verify.

2.6   Attitudes, Their Representations, and Manifestations

Conditional judgments of serious possibility and impossibility characterized by informational-value-based and consensus-based Ramsey tests have
useful applications.

Informational-value-based conditionals can be used in a practical deliberation to evaluate options as means to given ends. More generally, this
mode of conditional modal appraisal is appropriate in contexts where the
agent or inquirer is concerned to assess serious possibility relative to his or
her point of view modified by adding some supposition for the sake of the
argument.

Consensus-based conditionals are useful when an inquiring agent's opinions are confronted with alternative viewpoints recognized as meriting
respect sufficient to deserve considering a consensual point of view or when
the agent recognizes the need to engage in joint deliberation with other
agents and explores common ground on the basis of which joint decisions
may be taken. In this context, suppositional reasoning is to be based on the
consensual point of view.

The distinction being made is not a linguistic one. An agent's state of full
belief **K** at a given time is an attitudinal state of the agent. So is the state
constituted by the agent's judgments of serious possibility, the agent's state
of credal probability judgment, the agent's state of value judgment, and the
agent's states of judgments of serious possibility conditional on diverse
suppositions. Attitudinal states are to be distinguished from the linguistic
or other symbolic representations of them. Attitudinal states are also to be
distinguished from their linguistic manifestations.

The state of a mechanical system in classical mechanics can be
represented or described by a point in a phase space representing the states
a mechanical system of the kind in question is able to realize. One way of
(partially) representing a potential state of full belief **K** (i.e., a state of full
belief a deliberating agent is able to realize) is by means of a corpus or
theory in a suitable regimented language $\underline{L}$. Such a corpus or theory is a
set of sentences in $\underline{L}$ closed under the logical consequence relation for $\underline{L}$.
Due to limitations of the language $\underline{L}$, the representation of state of full
belief **K** by corpus $\underline{K}$ in $\underline{L}$ may be only partial. However, the limitations
will not raise any difficulties of relevance to our current discussion. I shall
continue to represent states of full belief by such corpora throughout this
book.

Suppose, for example, Mario says in English: "Dave Dinkins will win the
election." Mario's utterance may manifest a hope, belief, conjecture, or
other attitudinal disposition. Or it may be gibberish. Deciding this matter is
part of the task of interpreting Mario's linguistic behavior.

If we interpret Mario as expressing an attitude by his linguistic behavior, we are taking the linguistic behavior as manifesting an aspect of Mario's attitudinal state. But the task of interpretation is not done. Not only must it be settled whether Mario registers a belief, hope, conjecture, or whatever, but what the "content" of the attitude is: Is it that David Dinkins will win the mayoral election? In answering this question concerning the task of interpreting Mario's linguistic behavior, we are using language not to manifest or reveal Mario's attitude but to describe or represent it.

If the attitude is one of full belief, we are claiming that Mario is in a state of full belief that has as a consequence another state – to wit, the state represented by a corpus consisting of all and only logical consequences of "David Dinkins is the winner of the New York mayoral election of 1993." To put the matter slightly differently, Mario's utterance reveals Mario's state of full belief state to be represented linguistically by a corpus that contains the sentence "David Dinkins is the winner of the mayoral election of 1993."

Mechanical systems do not, in general, use the language of phase space or any other method to represent their mechanical states. Inquiring and deliberating agents use such linguistic representations and, in doing so, reveal commitments to attitudes of belief, doubt, supposition, or intentions to explain, describe, or teach. In such contexts, the linguistic behavior of the agent is, on the one hand, a token representation of a mechanical state and, on the other hand, a manifestation of various attitudinal commitments.

Deliberating agents, unlike mechanical systems, may represent aspects of their attitudinal states but the language $\underline{L}$ used for the purpose of systematic characterization of such states need not be one used by the deliberating agents to manifest their attitudinal states.

Deliberating agents sometimes reveal their full beliefs, however, by uttering or inscribing sentence tokens of sentence types belonging typically to natural languages supplemented occasionally by sentences from regimented language.

Thus, George Bush might firmly declare "Saddam Hussein will be removed from office." Bill Clinton might (perhaps foolishly) construe this utterance as revealing George's firm conviction that Saddam will be overthrown as ruler of Iraq.

However, Bill's interpretation of George's utterance does not require Bill's taking George's state of full belief as being represented by a regimented version of "Saddam will be overthrown" together with all its logical consequences. Bill would be too charitable to regard George as being as empty-minded as all that. George would be regarded as having a stronger state of full belief including, for example, information to the effect that he is married to Barbara and was defeated by Bill for a second term as well as other information that Bill might not be in a position to specify.

Bill's interpretation of George's utterance would amount to providing a partial specification of George's state of full belief by means of a partial specification of George's corpus. George's utterance, as understood by Bill, is, in this respect, a manifestation of an aspect of George's state of full belief.

A deliberating agent is in several different attitudinal states at one and the same time and these different states are represented differently in a systematic presentation. An agent's state of full belief is represented by a corpus $\underline{K}$ of sentences in $\underline{L}$. The state constituted by the agent's judgments of serious possibility is represented by a partition of the set of sentences in $\underline{L}$ into the set of sentences in $\underline{L}$ inconsistent with $\underline{K}$ (the impossible) and the set of sentences in $\underline{L}$ consistent with $\underline{K}$ (the possible). Given such a partition, the set $\underline{K}$ is uniquely determined just as the partition is uniquely determined by $\underline{K}$. Although the two modes of representation are different, we may take them as representing one attitudinal state rather than two, which we may call the state of full belief or the standard for serious possibility as we wish.

States of credal probability judgment are representable by nonempty, convex sets of probability functions defined over sentences in $\underline{L}$ assigning equal probabilities to hypotheses equivalent given $\underline{K}$ and 0 probability to all sentences inconsistent with $\underline{K}$ (Levi, 1974 and 1980a, ch. 4). If we supplement the credal state with a way of representing the distinction between certainly false and almost certainly false (i.e., carrying 0 probability while remaining seriously possible), the representation of the credal state uniquely determines a corpus $\underline{K}$ but not conversely. This is an indication of the fact that many different states of credal probability judgment cohere with the same state of full belief. Unlike states of serious possibility judgment, states of credal probability judgment cannot be identified with states of full belief.

Similar remarks apply mutatis mutandis to states of value judgment partially ordering the seriously possible hypotheses relative to $\underline{K}$ with respect to value.

Although states of unconditional serious possibility judgment may be equated with states of full belief, states of conditional serious possibility judgment cannot be so reduced to states of full belief when such states satisfy the requirements of informational-value-based or consensus-based Ramsey tests.

Consider judgments of serious possibility conditional on h. To determine how the distinction between the impossible and the possible is drawn, consensus based tests require appeal not only to $\underline{K}$ but to $\underline{K}'$ in order to determine the appropriate transformation of $\underline{K}$ relative to which judgments of serious possibility are made.

Informational-value-based tests require reference to a means for evaluating informational value – such as the use of the function M.

Appeal to the M-function is unnecessary in cases where neither h nor ~h is in $\underline{K}$. In those cases, the state of judgment of serious possibility conditional on h is uniquely determined by the state of full belief and vice versa. Even so, the two states are distinct in the sense that they are distinct standards for serious possibility and, hence, distinct potential states of full belief.

Consensus- and informational-value-based conditional appraisals of serious possibility are different types of attitudinal states. These types of attitudinal states are no more linguistic states than are states of full belief or states of unconditional modal judgment of serious possibility. Of course, in giving an account of the distinction between the informational-value-based and consensual types of conditional modal appraisal, some linguistic devices will be used, whether from natural languages or regimented languages, to represent the distinction. But these linguistic devices ought not to be confused with the aspects of attitudinal states represented. Moreover, the resources of natural languages for displaying such conditional modal appraisals ought to be kept separate from the aspects of attitudinal states manifested by their use and from the use of language in representing such states for the purpose of systematic discussion.

This book is not an essay in the use of natural language for the purpose of manifesting conditional modal appraisals. Yet, these two types of conditional modal appraisal are sufficiently different that it would not be too surprising if some natural languages provide different linguistic means for expressing them. A few brief remarks concerning the way in which English is used to register the distinction may be helpful for the purpose of indicating the distinction under discussion.

2.7 Conditionals and Hypotheticals

I conjecture that "if" sentences of the sort that V. H. Dudman calls "hypotheticals" have acceptability conditions given by consensus Ramsey tests (Dudman, 1983, 1984, 1985). "If Oswald did not kill Kennedy, somebody else did" is a hypothetical according to Dudman's grammatical classification. As was noted in section 2.5, individuals who disagree concerning Oswald's guilt tend to agree that somebody shot Kennedy. They would, therefore, assert the hypothetical because, so I contend, the appropriate test for acceptability is consensus based.

Hypotheticals contrast in important respects with a class of "if" sentences Dudman calls "conditionals." I suggest that Dudman's conditionals have acceptability conditions given by the Ramsey test according to

Ramsey subject to an additional constraint mandated by the considerations of tense adduced by Dudman (1983, 1984, 1985).

For example, a person (say, Jack Ruby) who, before the assassination, was convinced that Oswald would kill Kennedy on the visit to Dallas the next day could have said:

(1) If Oswald does not kill Kennedy in Dallas, Kennedy might escape from Dallas unscathed.

In uttering such a token of this sentence type, Ruby is manifesting an attitude of a certain kind – one I am calling a conditional judgment of serious possibility. To make such a judgment, Ruby would have to invoke an expansion of a contraction of his belief state. The contraction would be non-degenerate because Ruby's belief state is represented by a corpus containing the a sentence asserting that Oswald would kill Kennedy during the Dallas visit. This contraction of belief state is represented by a contraction of Ruby's corpus by giving up the claim that Oswald will kill Kennedy during the Dallas visit. An admissible contraction as I characterized it would give up the claim that Oswald kills Kennedy while minimizing loss of damped informational value. Dudman's account requires an additional proviso. No assumption in Ruby's corpus concerning events in the past or the present should be given up. This is because they do not occur after the "crossover point" – that is, the tense used in the "if" clause, which, in this case, is present. If Ruby were convinced before the event that there is no other backup or alternative plan to assassinate Kennedy in Dallas, he might well have asserted the conditional cited.

Notice that Dudman's construal of sentences like (1) is not intended to impose restrictions on contractions in general but only to indicate that when the speaker (in our case Ruby) uses a sentence token of (1) to manifest his judgment of conditional modality, the speaker has indicated a constraint on the contractions from which a choice of an optimal one is to be made.

After the assassination has occurred (which is, on our supposition, no news to Ruby), Ruby could register the same judgment by the conditional

(2) If Oswald had not killed Kennedy, Kennedy might have escaped from Dallas unscathed.

Here, as before, Oswald is required to give up the claim that Oswald killed Kennedy at the contraction step. As before, he must minimize loss of damped informational value. But now he is free to alter present and past events because the "crossover point" expressed by "had" is "past of the past." But this might not have altered his verdict at all.

Dudman points out that in sentences like (1) and (2) the tense figuring in the "if" clause precludes the clause from constituting a grammatical sentence conveying a coherent message. The first of the sentences begins with a clause "Oswald does not kill Kennedy (tomorrow)" which has a present tense but refers to a future event. The tense does not serve to date the occurrence of the event but to identify the crossover point or date imposing the constraint on what bits of known history may be revised in the fantasy at the contraction step.

The only difference between (1) and (2) in the two respects just mentioned is the date of the crossover point and the consequent constraint on the bits of known history that may be revised.

Consider by way of contrast

(3) If Oswald did not kill Kennedy, maybe Kennedy escaped from Dallas unscathed.

The "if" clause is a grammatical sentence carrying a coherent message. As has been so often pointed out, Ruby might have rejected this judgment after the assassination while affirming the two conditionals. Grammatically this sentence is a Dudman "hypothetical." I contend that it is subject to standards of acceptability given by the consensus Ramsey test. Ruby would utter such a sentence when discussing the events in Dallas with someone who shared his view that Kennedy was indeed shot by someone.

Many writers appeal to the differences in presystematic judgments concerning the acceptability of (2) and (3) to argue that Ramsey test accounts of acceptability can only do justice to one category of if sentences – the so-called "indicatives." These are normally taken to include sentences that are hypotheticals in Dudman's sense as well as many that are conditionals according to Dudman – namely those where the crossover point according to Dudman is present such as (1). According to this widely held view, Ramsey style tests do not apply to subjunctive conditionals like (2). Yet, (1) resembles (2) rather than (3) in all relevant respects except the time of utterance, which according to Dudman accounts for differences in tense.

Ruby would not only affirm (1) prior to the assassination knowing what we are supposing he knew, but he would refuse to affirm the following dual:

(1') If Oswald does not kill Kennedy (tomorrow), somebody else will.

After the fact, Ruby would not only affirm (2), he would refuse to affirm the following:

(2') If Oswald had not killed Kennedy, somebody else would have.

Again after the fact, Ruby would refuse to affirm (3) but would affirm the following:

(3') If Oswald did not kill Kennedy, somebody else did.

Thus, parallel to the comparison between (1), (2), and (3), (1′) resembles (2′) rather than (3′).

Alan Gibbard (1981) shares Dudman's view of how (1′), (2′), and (3′) should be classified. His indicatives coincide with Dudman's hypotheticals and his subjunctives are Dudman's conditionals. So the obvious affinity between (1′) and (2′) poses no problem for him. Even so, he too claims that Ramsey test criteria are applicable only to hypotheticals.[9]

I contend that Ramsey test criteria handle both conditionals and hypotheticals. The presystematic differences derive from differences between invoking consensus to do the job of contraction and relying on evaluations of losses of informational value with side constraints imposed by the crossover point.

I have found Dudman's arguments for his grammatical distinctions quite convincing; but my objective in this essay is not to settle this or any other linguistic issue. Whether one divides the domain of sentences under consideration into indicatives and subjunctives or into hypotheticals and conditionals, my contention is that all of them are amenable to analysis in conformity with Ramsey test standards. I favor the Dudman approach to the grammatical distinction because (a) his arguments are compelling and (b) the distinction between the domain of application of informational-value-based Ramsey tests and the domain of application of consensus-based Ramsey tests fits Dudman's classification better than it does the indicative–subjunctive dichotomy. Conditionals are informational value based whereas hypotheticals are consensus based. Grammatical distinctions correlate well with methods of interpretation – a point Dudman himself seems to regard as a desideratum.

It should, perhaps, be mentioned in passing that (as Dudman points out) there are English "if" sentences that fall outside the hypothetical-conditional dichotomy and many of these are not amenable to Ramsey test treatment.

Consider, for example, a sentence like "If Anne smokes, she chokes." This is an example of what Dudman calls a "generalization." I suggest that it asserts that Anne has, at the present moment, a dispositional property – to wit, the disposition to choke upon smoking. Attributing such a disposition to Anne suggests that one accepts the dispositional claim into one's corpus. If so, adding the claim that Anne smokes on some occasion to the corpus as a supposition will bring in its wake the claim that Anne chokes on that occasion. In this sense, the dispositional claim *supports* the conditional "If Anne had smoked, she would have choked" or "If she smokes, she will choke." But conditionals ought not to be confused with the generalizations that support them. Disposition statements can belong to the contents of a corpus standing for an inquirer's state of full belief. They can be true or false.

When neither a disposition statement nor its negation is in a corpus, the agent is in suspense as to its truth. Beginning in the next chapter, I shall argue that conditionals cannot be elements of the corpus representing the state of full belief except in an "epiphenomenal" sense. They are neither true nor false. An agent cannot coherently be in suspense concerning them.

I contend that consensus-based judgments expressed by hypotheticals also lack truth values. Because controversy over the scope of the Ramsey test focuses on conditionals, in the subsequent discussion, I shall restrict attention exclusively to conditionals ignoring the special qualifications appropriate for hypotheticals.

And there is controversy. The received view (corrected for the common practice of grouping (1′) together with (3′) rather than with (2′)) is to provide conditionals with truth conditions according to a semantics pioneered by R. C. Stalnaker (1968, 1984) and D. Lewis (1973) or to provide a belief revision account based on such a semantics according to a method of transforming states of full belief called "imaging" (see Collins, 1991; Katsuno and Mendelzon, 1992). The controversy does not primarily concern the method of classifying "if" sentences in English. Even if one accepts the method of classification elaborated with painstaking care and cogency by Dudman, as Gibbard appears to do, the issue concerns the kind of conditional modal appraisal expressed by conditionals in Dudman's sense. In sections 3.1–3.6, the main issue leading to doubts concerning the viability of the use of informational-value-based Ramsey tests will be considered and, in 3.7, the alternative interpretation of conditional "if" sentences as expressing imaging conditionals will be examined.

3 Modality without Modal Ontology

3.1 Truth, Information, and Full Belief

When an inquirer seeks to improve his current state of full belief, the legitimacy of the alteration made depends on the aims of the inquirer. There are many kinds of aims inquirers might and do have in altering their full beliefs. These aims need not be economic, political, moral, or aesthetic. Cognitive aims may be pursued as well. The kind of cognitive aim that, in my opinion, does best in rationalizing scientific practice is one that seeks, on the one hand, to avoid error and, on the other, to obtain valuable information. Whether inquirers always seek error-free information or not need not concern us here. I rest content for the present with making the claim that agents can coherently pursue cognitive aims of this kind.

A consequence of this view is that states of full belief should be classifiable as error free or erroneous. Otherwise it makes little sense for an inquirer to seek to avoid error in changing his or her state of full belief. Likewise states of full belief should be classifiable as stronger or weaker; for those who seek valuable information should never prefer weaker states of full belief to stronger ones.

The two classifications are interrelated. If state 1 is stronger than state 2 and is error free, so is state 2. If state 2 is erroneous, so is state 1.

I have argued elsewhere (Levi, 1991, ch. 2) that the space K of potential states of full belief carrying information and being error free (true) or erroneous (false) should, at a minimum, be partially ordered by the "stronger than" or consequence relation in a manner satisfying the requirements of a Boolean algebra.

This thesis is intended to make a claim about the states of full belief that are accessible to the inquiring agents. It contends that one should be as liberal as possible in allowing conceptual accessibility to belief states. Here is a brief sketch of the thinking elaborated previously (Levi, 1991).

If K_1 and K_2 are potential states of full belief an inquirer could settle into, the inquirer should be able to shift to a belief state that reflects the "common ground" or shared assumptions of these two states. There may be good reasons why the inquirer should or should not do so. But the issue should not be settled in advance by denying accessibility to such a belief state. I suggest that this requirement is reflected in the condition that the "join" $K_1 \vee K_2$ be available to the inquirer.

Similarly, if an inquirer could be in belief state K, its complement ~K should also be a potential state of full belief.

In general, views that do not guarantee the availability of joins, complements, and meets of available belief states as well as a weakest belief state 1 that is error free and its complement the strongest or inconsistent belief state 0 that is erroneous place roadblocks in the path of inquiry. There may be belief states that inquirers should avoid. But that needs argument and should not be precluded in advance of inquiry.

The agent's state of full belief K at a given time is one of the elements of that algebra. In virtue of being in that belief state, the agent is committed to judging K to be error free as well as to recognizing his or her location in that state. The agent is also committed to judging all consequences of K (all elements in the filter generated by K) to be error free.

The agent is also committed to judging the complement of K and all potential states that are stronger than the complement as erroneous. The remaining potential states are left in suspense with respect to whether they are error free or erroneous. Such belief states are judged to be serious possibilities relative to K.

An inquiring agent concerned to avoid error should avoid deliberately changing from his or her current belief state K to a potential state currently judged false. The agent contemplating a shift from K to a state not judged false but where the falsity of the state is a serious possibility should take the risk of error to be incurred by such a shift seriously where such a risk is assessed by means of the credal probability that the new belief state is erroneous. Incurring such a risk may be justified if the informational value gained can compensate for the risk taken by making the change in belief state. All such cases are cases of expansion. The details of justified expansion have been discussed previously (Levi, 1967; 1980; 1984, ch. 5; 1991), and some of them will be reviewed in Chapter 6.

In contraction, the inquirer shifts from the current belief state to a weaker one. The weaker belief state belongs to the filter generated by K and, hence, is judged error free. No risk of error is incurred by contraction. In general, however, loss of informational value results. To incur such a loss requires some compensating incentive. Previously (Levi, 1980, ch. 3; 1991, ch. 4), two kinds of incentive have been described. Sometimes an inquirer has expanded inadvertently into the inconsistent belief state. Contraction is then coerced. Sometimes the inquirer contracts to give a hearing to a point of view that, even though judged certainly false, carries sufficient informational value to warrant converting it to a conjecture considered seriously possible.

My concern here is not to enter into the details of this approach to justification of belief change but merely to emphasize that in the first instance it is not sentences or other linguistic entities that carry truth value and informational value but potential states of full belief.

I have, to be sure, proposed partially representing states of full belief by corpora or deductively closed theories in a suitably regimented language $\underline{L}$. When a corpus is finitely axiomatizable, the state represented by the corpus can also be represented by a single sentence in $\underline{L}$ – namely, any sentence whose consequences constitute the corpus in question. Thus, if $\mathbf{K}'$ is represented by $\underline{K}'$, which consists of all logical consequences in $\underline{L}$ of h, and if the current state of full belief $\mathbf{K}$ is stronger than $\mathbf{K}'$ so that $\mathbf{K}'$ is judged certainly true, we may express that information by saying that the inquirer believes that h.

To say this does not mean that the agent stands in a certain attitudinal relation to a sentence type or token, to the meaning of the sentence (a proposition) or a set of possible worlds. Rather it partially locates the agent's current belief state in the algebra of potential belief states by indicating that the agent's current belief state is as strong as the belief state represented by the potential state of full belief that h or by the corpus $\underline{K}'$ consisting of h and all of its consequences. If we want to speak of full beliefs as having "objects," these objects are, strictly speaking, potential states of full belief or "doxastic" propositions as I sometimes call them (Levi, 1991, ch. 2).

A sentence h has a truth value if the logical consequences of h constitute a corpus or theory representing a potential state of full belief in the given language $\underline{L}$. h is true (false) if and only if the state of full belief represented is error free (erroneous). h is judged true (false) if and only if the state of full belief represented is judged error free (erroneous). Once a decision is reached as to which sentences in $\underline{L}$ represent potential belief states in the conceptual framework, one can then proceed to provide a Tarski-like truth definition for those sentences.

Inquirers concerned to avoid error are concerned, in the first instance, with the truth values of potential states of full belief. Interest in the truth values of sentences arises only insofar as such sentences represent potential states of full belief or, equivalently, determine potential corpora or theories representing potential states of full belief.

3.2   Are Judgments of Serious Possibility Truth Valued?

There are, as already emphasized, other attitudinal states besides states of full belief.

Consider first the state of judgments of serious possibility. Given the state of full belief **K**, the set of potential states of full belief judged seriously possible is uniquely determined. All the rest are judged impossible. Conversely given the partition of potential states of full belief into the possible and the impossible, **K** is uniquely determined. The attitudinal state constituted by judgments of serious possibility and impossibility can be said, therefore, to have a truth value – namely, the truth value of the potential state of full belief associated with it. In this sense, an inquirer who judges the doxastic proposition (i.e., potential state of full belief) that h to be seriously possible "accepts" the judgment if and only if the agent is in a state of full belief determining that judgment. But the doxastic objects judged seriously possible according to that state are not, in general, consequences of the agent's state of full belief and are not judged error free.

More crucially, the judgments of serious possibility themselves are not potential states of full belief or doxastic objects. When the inquiring agent expands his state of full belief, doxastic propositions initially judged seriously possible will be judged impossible. Suppose that initially the inquirer judges that h is a serious possibility and that this judgment is a doxastic proposition or state of full belief. If avoiding false judgments of serious possibility were a feature of the aim to avoid error, expansion would inevitably import error into the evolving doctrine. Indeed, all expansion would inevitably yield the inconsistent state of full belief. In contraction, a similar result would ensue. The inquirer's goals in expansion cannot include a concern to avoid false judgments of serious possibility. The agent makes such judgments and, in this sense "accepts" them, without full belief.

Saying this does not preclude the use, if desirable, of linguistic representations of judgments of serious possibility. Indeed, starting with a non-modal language $\underline{L}$ we can introduce a language $\underline{LP}$, which includes all sentences in $\underline{L}$ and sentences of the form $\Diamond$h and $\sim\Diamond$h as well as Boolean combinations of such sentences and sentences in $\underline{L}$. Given a corpus $\underline{K}$ in $\underline{L}$ representing the agent's state of full belief **K**, we can determine a theory or corpus $LP(\underline{K})$ in $\underline{LP}$ consisting of sentences in $\underline{K}$ together with judgments of serious possibility and impossibility according to $\underline{K}$ and logical consequences of these.

There can be no objection to constructing such a theory. If one likes to engage in such exercises, one can even construct a semantics supplying truth conditions of the modal sentences in such corpora. My contention is that when the inquirer is concerned to avoid error in changing his or her belief state, the inquirer takes for granted that the set $\underline{K}$ is free of false sentences. The inquirer concerned to avoid error should not deliberately shift from $\underline{K}$ to a corpus containing false sentences on the assumption that all elements of $\underline{K}$ are true. Such an inquirer can contract $\underline{K}$ and respect the concern to avoid error. The inquirer may consistently expand $\underline{K}$ and show such respect. But the inquirer will betray such respect if he or she deliberately shifts from $\underline{K}$ to a theory inconsistent with it.

The truth values of sentences in $LP(\underline{K})$ cannot coherently be of concern to the inquirer seeking to avoid deliberate importation of falsity or error into his belief state. $LP(\underline{K})$ can neither be expanded to a consistent corpus by adding consistent h nor can it be contracted to a consistent corpus by removing consistent h. The constitution of the nonmodal corpus $\underline{K}$ uniquely determines $LP(\underline{K})$. More crucially, for every h in nonmodal $\underline{L}$ either $\Diamond$h or ~$\Diamond$h is in $LP(\underline{K})$. In particular, if $\underline{K}$ contains neither h nor ~h so that both $\Diamond$h and $\Diamond$~h are in $LP(\underline{K})$, expansion of $\underline{K}$ by adding h (~h) leads to the removal of $\Diamond$~h ($\Diamond$h) and its replacement by ~$\Diamond$~h (~$\Diamond$h). Hence, $LP(\underline{K})$ cannot be expanded. It cannot be contracted either. If all the sentences in $LP(\underline{K})$ are judged true by the inquirer, changing this theory must perforce deliberately import error. An inquirer cannot coherently regard the modal sentences in LP to be truth value bearing and be concerned to avoid error in the sense that he or she is concerned to avoid having false sentences in the $LP(\underline{K})$ representing his modal and nonmodal belief state. An inquirer who does so will be prohibited from making any change in his or her belief state.

Someone who does not think avoidance of error in a sense that prohibits deliberate replacement of sentences the agent is certain are true by sentences the agent is certain are false to be a desideratum in inquiry need not be distressed by this observation and may be prepared to take modal judgments as carrying truth values. Once the importance of truth has been downgraded, one can afford to be more generous in deciding what is to carry truth value.

If avoidance of error is a desideratum in inquiry, sentences of the form "It is seriously possible that h" and "It is not seriously possible that h" should not be considered to carry truth values, and the judgments of serious possibility they represent should not carry truth values either.

3.3   Truth and Probability

States of credal probability judgment, as we have already seen, are not uniquely determined by the agent's state of full belief. They supplement the structure imposed on potential states of full belief by the current state of full belief by making fine-grained distinctions between serious possibilities with respect to credal probability.

Just as the language $\underline{L}$ can be extended to $\underline{LP}$ to represent judgments of serious possibility, it can be enriched by extending the language to include sentences reporting the credal or subjective probabilities of agents.

Can these sentences and the judgments of credal probability they represent be taken to be truth value bearing?

The argument of section 3.2 applies almost immediately to yield a negative answer for those who think avoidance of error is a desideratum in changing full beliefs. If the nonmodal nonprobabilistic corpus $\underline{K}$ expressible in $\underline{L}$ contains neither h nor ~h, then neither h nor ~h carries 0 credal probability in the sense in which 0 credal probability means absolutely certainly false as opposed to "almost" certainly so. If there were a truth of the matter concerning the probability judgments, the agent would be committed to denying that either h or ~h is absolutely certainly false. However, expansion by adding h (~h) would require deliberately betraying this commitment and deliberately importing error into the state of full belief.

Even so, someone might insist that as long as the issue is not one of changing the corpus $\underline{K}$ but changing the credal state relative to $\underline{K}$ so that the distinction between what is and what is not seriously possible remains constant, a fact of the matter is involved. Both h and ~h are counted as serious possibilities. The initial credal state assigns 0.1 to h and 0.9 to ~h. The credal probabilities might have been 0.9 and 0.1. Is there a question of truth or falsity involved here?

Again the answer is negative. If there were a truth of the matter, it should be open to the inquirer concerned to avoid deliberate importation of error to move to a position of suspense as to the truth of the the two credal probabilities:

(a) $p_1(h) = 0.1$ and $p_1(\sim h) = 0.9$.
(b) $p_2(h) = 0.9$ and $p_2(\sim h) = 0.1$.

But then the agent should be in a position of assigning numerically definite credal probabilities to (a) and (b).

It is impossible to do so, however, without either assigning probability 1 to p_1, assigning probability 1 to p_2, or assigning probability 1 to the judgment that the probability distribution is other than p_1 or p_2. The reason is that the credal probability for h will equal $xp_1 + (1 - x)p_2$ where x is the probability assigned to p_1.[1]

The initial credal state is represented by p_1. If credal states bear truth values, shifting to p_2 or to some other state p_i is to betray the injunction to deliberately avoid error.[2]

Thus, if credal probability judgments are truth value bearing and the inquirer is concerned to avoid deliberate importation of error, the inquirer cannot coherently alter his nonmodal, nonprobabilistic corpus or his credal state.

3.4   Conditional Judgments of Possibility

Remarks similar to those covering judgments of serious possibility and credal probability apply mutatis mutandis to attitudinal states constituted by judgments of value or of desire. These too lack truth values from the point of view of someone who is concerned to avoid error and thinks that both states of full belief and other attitudinal states are open to legitimate change. I will not elaborate on these points here but turn finally to the conditional judgments of serious possibility that are the product of suppositional reasoning.

According to the three versions of the Ramsey test introduced in section 2.2, given a nonmodal consistent corpus $\underline{K}$, either h > g is acceptable relative to $\underline{K}$ or ~(h > g) is acceptable relative to $\underline{K}$. h > g is the judgment that the falsity of g is not a serious possibility on the supposition that h. ~(h > g) is the judgment that the falsity of g is a serious possibility on the supposition that h.

Just as we can extend $\underline{L}$ to include judgments of serious possibility, credal probability, utility, and the like, so too we can encompass conditional judgments of serious possibility, credal probability, and so forth. Indeed, in section 3.6, we shall utilize such an extended language in a systematic manner. For the present, however, it is sufficient for us to notice that if the denizens of a language encompassing conditionals are truth value bearing, consistent expansion of a corpus becomes impossible as does contraction. The inquirer who seeks to avoid deliberately importing error into his or her belief state is driven to avoiding any change in belief state altogether.

From the perspective just sketched, sentences representing judgments of serious possibility, credal probability, value judgment, and conditional serious possibility should not be members of theories or corpora representing states of full belief.

Judgments of serious possibility, credal probability, value judgment, and conditional serious possibility are, therefore, neither true nor false in a sense of concern to the inquirer seeking to avoid error when engaged in modifying belief states. Evaluating such judgments with respect to serious possibility, credal probability, value judgment, or conditional possibility is incoherent. In particular, the probability of a conditional is not a conditional probability because conditionals cannot be assessed with respect to credal probability. And conditions for the acceptability of conditionals furnished by Ramsey tests are not conditions for fully believing such conditionals.

3.5 Reduction of Acceptance of Conditionals to Full Belief

I have just argued that the acceptability conditions for conditionals imposed by the Ramsey test – that is, the conditions for making conditional judgments of serious possibility – are not conditions for fully believing such conditionals. Gärdenfors (1988, pp. 164–5) has embraced a point of view contravening this claim. He endorses the following condition:

Reduction Condition for Conditionals:

$h > g$ ($\sim$($h > g$)) is acceptable relative to $\underline{K}$ if and only if
$h > g$ ($\sim$($h > g$)) is in $\underline{K}$.

I have already alluded to one serious objection to this requirement if the conditional judgments of serious possibility are derived from $\underline{K}$ via some version of a Ramsey test. Ramsey test conditionals cannot, in general, be considered immune to change. If they are represented by sentences in the agent's corpus, such sentences should be eligible for removal by contraction. And when they are not accepted in the agent's corpus, it should be possible to come to accept them by consistent expansion.

The reduction condition together with the Ramsey test precludes both contraction and consistent expansion (Levi, 1988; Rott, 1989). Whether one uses RTGL or RTR, this reduction condition implies that either $h > g$ or $\sim$($h > g$) must be in the corpus; for conditional on h, g must be judged seriously possible (according to the negative version of the given Ramsey test) or judged seriously impossible (according to the positive version). One cannot, therefore, remove $h > g$ from $\underline{K}$ without replacing it by its negation. More specifically, one cannot remove $T > h$ ($= \sim\diamondsuit\sim h$) without replacing it with its negation $\sim$($T > h$) ($= \diamondsuit\sim h$) and vice versa. This maneuver precludes contraction removing h from $\underline{K}$ that contains h, which requires that the contraction of $\underline{K}$ be a subset of $\underline{K}$. It also precludes consistent expansion by adding h to $\underline{K}$ that does not have h as member; for the expansion of $\underline{K}$ by adding h (i.e., the smallest corpus containing the logical consequences of $\underline{K}$ and h, which also satisfies RTGL or RTR and, hence, represents a potential state of full belief when reduction is adopted) must be the inconsistent corpus. The only kind of change allowed will be replacement of a claim initially fully believed to be true by one initially fully believed to be false. Anyone concerned to avoid error in changing his or her state of full belief should regard this result as untenable.

One might seek to finesse this result by following Gärdenfors's view (1988, sec.1.6) according to which avoidance of error is not a desideratum in belief change. According to this view, there is no need to object to the impossibility of coherently contracting or expanding. One can deliberately replace sure truth with sure error because avoidance of error does not matter.

For Gärdenfors, avoidance of error is not a matter for concern. But avoidance of inconsistency is. So one might retain the reduction condition by requiring that the test for the acceptability of conditionals requires that g be in some potential corpus that is a consistent transformation of $\underline{K}$ by adding h.

Suppose, however, that neither h nor ~h is in $\underline{K}$. Both ~(T > h) and ~(T > ~h) are acceptable by negative Ramsey tests. The condition of adequacy for Ramsey tests requires in that case that h > g be acceptable if and only if g is in the the expansion of $\underline{K}$. Moreover, to guarantee the coherent applicability of this condition of adequacy, the consistent expandability of $\underline{K}$ by adding h is required. Such expansion, however, requires adding T > h without removing ~(T > h). So consistent expandability fails. Hence, even those who are indifferent to whether error is avoided but care about avoiding inconsistency still have reason to worry.

Gärdenfors could reply that he does not endorse the negative version of RTGL and, hence, is not saddled with the result that either h > g or ~(h > g) must be accepted relative to a belief state. When neither h nor ~h is in $\underline{K}$, there is no requirement that ~(T > h) and ~(T > ~h) be in $\underline{K}$ as well. Neither may be present.

This cannot be satisfactory. Expand $\underline{K}$ to $\underline{K}'$ by adding ~(T > h). By consistent expandability, $\underline{K}'$ must be consistent. However, neither h nor ~h is in $\underline{K}'$. Hence, expanding $\underline{K}'$ by adding h should be consistent. But it cannot be; for by the positive Ramsey test, T > h will be in that expansion. Jettisoning the negative Ramsey test does not avoid the conflict with consistent expandability and, hence, with the coherent applicability of positive Ramsey tests.

Furthermore, jettisoning the negative Ramsey test is abandoning the use of the current belief state as a standard for serious possibility and suppositional belief states as conditional standards for serious possibility. Doing that deprives states of full belief of their main function in deliberation and inquiry without affording any benefit.

Gärdenfors does not appear to appreciate that Ramsey tests, including the positive RTGL he favors, presuppose consistent expandability. Consistent expandability is tantamount to (K*r4). To abandon it is to abandon a key presupposition of the coherent applicability of RTGL. Gärdenfors (1986) contends that the Ramsey test together with its reduction condition is inconsistent with the preservation condition or more specifically with that part of preservation expressed by (K*r4).

My disagreement with Gärdenfors does not concern the existence of an inconsistency. It concerns his vision of how one should retreat from inconsistency. He holds the reduction condition (or his condition O) sacrosanct and contends that we have a choice of retreating from the positive Ramsey test (whether as RTGL or RTR) or from preservation. I have argued that retreating from preservation by abandoning (K*r4) is retreating from the positive Ramsey test. Given that one is going to insist on condition O or the reduction condition, the only option is to abandon the Ramsey test. One can, to be sure, abandon it by abandoning preservation or abandon it while retaining preservation; but, either way, positive informational-value-based Ramsey tests satisfying the condition of adequacy endorsed by Ramsey are given up.

Gärdenfors thinks it desirable to retain the reduction condition in order to account for iterated conditionals. No doubt, natural languages do have the resources to construct "if" sentences nested within other "if" sentences. When such sentences are used to manifest attitudes, it is not clear that they express conditional judgments of serious possibility requiring conditional sentences to be elements of potential corpora representing conditional doxastic propositions. In section 3.7, I shall argue that they do not. If this view is right, the chief motive Gärdenfors offers for embracing the reduction condition will be removed.

There may be other motives than Gärdenfors's for retaining the reduction condition and abandoning Ramsey test criteria for the acceptability of conditionals. Advocates of imaging tests for the acceptability of conditionals abandon Ramsey tests while insisting on the reduction condition. In section 3.8, I shall summarize briefly some of the main features of imaging that allow for those who favor this approach to embrace the reduction condition. My worry about such approaches is that they lack clear application either to practical deliberation or scientific inquiry. Ramsey test conditionals, on the other hand, do have clear and useful applications.

Sections 3.1–3.4 argued against the reduction condition on the grounds that judgments of serious possibility should not be assessed for truth or falsity in the sense relevant to the injunction to avoid error in inquiry. This argument was supplemented in this section with the observation that the reduction condition is incompatible with the satisfaction of the condition of adequacy for Ramsey tests and, hence, with the coherent applicability of either RTGL or RTR. Gärdenfors and others do not appreciate that (K*r4) is presupposed as a condition of adequacy vital to the applicability of Ramsey test conditionals. For this reason, section 3.6 rehearses the argument in greater detail.

3.6   Trivializing the Ramsey Test

In section 3.5, the bare bones of the conflict between RTGL or RTR and the reduction condition were brought out. Since neither Gärdenfors nor others who have worked on this subject have been prepared to recognize the issue in this light, at the cost of considerable repetition, the matter will be rehearsed once more in some greater detail.

The main point that needs to be established is that adoption of the reduction condition undermines the Ramsey test criterion precisely for the case where neither h nor ~h is in $\underline{K}$. If, for every potential $\underline{K}$ in K to which the revision operator * ingredient in the Ramsey test applies and for every sentence h in $\underline{L}$, either h or ~h is in $\underline{K}$, the "belief revision system" consisting of K and * has been rendered *weakly trivial* in the sense of Rott (1989, p. 93). If we are going to permit belief revision systems to be nontrivial so that agents can be in doubt concerning some substantive claims, the conflict between the reduction condition and Ramsey test becomes inescapable.

Consider a language $\underline{L}^1$ with an underlying truth functional or first order logic but with no modal or conditional connectives. $\underline{L}^2$ contains all sentences in $\underline{L}^1$ together with sentences of the form h > g where h and g are in $\underline{L}$ and all truth functional compounds of these. In general, $\underline{L}^{i+1}$ is the language consisting of all sentences in $\underline{L}^i$, sentences of the form h > g where h and g are in $\underline{L}^i$, and truth functional compounds of these. Let $\underline{ML}$ be the language consisting of all sentences in the union of the sets of sentences in the languages $\underline{L}^i$.

K^1 is the set of corpora or theories closed under consequence in $\underline{L}^1$. K^i is the set of corpora or theories closed under consequence in $\underline{L}^i$ and MK is the set of corpora or theories closed under consequence in $\underline{ML}$.

Consider an informational-value-based Ramsey test – in particular, either RTGL or RTR. For $\underline{K}^1$ in K^1 and sentences in $\underline{L}^1$ (i.e., sentences in the nonmodal language), let $\mathbf{RT}(\underline{K}^1)$ be the logical consequences of the set of sentences acceptable according to that Ramsey test relative to $\underline{K}^1$ together with the sentences in $\underline{K}^1$. The result is a corpus or theory in K^2 but not in K^1.

If the sentences in $\mathbf{RT}^1(\underline{K}^1)$ that are not in $\underline{L}^1$ are not merely sentences acceptable at $\underline{K}^1$ but fully believed as the reduction condition requires, then the corpus of full belief on which the Ramsey test operates is not $\underline{K}^1$ but includes all the sentences in $\mathbf{RT}^1(\underline{K}^1)$.

Consider then the set $\mathbf{RT}^2(\mathbf{RT}^1(\underline{K}^1))$ of sentences in $\underline{L}^3$ that is the deductive closure of $\mathbf{RT}^1(\underline{K}^1)$ and the acceptable sentences in $\underline{L}^3$ according to the Ramsey test. The reduction condition requires this theory also to be contained in the corpus representing the belief set. Indeed, the full reduction condition stipulates that the corpus $\underline{MK}$ of full belief is the union of all corpora obtained by reiterating application of the Ramsey test in this way.

We can also think of partial reduction conditions to various levels. The 0th level is where $\underline{K}$ is simply $\underline{K}^1$ and no sentence containing a conditional connective appears in a corpus representing the state of full belief. ith-level reduction stipulates that corpora obtainable from $\underline{K}^1$ in K^1 by i iterations of the Ramsey test represent the state of full belief *when restricted to the members of the set belonging to* $\underline{L}^1$. Thus, first level reduction declares the set of sentences in the intersection of $\mathbf{RT}^1(\underline{K}^1)$ and $\underline{L}^1$ to represent the corpus of full belief.

Consider the first-level partial reduction condition according to which the theories representing potential states of full belief are expressible in $\underline{L}^2$ but not in $\underline{L}^1$. Start with either of the informational-value-based Ramsey tests. Let $\underline{K}^2$ be the set of sentences in the language of noniterated conditionals and modality representing the state of full belief. Consider then $\mathbf{RT}^2(\underline{K}^2)$. $\underline{K}^2$ must be identical with the intersection of this corpus with $\underline{L}^2$.

Consider then the intersection $\underline{K}^1$ of $\underline{K}^2$ with $\underline{L}^1$. This corpus must itself be the nonmodal part of the corpus of full belief and should be identical with the intersection of $\mathbf{RT}^1(\underline{K}^1)$ and $\underline{L}^1$.

Hence, if we are given the Ramsey test and the corpus $\underline{K}^1$, the corpus $\underline{K}^2$ representable in $\underline{L}^2$ is uniquely determined and vice versa. Both corpora must represent the same state of full belief. We may reasonably ask in the light of this whether the first-level reduction condition raises any interesting question. More generally, what is at stake in defending or rejecting ith level reduction or full reduction?

The answer is that partial reduction, to various degrees, and full reduction impose restrictions on what constitutes a potential corpus or a potential state of full belief. Without reduction conditions (or, equivalently, with 0-level partial reduction), nothing prevents an inquiring agent from suspending judgment between two extralogical, and nonmodal hypotheses. With reduction, suspension of judgment is prohibited in such cases. One must be opinionated concerning any logically consistent, extralogical, and nonmodal claim.

Suppose that the agent's corpus in $\underline{K}^2$ contains neither h nor ~h. The negative Ramsey test requires that $\sim(T > h)$ and $\sim(T > \sim h)$ will be in $\mathbf{RT}^1(\underline{K}^1)$. This corpus should be identical with the intersection of $\underline{L}^2$ and $\mathbf{RT}^2(\underline{K}^2)$ which, in turn, must be identical with $\underline{K}^2$ itself.

No difficulty arises if the first-level partial reduction condition does not obtain. The expansion of $\underline{K}^1$ by adding h (or ~h) in order to determine the acceptability of a conditional h > g or its negation can be done consistently as the coherent applicability of the Ramsey test requires. If first-level partial reduction does obtain, the Ramsey test for h > g must consider expanding $\underline{K}^2$. Consequently, $\sim(T > h)$ remains in the expanded corpus and the Ramsey test requires that $T > h$ also be present. $(\underline{K}^2)_h^+$ is inconsistent.

One can escape from such embarrassment by prohibiting the agent from suspending judgment between h and ~h. Weak triviality then sets in.

Relief from weak triviality might be sought by denying that when the agent is in suspense between h and ~h, $\underline{K}_h^* = \underline{K}_h^+$. To do this is to reject the part of preservation corresponding to (K*r4). This strategy *abandons*

Ramsey test conditionals. $K*r4$ is a condition of adequacy on Ramsey tests for conditionals in the sense relevant to practical deliberation. In "practical" applications, the belief revision that takes place in open situations should be an expansion.

Consequently, either (1) there is no corpus in $\underline{L}^2$ containing neither h nor ~h for some extralogical h in $\underline{L}^1$ so that weak triviality applies, (2) the condition of adequacy for Ramsey tests fails, or (3) the partial reduction condition and a fortiori the full reduction condition fail.[3]

The key step in this argument is the point that use of the Ramsey test requires that for every $\underline{K}$ in K either $T > h$ is acceptable relative to $\underline{K}$ or $\sim(T > h)$ is acceptable relative to $\underline{K}$. This implication cannot be removed without abandoning the idea that a potential corpus representing a potential state of full belief is a standard for serious possibility distinguishing what is seriously possible from what is not. Moreover, for any consistent h and g and consistent $\underline{K}$, we should expect that either $h > g$ is acceptable relative to $\underline{K}$ or $\sim(h > g)$ is acceptable. The reason is that conditionals represent conditional judgments of serious possibility. To abandon these ideas is to abandon the raison d'être of states of full belief and supposition for the sake of the argument in inquiry and deliberation. Ramsey tests insure that these functions are preserved. But it is precisely these functions that get us into trouble with the partial reduction condition.

It is fairly obvious that shifting to the full reduction condition or partial reduction conditions at higher levels will not alleviate the situation with respect to Ramsey tests. We have a choice: either to abandon Ramsey tests for the acceptability of conditionals or abandon partial or full reduction conditions.

Peter Gärdenfors (1986; 1988, Secs. 7.4–7.7) does not see matters in this way. He shows that the positive Ramsey test clashes with (i) condition 0 (the reduction condition); (ii) nontriviality of belief revision in the sense that the domain of potential corpora on which it operates includes a corpus where the agent suspends judgment between three pairwise propositions; (iii) $(K*4)$ or preservation (which says that if g is in $\underline{K}$ and ~h is not, then g is in $\underline{K}^*_h$), $(K*2)$, and the portion of $(K*5)$ requiring that revisions, by adding logically consistent sentences, should be consistent. He holds the reduction condition sacrosanct and rightly finds nontriviality, $(K*2)$, and $(K*5)$ to be noncontroversial. So he ends up seeing the result as posing a tension between the positive Ramsey test and $(K*4)$ or, more specifically, $(K*r4)$.[4]

As mentioned in section 3.2, Gärdenfors fails to take into account the fact that the Ramsey test *presupposes* $K*r4$ so that giving $(K*r4)$ up cannot save the positive Ramsey test. He suggests to the contrary that the positive RTGL can be saved by abandoning $(K*4)$ or the preservation condition through jettisoning $(K*r4)$ while retaining $(K*4w)$ = weak preservation.

But no test can count as a Ramsey test in Ramsey's sense unless it is coherently applicable in contexts where the inquirer is in doubt with respect to the truth of extralogical supposition h where h contains no modal connectives. Ramsey explicitly formulated his test for such cases so that revisions are expansions. But if we set aside the point that Ramsey tests so construed are more in keeping with the ideas of the historical Ramsey, there are good reasons for being interested in conditionals whose acceptability is controlled by such tests no matter what their name.

Ramsey test conditionals are of interest because they model quite plausibly those conditionals relevant to the deliberating agent seeking to make determinations of what would happen if he were to make such and such a choice. Moreover, as we have seen, conditionals furnished with Ramsey tests for acceptability free us from having to take unconditional and conditional modal judgments into account when examining changes in belief states through inquiry. Given the changes in the nonmodal corpora representing belief states, changes in judgments of conditional and unconditional serious possibility are determined.

As we shall see, imaging criteria for the acceptability of conditionals do not, in general, agree with Ramsey's proposal in the core cases of open conditionals. Whatever one's view of such criteria might be, the clarity of the contrast between Ramsey tests and imaging tests would be obscured considerably if we took the position that when an inquirer is in suspense between h and ~h, he cannot consistently fix on one of these answers without lapsing into contradiction.

We should, therefore, recognize that the conflict generated by the argument offered here and the different argument advanced by Gärdenfors (1986; 1988, pp. 158ff.) is between the Ramsey test and the reduction condition and not between the Ramsey test and that portion of (K*4) equivalent according to the core requirement for Ramsey tests to consistent expandability.

Gärdenfors claims (1988, pp. 158–9) that if we endorse the reduction condition, (K*8) cannot be satisfied together with the positive Ramsey test (whether it is PRTGL or PRTR) even if the corresponding negative Ramsey test is not assumed. Gärdenfors (1988, pp. 154–6) distinguishes between ~(h > g) and the "might" conditional "if h were the case, g might not be the case." He contends that when g is not in $\underline{K}^*_h$, the "might" conditional is acceptable but denies that it is equivalent to ~(h > g). And he defends proceeding in this manner as a means of allowing the positive Ramsey test to apply coherently to corpora in MK by insuring satisfaction of the consistent expandability condition. This is a somewhat strange argument since, in the final analysis Gärdenfors wants to show that the positive Ramsey test is in tension with consistent expandability.

Gärdenfors's approach implies that when an inquirer is in suspense between h and ~h, $\Diamond$h and $\Diamond$~h are in $\underline{K}$. But even if $\Diamond$~h (= if T, then ~h might be true) is not equivalent to ~(T > h), it implies it. It would be incoherent to judge ~h a serious possibility and yet refuse to reject T > h. So the negative Ramsey test as a rule for "might" conditionals is sufficient to secure the acceptability of both ~(T > h) and ~(T > ~h) in the case where the inquirer is in suspense between h and not h. In spite of what Gärdenfors claims, the consistent expandability condition is violated.

To avoid this result, one might consider abandoning the negative Ramsey test as a criterion for the acceptability of "might" conditionals. This alone cannot save the consistent expandability condition. One needs to *prohibit* the acceptance of ~(T >h) and ~(T >~h) in $\mathbf{RT}(\underline{K}^1)$ in the case where neither h nor ~h is in $\underline{K}$. And this requires prohibiting acceptance of both $\Diamond$h and $\Diamond$~h in such cases. But the corpus $\underline{K}$ is intended to serve as the standard for serious possibility distinguishing between hypotheses eligible for assignment of positive subjective or credal probability (the seriously possible hypotheses) and hypotheses that are ineligible (the impossible). Abandoning the negative Ramsey test in this case prevents the corpus $\underline{K}$ from rendering this service. This untoward consequence is made worse by prohibiting the judgment that h and ~h are both seriously possible in these cases.

The dialectic can be reiterated for the case where the negative Ramsey test delivers the verdict relative to $\underline{K}$ that both ~(h > g) and ~(h > ~g) are acceptable in $\mathbf{RT}(\underline{K}^1)$. Gärdenfors prohibits the use of the test as securing the acceptability of these conditionals but endorses the use of the test to secure the acceptability of "If h, then maybe ~g" and "If h, then maybe g." But the "might" conditionals presuppose the negations of the corresponding "would not" conditionals. $(\underline{K}^1)^+_h$ cannot be consistently expanded by adding g or by adding ~g.[5]

But even if we rest content with the positive Ramsey test, Gärdenfors showed that trouble arises when this test is applied to potential corpora in K consisting of all theories in $\underline{L}^2$ derivable from theories in K^1 with the aid of the positive but not the negative Ramsey test.[5]

Gärdenfors thinks that a champion of Ramsey tests could avoid trivializing his or her account of belief change by abandoning that part of K*4 equivalent to consistent expandability and retaining K*4w that endorses recovery.

This conclusion gets things backward. If RTR is adopted rather than RTGL, as I now think should be done, (K*4w) should be abandoned as a condition on revision because of the dubiety of recovery as a condition on admissible contraction. (K*r4), on the other hand, should be retained because RTR like RTGL presupposes and should presuppose it.

In all fairness to Gärdenfors, he does not think that his efforts to rescue the positive Ramsey test are satisfactory and that in the end the applicability of the Ramsey test to theories belonging to *MK* may have to be abandoned for the sake of preservation.

When he is in this frame of mind, Gärdenfors agrees with me that neither the positive nor the negative Ramsey test should be applied to potential corpora in K^i (i > 1) or *MK*. He also does not question that both the positive and negative Ramsey tests are applicable to potential corpora in K^1 without triviality. But he thinks this strategy is unsatisfactory because, so he alleges, an account of iterated conditionals is then unavailable. I disagree; but the more serious disagreement concerns his conviction that one can retain Ramsey tests while giving up preservation or (K*r4).

I do not wish to become embroiled in a logomachy over the correct use of the expression "Ramsey test." Ever since Stalnaker took to characterizing Ramsey test conditionals in terms of minimal revisions, many writers have understood minimal revision so broadly that it includes the product of imaging as a species of minimal revision. (Recent examples include Collins, 1991, and Katsuno and Mendelzon 1992.) Construed so broadly, Ramsey tests need not presuppose the condition of adequacy we have required for Ramsey test conditionals stipulating that, for open conditionals, revisions used in Ramsey tests should be expansions. However, according to the understanding of Ramsey tests deriving from Ramsey himself and worthy of serious consideration because of its relation among other things to decision theory, Ramsey tests become incoherent if the reduction postulate is endorsed. They do so precisely because the condition of adequacy is violated.

Conditional assessments of serious possibility invite us to transform the current state of full belief by stipulating for the sake of the argument what the input sentences to be added or removed are to be. In inquiry (in contrast to fantasy), we require the addition or removal of inputs to be legitimate.

Perhaps we should think of Ramsey tests meeting the condition of adequacy as appropriate for assessments of serious possibility relative to legitimate addition or removal of inputs and we should expect a criterion satisfying reduction to apply to hypothetical changes. (See Rott, 1989, pp. 105–6; and Collins, 1991.) If the intended contrast is between two kinds of conditional judgments of serious possibility to be made when the content h of the "if" clause is open relative to the initial belief state, I do not see the difference. The account of conditional judgments of serious possibility ought to be the same for changes "for the sake of the argument" and for changes made in earnest – modulo the differences in the way inputs are introduced. They cannot be the same if the reduction condition is endorsed for conditional judgments of serious possibility for the sake of the argument and follows the requirements of the Ramsey test for genuine belief changes.

Thus, when it comes to giving an account of conditional judgments of serious possibility for the open case, the issue cannot be ducked by consigning one approach (the Ramsey test) to the domain of genuine belief changes and another approach (favoring a reduction condition) to hypothetical belief changes. One approach needs to be adopted for both domains. Should it be the Ramsey test approach? Or should an approach satisfying the reduction condition be adopted?

Rejecting the reduction condition or various partial reduction conditions does not imply that we must be deprived of the benefit of representing the set of conditionals acceptable relative to $\underline{K}^1$ by $\mathbf{RT}^1(\underline{K}^1)$. $\underline{K}^1$ uniquely determines $\mathbf{RT}^1(\underline{K}^1)$ and vice versa. In providing a linguistic representation of the state of full belief, we may use either theory. However, when judging whether the belief state is error free or not, only the truth values of elements of $\underline{K}^1$ ought to be of concern.

According to the view I favor, justification of changes in belief state should show that the change optimizes expected yield of valuable and error-free information as judged by the inquirer relative to $\underline{K}$. Combine that with the assumption that states of full belief are standards for serious possibility. If judgments of serious possibility and impossibility are included in the belief state, there can be no alteration of state of full belief that does not replace a claim judged certainly true by one judged certainly false relative to the initial state. To avoid this unfortunate result, either we must give up the concern to avoid importing falsity into the belief state, deny the role of the belief state as a standard for serious possibility, or deny that judgments of serious possibility are doxastic propositions carrying truth values. I favor the third option. And if that option is followed, there is no incentive for regarding conditional judgments of serious possibility to be doxastic propositions.

Hence, the belief states subject to such change are represented by theories in K^1 that contain no modal judgments. The concern to avoid error relevant to inquiry is represented as a concern to avoid error in the contents of the corpus. Shifting to a corpus $\underline{K}^1$ consisting exclusively of true sentences in nonmodal $\underline{L}^1$ represents moving to an error-free potential state of full belief. Doing so exhausts the concern to avoid error. There is no further concern to avoid error in the judgments of serious possibility relative to $\underline{K}^1$ contained in the corpus $\underline{K}^2$ in $\underline{L}^2$. Even if one can construct a semantics for $\underline{L}^2$ (or $\underline{ML}$), the risk of error involved in shifting from an initial corpus to $\underline{K}^2$ (or $\underline{MK}$ generated by $\underline{K}^1$) just is the risk of error incurred in shifting to $\underline{K}^1$.

This is the sense in which both unconditional and conditional judgments of possibility and impossibility lack truth values. Whether or not such judgments are assigned truth values according to some modal semantics does not matter in inquiry concerned to change beliefs in a manner that seeks to promote error-free informational value.[6]

Thus, there are two kinds of consideration arguing against the reduction condition:

(1) Giving up the reduction condition allows Ramsey tests to satisfy coherently the condition of adequacy.

(2) Giving up the reduction condition allows the Ramsey test (which implies that the corpus is a standard for serious possibility) and the thesis that in inquiry one ought to seek error-free information to be endorsed coherently.

Authors like Gärdenfors (1988, pp. 164–5) and S. O. Hansson (1992, p. 15) insist that my preoccupation with avoidance of error in the revision of corpora in $\underline{L}$ is irrelevant. Preoccupation with distinguishing the domain of the truth valued from the domain of the nontruth valued is pointless. This view discounts avoidance of error as a desideratum in inquiry. Insofar as there is a disagreement between Gärdenfors, Hansson, and myself on this score, it concerns the aims of inquiry. But surely when discussing belief change, the topic of what the aims of such change are is not irrelevant. Gärdenfors and Hansson may disagree with my view of the aims of inquiry concerned with belief change. Given their view that avoidance of error is not a desideratum in inquiry, my preoccupation with truth value *is* irrelevant. But their view of the aims of inquiry is precisely the issue under dispute.

Instrumentalists disavow a concern with avoidance of error at least with respect to certain classes of beliefs. Sometimes they are happy to proceed with these beliefs "as if" they were realists. Sometimes they are not. Thus, both Hansson and Gärdenfors agree that the reduction condition is in tension with the Ramsey test. Gärdenfors is prepared to sacrifice the Ramsey test to preserve the reduction condition. As we shall see, if Gärdenfor's strategy is adopted, it is possible to resort to alternative criteria for acceptability such as imaging according to which a corpus in K need not be uniquely determined by its nonmodal part in K^1. Such views leave us free to take avoidance of error in conditional and modal judgments as desiderata in inquiry, as good modal realists would do. Gärdenfors may disavow modal realism; but his position is a reasonable facsimile of a modal realist view.

Hansson, though concerned to give an account of the acceptability of iterated conditionals, is also anxious to preserve the Ramsey test and seems prepared to do so by giving up reduction. I take it, he is persuaded by the first kind of consideration I mentioned earlier that reduction should go. His view is a reasonable facsimile of an antimodal realist view.

Gärdenfors's main reason for insisting on reduction seems to be his concern for giving an account of iterated or nested conditionals. I am not myself convinced that the concern is urgent. Nonetheless, the topic will be discussed in the next section.

3.7   Iteration

Gärdenfors defends his adherence to the reduction condition on the grounds that acceptability conditions for iterated conditionals expressible in ML are needed to accommodate presystematic linguistic practice permitting the use of iterated conditionals.

   "If" sentences undoubtedly do appear in English and some other natural languages where either the "if" clause or the "then" clause contains "if" sentences. And the resources are available for iterating the constructions ad nauseam. It is desirable, therefore, to identify the attitudes the use of such iterated "if" sentences are intended to reveal or communicate.

   I have suggested that "if" sentences that Dudman denominates "conditionals" express the kind of suppositional reasoning characterized by informational-value-based Ramsey test conditionals. But that applies only to Dudman conditionals that lack iteration. I have also suggested elsewhere (Levi, 1988) that the occurrence of iterated Dudman conditionals in English is restricted. More to the point, when such conditionals are deployed, I denied that their acceptability required the application of a Ramsey test subject to a partial or full reduction condition (Gärdenfors's condition 0).

   I also registered skepticism as to the feasibility of providing a systematic account of the acceptability conditions for the conditional judgments expressed by such iterated Dudman conditionals for arbitrary iterations. This skepticism was excessive. Thanks to the prompting of Hansson (1992), I now think the ideas I advanced earlier (Levi, 1988) can be captured in a more systematic manner. If this is right, it is feasible to provide acceptability conditions for iterated conditionals without endorsing either a full or partial reduction condition.

   To be sure, no Ramsey test can yield acceptability conditions for iterated conditionals. Applying a Ramsey test to iterated conditionals requires adoption of the reduction condition. Since, however, the Ramsey test presupposes the consistent expandability condition and this condition conflicts with the reduction condition and, indeed, with the minimal partial reduction condition we considered in the previous section, the net result is incoherent.

   Ramsey tests allow for judgments of serious possibility and impossibility relative to transformations of the agent's current nonmodal corpus belonging to K^1. The transformations are of a special kind. They add suppositions endorsed for the sake of the argument to modifications of the current corpus K. Such modifications "prepare" K by removing items standing in the way of consistently expanding by adding the supposition in the case of RTGL or precluding adding the supposition to a corpus that holds the supposition in suspense in the case of RTR. The preparation is a contraction of $\underline{K}$ that is "admissible" in the sense that it minimizes loss of damped informational value. Sometimes, of course, $\underline{K}$ needs no such massaging by contraction. The supposition can be added by expansion. Thus, $\underline{K}$ is either transformed directly or by expanding an admissible contraction. The acceptability conditions are intended to express the results of suppositional reasoning – that is, the judgments of serious possibility relative to the addition of the supposition to the prepared corpus.

As expressions of suppositional reasoning, Ramsey test conditionals are limited to supposing that nonmodal judgments are true. And the modal judgments based on the suppositions are judgments of serious possibility.

The second restriction can be lifted at least partially with little difficulty. Given the current corpus, judgments of conditional serious possibility can be made. These judgments are tantamount to judgments of serious possibility based on suppositions made for the sake of the argument. There seems to be no difficulty in allowing judgments of conditional serious possibility to be made based on the same suppositions. That is to say, there seems to be no obstacle to providing acceptability conditions for h > (f > g) or h > ~(f > g) relative to $\underline{K}$. One can identify $\underline{K}^{*r}_{h}$ (or $\underline{K}^{*}_{h}$) and determine whether f > g or ~(f > g) is acceptable relative to the revision. Nor is there any obstacle to allowing for cases where the consequent g of the nested conditional is a conditional. Iteration of this limited type can be carried on indefinitely.

Furthermore, there is some small advantage in giving an account of such cases. In ordinary language one can say things like: "If Robert Kennedy had not been assassinated, then, conditional on Kennedy beating Hubert Humphrey for the Democratic nomination, Nixon would not have won the 1968 presidential election."

The test for the acceptability of these conditionals is not strictly speaking according to a Ramsey test. For the "consequent" is not representable as a member of a corpus of full belief. It is a judgment of conditional serious possibility or impossibility relative to the suppositional corpus, but is not a member of that corpus. No reduction condition partial or full, is assumed.

Yet, the proposed test remains in the spirit of the Ramsey test. It construes such conditionals as expressions of suppositional reasoning. The inquirer adopts the nonmodal supposition h for the sake of the argument. Instead of determining whether some nonmodal proposition k or its negation is in the revision of his or her corpus so that k may be judged possible or impossible as the case may be, the concern is to make judgments of conditional possibility and impossibility relative to the suppositional corpus. Such judgments require further transformations of the suppositional corpus; but all the transformations involved are transformations of potential corpora in the nonmodal language $\underline{L}^{1}$.

The iterated "if" sentences in English meeting Dudman's requirements for conditionals that trip relatively lightly off the tongue appear to nest "if" sentences in the "then" clauses but not in the "if" clauses. It may be enough, therefore, to provide an account of such iterations without reduction to respond to Gärdenfors worry about iteration.

Even so, it may be worthwhile attending to nesting in "if" clauses. This calls for providing acceptability conditions of $(h > f) > g$ and $\sim(h > f) > g$ where h, f, and g are sentences in nonmodal $\underline{L}^1$.

We cannot treat $h > f$ and $\sim(h > f)$ as suppositions; for what is supposed is supposed true and should be incorporated into a belief state. That implies commitment to at least a partial reduction condition.

Supposing a proposition true for the sake of the argument, however, is but a species of a broader category of fantasizing. Anytime the current belief state is modified for the sake of the argument, hypothetical thinking is undertaken. One might contemplate characterizing such fantasizing by stipulating that one shift is to be implemented from the current corpus $\underline{K}$ to one relative to which $h > f$ ($\sim(h > f)$) is acceptable according to the Ramsey test.

Doing this does not require somehow adding $h > f$ or its negation to a suitably prepared modification of the current corpus.

Consider, for example, the task of "supposing" that $h > f$ is acceptable. Either $h > f$ is already acceptable relative to $\underline{K}$ according to the Ramsey test of choice or its negation is.

Case 1: $h > f$ is acceptable relative to $\underline{K}$. $(h > f) > g$ is acceptable at $\underline{K}$ if and only if g is in $\underline{K}$ – provided no conditionals are embedded in g. Otherwise, the negation is.

Case 2: $\sim(h > f)$ is acceptable relative to $\underline{K}$. There are three cases to consider:

    (a) Both $h \supset f$ and $h \supset \sim f$ are in $\underline{K}$ and the *M*-function is such that removing $\sim h$ (which must be in $\underline{K}$) favors removing the first of these material conditionals. In this case, $\underline{K}$ should be contracted with a minimum loss of information by removing $h \supset \sim f$.

        (i) If $h \supset f$ is retained in the admissible contraction, that contraction is the relevant suppositional corpus and $(h > f) > g$ is acceptable relative to $\underline{K}$ if and only if g is in this contraction.

        (ii) If $h \supset f$ is removed in the admissible contraction removing $h \supset \sim f$, expand the contraction by adding $h \supset f$ and use that corpus.

    (b) Neither $h \supset f$ nor $h \supset \sim f$ is in $\underline{K}$. Expand by adding the former.

    (c) $\sim[h \supset f]$ is in $\underline{K}$. Remove it with minimal loss of damped informational value and then expand by adding $h \supset f$.

One of these recipes can be used to determine the transformation of $\underline{K}$ used in appraising the acceptability of $(h > f) > g$ relative to $\underline{K}$. However, it is worth mentioning that such recipes may carry fantasy to very dramatic lengths. Consider case (c). Let h state that Levi throws a baseball and f state that it travels faster than the speed of light. To remove $\sim$[h $\supset$ f] from $\underline{K}$, a very substantial portion of physics needs to be given up in a manner minimizing loss of informational value. The fantasy is so extreme that even a sophisticated physicist and mathematician may need to think carefully before reasoning hypothetically from a corpus relative to which h > f is acceptable.

One might think of mitigating the problem by allowing the agent to change his or her assessment of damped informational value (the M-function and basic partition V as in section. 2.1) in the course of fantasizing. In my judgment, however, any such change should be expressed as part of the supposition. Otherwise in suppositional reasoning, assessments of informational value (and, hence, of entrenchment or incorrigibility) are understood to be held fixed relative to the initial corpus $\underline{K}$.[7]

Consider now $\sim$(h > f) > g.

Case 1: $\sim$(h > f) is acceptable relative to $\underline{K}$. $\sim$(h > f) > g if and only if g is in $\underline{K}$.

Case 2: (h > f) is acceptable relative to $\underline{K}$. Contract by removing h $\supset$ f.

Again in many cases, implementing the instruction in case 2 might require extreme fantasy.

Observe that if h = T, h > f = T > f = $\sim\Diamond\sim$f, $\sim$(h > f) = $\sim$(T > f) = $\Diamond\sim$f.

The principles proposed here imply that $\sim\Diamond\sim$f > g is acceptable relative to $\underline{K}$ if and only if g is in $\underline{K}^*_f$. Notice that AGM* and not Ramsey revision *r is to be used even if one endorses the RTR for determining the acceptability of f > g. The reason is that either $\Diamond\sim$f or $\sim\Diamond\sim$f is acceptable relative to $\underline{K}$. Accepting neither is not optional.

Similarly $\Diamond\sim$f > g is acceptable relative to $\underline{K}$ if and only if g is in the contraction of $\underline{K}$ removing f. This corresponds to what Gärdenfors (1988, pp. 151–2) calls an "even if" conditional.[8]

Once one has contemplated fantasizing in a manner that is controlled by determining which conditionals or negations of conditionals are acceptable in the fantasized belief state, the possibility of providing for the acceptability of nested conditionals of arbitrary complexity opens up.

Can this modest proposal for a single iteration be extended to cover multiple iterations? The proposal, as it stands is well defined for arbitrarily many iterations. Whether it conforms to presystematic judgment, however, is unclear – partially because presystematic judgment is itself not clear concerning multiple iterations.

S. O. Hansson (1992) appreciates the nature of the problem of providing a systematic extension of the Ramsey test criterion to cover all forms of iteration and has prompted me to rethink my own view of the subject. To address the problem, Hansson abandons the requirement that the acceptability of conditionals be determined by the corpus or theory closed under deduction in $\underline{L}^1$ and the information-determining probability M defined over the domain K of such theories. He substitutes for the corpus $\underline{K}^1$ a basis for the corpus – that is, a set $\underline{B}^1$ of sentences not necessarily closed under deduction whose closure is identical with $\underline{K}^1$. Revisions (expansions of contractions) are restricted to expansions of contractions of $\underline{B}^1$. The idea is to choose an optimal contraction from among those allowed by the base $\underline{B}^1$. By restricting the set of available contraction strategies in this way, one can restrict also the strategies for revision. Consonant with his approach, the M function is replaced by a relation of comparative similarity among bases.

Thus, suppose that $\underline{K}^1$ is the closure of 1 = "All the coins in my pocket on V-E day were silver", ~h = "This coin was not in my pocket on V-E day" and ~g = "this coin is not silver." One basis might be $\underline{B}_1^1$ = {1, ~h&~g}. Removing ~h from $\underline{K}^1$ requires removing ~h&~g alone from the basis. Adding h to the result and taking the closure yields g. h > g is acceptable relative to $\underline{B}_1^1$. Another basis might be {1 <–> ~h&~g,~h ~g} = $\underline{B}_2^1$. Here, it is mandatory to remove ~h. Removing ~h requires removing 1 from the closure of the result. The closure of the result by adding h does not yield g. h > g is unacceptable.

Hansson's approach violates the requirement that the acceptability of conditionals should be uniquely determined by the deductively closed corpus $\underline{K}^1$ and M defined over potential corpora.

Suppose that an inquirer changes from one basis $\underline{B}^1$ to another $\underline{B}^{1'}$ for the same theory $\underline{K}^1$. According to Hansson's view, this is a change in the available strategies for changing beliefs and a change in the informational value M = function applicable to corpora (as opposed to bases). Insofar as the question of logical omniscience is not an issue, I do not understand why differences in bases should make a difference here.

Hansson and I agree, however, that the reduction condition should be replaced by an account of iterated conditionals that allows for appeal to a wider variety of transformations of nonmodal $\underline{K}^1$ than the Ramsey test recognizes. We differ with respect to the role of bases and with respect to the need to keep the demands for informational value assigned to theories fixed.

I have undertaken this exploration of iterated conditionals in order to respond to Gärdenfors's complaint that in the absence of the reduction condition, we lack any sensible account of iterated conditionals. I remain unconvinced that giving such an account is of importance although I am prepared to be proved wrong by future developments. My chief concern is to show that abandoning the reduction condition whose presence seems an obstacle to giving such an account according to Gärdenfors does not preclude systematic treatment of iteration should that prove desirable.

In Chapter 4, some ramifications of the proposals made here for a logic of iterated conditionals will be explored in a severely limited fashion.

3.8   Imaging

The view of modal judgment advanced in the previous discussion is decidedly antirealist. Judgments of neither unconditional nor conditional serious possibility and impossibility carry truth values.

Opposition to modal realism does not deny usefulness to all realistically construed notions of modality. I myself have endorsed interpretations of dispositions and abilities as well as of statistical probability that allows that attributions of dispositions, abilities, and chances are true or false of systems or objects even though the chief function of such attributions in inquiry is to serve as "placeholders" in stopgap explanations (Levi and Morgenbesser, 1964; Levi, 1967, 1984a, ch. 12, 1980).

Many authors have followed C. S. Peirce in mistakenly thinking that attributions of dispositions and abilities may be made by using conditionals.

Those "if" sentences of English Dudman calls "generalizations" do, indeed, make attributions of dispositions and abilities. "If Joe was angry, he sulked" states that in the past, Joe had the disposition to sulk when angry. Attribution of this disposition to Joe supports the acceptability of the conditional "If Joe had been angry, he would have sulked." This latter sentence expresses a conditional judgment of serious possibility. It rules out "Joe did not sulk" as being a serious possibility relative to a revision of the current corpus by adding "Joe was angry." The presence of the disposition statement in the corpus warrants the judgment, provided it is not removed at the contraction step of the revision. The disposition statement is often well entrenched due to its usefulness in explanatory tasks. Because of this, full belief that the disposition statement is true "supports" the acceptability of the conditional. The disposition statement does not, however, deductively entail the conditional. The disposition statement carries a truth value (hence, the modal realism). The conditional does not (hence, the modal antirealism).

The relation between belief that objects have dispositions and abilities and judgments of conditional serious possibility is analogous to the relation between belief that systems have statistical properties indicating chances of outcomes on trials and judgments of conditional credal probability. Statistical predicates are attributed truly or erroneously. Judgements of conditional credal probability, on the other hand, lack truth values. Yet knowledge of chances (statistical properties) can warrant, ground, or support judgments of conditional credal probability just as knowledge of dispositions and abilities can warrant, ground, or support judgments of conditional serious possibility (Levi, 1984a, ch. 12; 1980).

Stalnaker (1968) and Lewis (1973) provide a truth conditional semantics for conditionals that appear to express possibilities and impossibilities conditional on suppositions. They both exploit the framework of a possible-worlds semantics. Stalnaker is alleged to be less of a realist than Lewis; but the differences are not of great concern here. They are both excessively realist in their contention that conditionals have truth values.

Lewis's approach begins by introducing a relation of comparative similarity between possible worlds relative to a given world (Lewis, 1973, p. 48). A truth theoretic semantics for a language like ML is constructed including a specification of truth conditions for h > g and the "might" conditional h –◇–> g. h > g (h –◇– >g) is true at a given world w if and only if g is true in every (some) world in which h is true that is most similar to w among worlds in which h is true.

It is possible on this view to provide acceptability conditions for conditionals without appealing to suppositional transformations of belief states at all. Because conditionals carry truth values, they are objects of full belief. Instead of specifying conditions of acceptability for conditionals, we may specify conditions under which conditionals belong to the content of states of full belief. States of full belief can be represented by the sets of possible worlds in which such states are true. Conditional h > g is believed in a state of full belief represented by corpus K if and only if the state of full belief is also represented by a set of possible worlds such that h > g is true at all worlds in the set.

Thus, it is unnecessary to supply a suppositional account of the acceptability of conditionals once one has embraced a specification of truth conditions for conditionals within the framework of a possible worlds semantics. An account of belief conditions is secured without appeal to transformations of belief states of any kind.

Even though belief conditions for conditionals can be provided without appeal to transformations of belief states, there is an alternative (equivalent) characterization of belief conditions for conditionals that does explicitly appeal to transformations of belief states. The belief conditions for conditionals can be furnished by appealing to a special kind of transformation of K through imaging (in a sense to be explained shortly).

It is important to attend to imaging transformations to this extent. If judgments of conditional possibility understood along lines conforming to possible-worlds semantics of the sort pioneered by Stalnaker and Lewis are to be relevant to the use of suppositional reasoning in practical deliberation and inquiry, a person who fully believes that h > g and who changes his or her belief state suppositionally by adding h for the sake of the argument should be committed to judging true (for the sake of the argument) that g. If this condition is not satisfied, the realistic construal of conditionals has lost its claim to have bearing on human conduct and experience. Conditionals understood in terms of imaging become mere artifacts of the metaphysician's fevered imagination.

Stalnaker originally thought that conditionals with this kind of truth theoretic semantics are Ramsey test conditionals and, hence, could have the application that Ramsey test conditionals have in practical deliberation. This cannot be so.

Consider a situation where agent X is in suspense between h and ~h but is sure that h ⊃ g is true. Any criterion for the acceptability of h > g meeting the requirements of the condition of adequacy for Ramsey tests must insist that h > g is acceptable relative to the agent's state of full belief $\underline{K}$. However, it is easy to describe a situation where $\underline{K}$ holds h > g and ~(h > g) in suspense. Suppose all the possible worlds in which h is true in the representation of $\underline{K}$ are worlds in which g is true. Suppose, however, that some of the nearest h worlds to some of the ~h worlds in this representation are ~g worlds. In that case, X should hold h > g and ~(h > g) in suspense.

Consequently (*pace* Stalnaker), it appears that in those practical or theoretical contexts where suppositional reasoning seems relevant, judgments about the truth values of conditionals realistically construed are irrelevant – provided that suppositional reasoning is understood along the lines meeting Ramsey's requirements.

To render realistically construed conditionals relevant to practical deliberation and scientific inquiry through suppositional reasoning, apologists for realistic interpretations have at least tacitly abandoned views of suppositional reasoning meeting the condition of adequacy built into the Ramsey approach. This strategy has clearly and explicitly been elaborated by J. Collins (1991).[9]

According to the new form of suppositional reasoning, when an agent supposes that h is true for the sake of the argument, the agent is supposed to modify $\underline{K}$ by shifting to the (generalized) *image* $\underline{K}_h^\#$ of $\underline{K}$ by adding h.[10] For each possible world w in the proposition interpreting $\underline{K}$, determine the set w#h of nearest h worlds to w. The corpus interpreted by w#h is the image of the corpus interpreted by {w} by adding h. And the corpus $\underline{K}_h^\#$ that is the intersection of all image revisions of elements of the set of worlds interpreting $\underline{K}$ is the image of $\underline{K}$ by adding h.

Clearly h > g is in $\underline{K}$ if and only if h > g is true at every world in $\underline{K}$ and, hence, g is in $\underline{K}_h^\#$.

Thus, there is a belief revisional test for whether the agent should fully believe h > g that agrees with the verdict delivered by checking whether h > g is true at every world in the set representing the belief state. Modal realists can endorse a positive imaging test analogous to a positive Ramsey test. But there the resemblance ends.

The positive imaging test is derived from an account of truth conditions for the conditional and the possible-worlds representation of belief states. The positive Ramsey test is not derived from a semantic interpretation of conditionals.

The force of this point comes out with respect to cases where conditionals flunk the positive imaging test. g is not in $\underline{K}^{\#}_h$ and h > g is not fully believed.

In particular, let h > g be true according to some possible worlds according to $\underline{K}$ and false according to others. g is true at some worlds according to $\underline{K}^{\#}_h$ and false at others. That is to say, the agent can be in suspense as to the truth of h > g, while, in the suppositional reasoning adding h for the sake of the argument, g and ~g will both be judged possibly true. If a negative imaging test were introduced, it would stipulate that ~(h > g) is in $\underline{K}$ because g is not in $\underline{K}^{\#}_h$. Negative imaging is ruled out, therefore, for belief states like the one just constructed.

Consider, however, a situation where the agent is in a state K of full belief where for every possible world w in the set representing $\underline{K}$, there are several nearest h worlds to w and some of these are worlds where g is true while others are worlds where g is false.

In this case, like the previous one, adding h for the sake of the argument brings suspense between g and ~g. Yet, here the agent is not in suspense between h > g and ~(h > g). The agent should be certain that ~(h > g); for it is true at every world representing $\underline{K}$. Negative imaging works in this case.

The upshot of this is that the negative imaging test will work for some states of full belief but not for others. Moreover, the difference in the cases where it works and where it fails is not manifested in the judgments of possibility and impossibility conditional on suppositions made in suppositional reasoning. Whether the agent is in suspense concerning the truth of h > g and ~(h > g) or fully believes ~(h > g) and also ~(h > ~g) is not revealed by his or her judgment as to the possibility that g as a result of supposing that h is true for the sake of the argument. Both g and ~g count as possibilities in both situations. The implication of this is that the use of imaging tests does not fully succeed in making conditionals interpreted according to possible-worlds semantics relevant to suppositional reasoning even as conceived according to advocates of imaging.

Imaging methods of changing belief states by adding suppositions are clearly different from the use of AGM or Ramsey revisions. Suppose, for example, that $\underline{K}$ is represented by a set of possible worlds, some of which are worlds in which h is true and others in which it is false. All the h worlds are worlds in which g is true. Hence, h $\supset$ g is true. But g is true in all of the most similar h worlds to some ~h worlds, false in all of the most similar worlds to some other ~h worlds, and true in some and false in other h worlds most similar to still another category of ~h worlds. Hence, relative to $\underline{K}$, the agent is in suspense between h > g, h > ~g, and the conjunction of ~(h > g) and ~(h > ~g).[11] Hence, the agent is in suspense between h > g and ~(h > g). Because g is not in $\underline{K}_h^{\#}$, the positive image test does not recommend accepting h > g.[12]

The Ramsey test, whether RTGL or RTR, recommends accepting h > g. Indeed, it must do so if the condition of adequacy for the Ramsey test is to be satisfied. Clearly, positive imaging does not recommend believing h > g in this example in violation of the condition of adequacy for Ramsey tests. In spite of the fact that, beginning with Stalnaker, many philosophers seem to have taken positive imaging to be a Ramsey test, it is not.[13]

Even though the semantic interpretation of conditionals used here and the positive imaging test based on it presuppose the satisfaction of the reduction condition, no triviality result ensues. Gärdenfors and Ramsey tests presuppose that AGM and Ramsey revisions satisfy (K*r4) and consistent expandability. These revision transformations are monotonic in the presence of the reduction condition (in the sense that if $\underline{K} \subseteq \underline{J}$, $\underline{K}_h^* \subseteq \underline{J}_h^*$). They are not in the absence of the reduction condition. Gärdenfors has shown, in effect, that consistent expandability, (K*r4), and monotonicity cannot be consistently satisfied without triviality. Monotonicity has to go if the Ramsey test is to be saved. This means that reduction has to go.

Imaging revisions are also monotonic. Unlike revisions used in Ramsey tests, the monotonicity does not depend upon the assumption of reduction. Moreover, they do not require satisfaction of (K*r4). As Collins (1991) and Katsuno and Mendelzon (1992) recognize, reduction, consistent expandability, and monotonicity can live together consistently and nontrivially.[14]

Of course, imaging tests purchase this result at the cost of violating a presupposition of adequate Ramsey tests for the acceptability of conditionals.

These observations bring out how strikingly different positive imaging tests are from Ramsey tests as accounts of suppositional reasoning and tests for the acceptability of conditionals. But the two tests are not only different, they have been used to advance incompatible views concerning suppositional reasoning. Collins suggests that positive imaging tests are relevant to practical deliberation and scientific inquiry because of the role they play in suppositional reasoning. Ramsey and others indicate that Ramsey tests are alleged to be relevant for the same purpose. The difference in the tests now becomes a clear conflict between the perspectives of those who embrace the Ramsey test approach to conditionals and those who favor imaging test approaches.

In my judgment, advocates of Ramsey tests have the clear upper hand in this confrontation, even though one cannot demonstrate this from first principles that advocates of imaging are not prepared to question.

Imaging accounts of the kind of transformation of belief state appropriate in suppositional reasoning violate the idea that when the supposition is held in suspense relative to $\underline{K}$, the result of adding the supposition for the sake of the argument should be the expansion of $\underline{K}$ by adding the expansion. The corresponding imaging test for the acceptability of conditionals violates the condition of adequacy for tests of the acceptability of conditionals laid down by Ramsey. This circumstance itself ought to be sufficient to bring into question the propriety of imaging in an account of suppositional reasoning and the acceptability of conditionals. And this worry undermines the claimed relevance of imaging to deliberation and inquiry.

The introduction of possible-worlds semantics for conditionals was followed by the introduction of a so-called causal decision theory, which has built into it an imaging interpretation of the conditional judgments expressing the results of suppositional reasoning from the implementation of an option to judgments about outcomes. Collins has suggested that the product of such suppositional reasoning need not rely on causal judgments. If one could take suppositional or causal decision theory seriously, it would constitute a substantial response to the worry just registered.

However, no one has demonstrated a clear motivation for introducing such a decision theory *independent* of the wish to preempt suppositional reasoning for belief revision of the imaging type. To the outsider unconvinced by the metaphysics of modal realism, there is no independent motivation for proposing imaging revision combined with causal (suppositional) decision theory other than to supply conditionals realistically construed with an application. By combining the lack of motivation for causal decision theory independent of a love affair with modal realism with the remarkable implications of causal decision theory, it seems doubtful that one can justify the applicability of conditionals understood along the lines of Stalnaker or Lewis to serious matters by an appeal to causal decision theory.

These claims do, of course, need substantiation. I have supplied arguments supporting them elsewhere and will not repeat them here. (See Chapter 1, note 8, for references.)

Laws are commonly held to support conditional judgments and this is alleged to call for a realistic construal of conditionals. This would be so if the support proffered were in virtue of the fact that laws entail conditionals. But, if the universal generalization *l* belongs to an inquirer's corpus $\underline{K}$ and is judged by the inquirer to be a law, this does not mean that, in addition to taking for granted the truth of l = "All A's are B's," the inquirer is convinced that l has an objective nomic necessity entailing the truth of conditionals of the type "If x were an A, it would be a B." On the view I am advocating, the lawlikeness of l is a result of the inquirer's imputing to l explanatory and, hence, informational value relative to the inquirer's research program or demands for information. (Such demands are neither true nor false. The inquirer fully believes that l is true but does not fully believe in addition that it is a law.) As a consequence, if, in considering, for the sake of the argument, the supposition that x is an A in contravention of his or her current full beliefs, the inquirer needs to give up some of his or her full beliefs, l will tend to be retained at the contraction step. The Ramsey test then secures the claim that x is a B upon expansion by adding the claim that x is an A. No doubt the inquirer's full belief that l together with his or her demands for explanation support the conditional judgment; but this is a far cry from entailment of one truth value bearing claim by another.

Of course, this does not utterly refute the realist construal of the support relation between l and conditionals; but it does provide a nonrealist alternative construal permitting a distinction between laws supporting conditionals and accidental generalizations that do not. Realist construals are not indispensable, therefore, for accounting for the "support" in question.

Katsuno and Mendelzon (1992) and Morreau (1992) favor a distinction between two kinds of belief change. They claim that when one obtains new information about a "static" world, "revision" should proceed along the lines of Alchourrón, Gärdenfors, and Makinson (1985) or Gärdenfors (1988). But sometimes they say the "world" or situation changes and we must bring our "knowledge base" up to date. In that case, Katsuno and Mendelzon claim that the belief change should be in conformity with imaging. They call this kind of change "updating" (1992, p. 183) in contrast to Collins's usage of the same term for what Katsuno and Mendelzon call "revision."

The puzzling feature of the Katsuno and Mendelzon or Morreau view is that none of the examples they offer warrants the conclusion that two forms of belief change are involved. Suppose an inquiring agent is in a belief state $\underline{K}$ that contains laws and theory specifying possible transitions from states at t to states at t'. For simplicity represent this information as the consequences of the "law" l. In addition, the agent has some information (often very partial and meager) concerning the state at t and the state at t'. Taken together with l, the corpus $\underline{K}$ countenances some of the transitions consistent with l as seriously possible and, perhaps, others as not seriously possible. Finally, let h be new information (I shall suppose it is consistent with $\underline{K}$), adding additional detail to what is known about the transition from the state at t to the state at t'. Take, for example, an illustration used by Katsuno and Mendelzon. Suppose the inquirer fully believes that at time t, there is a table and five objects, A, B, C, D, and E, in a certain room, some of which are on the table and others not. In particular, the inquirer fully believes that at time t B is not on the table but that either A alone is on the table or objects C, D, and E alone are. The inquirer does not know what will be the case in time period t'. But he does know that whenever an alpha robot is present at a given time, it either removes all objects from the table or places them all on the table subject to the constraint that the number of objects whose position is to change is minimized. Call this transition law *Con*. And let R be the hypothesis that an alpha robot will operate at time t'.

Initially the agent is in suspense concerning R although he or she knows *Con*. So adding R to $\underline{K}$ is an instance of expansion. The change is from $\underline{K}$ to $\underline{K}_R^+$. Hence, the change in belief state must be a revision in the sense of Gärdenfors and also in the sense of Ramsey.

On the other hand, if we compare the information contained in both $\underline{K}$ and in $\underline{K}_R^+$ concerning the state at t with the information in $\underline{K}_R^+$ concerning the state at t' and suppress the difference in dates of the two state descriptions, we have what looks like a revision according to imaging. According to the information about the location of the objects at time t, there are two possible states of the five objects at that time. State 1 has A alone on the table and the rest off the table. State 2 has C, D, and E on the table. If state 1 holds at t, then R and *Con* imply that at the state at time t', all objects are off the table. If state 2 holds at t, all objects are on the table at t'. If one suppresses the dates and takes the disjunction of states 1 and 2 to represent a theory $\underline{T}$, the transformation of the theory by shifting to the disjunction of the claims that either all objects are on the table or all are off is an imaging transformation of $\underline{T}$. But this transformation does not represent a transformation *of the belief state* $\underline{K}$, by adding R. In the example, the change in belief state $\underline{K}$ is an expansion by adding R and, hence, is a revision in the sense of AGM and also in the sense of Ramsey. It is not an example of updating in the sense of Katsuno and Mendelzon.

The application of imaging is to the representation of predictions about future states from information about initial states and constraints on transitions. That is to say, the predictions about future states are deducible from information about the initial states and constraints. Since deduction from information in the linguistic representation of a belief state does not represent a change in belief state (on this point, there seems to be no dispute with Katsuno and Mendelzon), the imaging transformations used by Katsuno and Mendelzon cannot represent transformations of belief states or corpora.

The kinds of transformations of interest to Katsuno and Mendelzon are well illustrated by specifying the initial mechanical state of a statistical mechanical system by a set of points in phase space. The mechanical laws of the system together with initial set of points specify the possible trajectories of the system over time. Formally, they determine the set of points in phase space at t′ to be a transformation of the set of points at t. If we think of a point in phase space as a possible world p at t, the world p′ at t′ connected by the trajectory from p at t′ represents the nearest world at t′ to p at t. The imaging transformation does not itself represent a change in belief state. Rather it characterizes aspects of the dynamic principles regulating the statistical mechanical system.

There is much more to say about the views of Katsuno and Mendelzon and of Morreau concerning the relevance of their views for planning and action. My contention is not that they have failed to identify some contexts where an imaging formalism is applicable. They do, indeed, seem to have succeeded in doing so. Sometimes it may be useful to represent information *contained in a belief state* concerning how states of a system change over time by imaging transformations. And such information can exercise control over how the belief state may be modified. But whatever the influence may be, it will be due to the content of the information and how well it is entrenched in the belief state. It does not represent a type of belief change distinct from the belief changes covered by theories based either on Gärdenfors revision or Ramsey revision. Katsuno and Mendelzon may have identified an application of imaging; but they have not identified an application as an account of belief change.

There may, perhaps, be other proposed applications of imaging as a method of belief change that are being pursued. One should not be dogmatic and conclude that accounts of conditionals utilizing similarity between possible worlds will never be useful. But as matters now stand, realistically construed accounts of conditionals and imaging accounts of belief change are formalisms in search for an as yet undiscovered application.

3.9   Conclusion

In Chapter 2, I gave two distinct versions of the Ramsey test for conditionals: RTGL and RTR. Both of them construe the acceptance of a conditional as the making of a judgment of serious possibility conditional on a supposition.

In this chapter, I have argued that, so construed, the acceptance of a conditional cannot be understood as full belief in the truth of a conditional judgment as partial or complete reduction conditions require.

One step in the argument aimed to show that the various so called triviality results are best understood as showing that either reduction or Ramsey tests should be abandoned. The abandonment of Ramsey tests that is necessary requires not only giving up the idea of conditionals as conditional assessments of serious possibility for the sake of the argument but the view that belief states serve as standards for serious possibility. This is a cost too great, in my judgment, to bear.

Giving up the reduction condition precludes giving a certain kind of account of iterated conditionals; but, as I have also argued, there are alternative accounts of iteration that do not require the reduction condition and accommodate presystematic judgment quite well. Appearances to the contrary notwithstanding, giving up the reduction condition does not entail giving up the use of iterated conditionals but only of a certain kind of analysis of iterated conditionals.

Adding up the costs and the benefits of both options, I argue that retaining a Ramsey test approach to conditionals is preferable to abandoning Ramsey tests for the sake of the reduction condition.

There is, however, more to be said about conditionals. In later chapters, the question of extending Ramsey tests by converting expansions into inductively extended expansions will be considered. In such cases, it is of interest to identify the conclusions that may be drawn as a result of making a supposition for the sake of the argument not only by deduction but also by nondeductive or inductive inference.

Prior to that, however, some attention will be paid to what can be said about conditional logic when conditionals are Ramsey test conditionals and conditionals cannot be objects of full belief, probability judgment, or desire. The account of conditional logic will be accompanied by a discussion of the forms of so-called nonmonotonic reasoning.

Having discussed these matters, I shall at last turn to a discussion of inductively extended Ramsey test conditionals and the kinds of conditional logic related to these kinds of conditionals. Then a consideration of the types of nonmonotonic reasoning associated with inductively extended Ramsey test conditionals will be undertaken. This, once more, will be associated with new versions of nonmonotonic reasoning. Finally, so-called default reasoning will be examined in the light of the previous results.

4 Aspects of Conditional Logic

4.1 Logical Principles

A central aim of studies of deductive logic is to characterize formally the trichotomous distinction between logical truths, logical falsities, and truth-value-bearing sentences that are neither logical truths nor logical falsities. Another such aim is to formalize the relation of logical or deductive consequence between a set of sentences and a sentence.

When a formal language $\underline{L}$ is used to represent potential states of full belief, one might characterize the logical truths as those that represent the potential state that is weaker than every other state, the logical falsities as those that represent the potential state that is stronger than every other state, and sentences that are neither logical truths nor logical falsities as sentences representing potential states that are neither the strongest nor the weakest. One might characterize the relation of logical consequence by saying that set $\underline{H}$ has h as a consequence ($\underline{H} \vdash h$) if and only if any potential state that is as strong as all of the states represented by sentences in $\underline{H}$ is as strong as the state represented by h.

Some may complain that formal languages are worth studying without regard to their intended applications. A characterization of logical consequence and logical truth may be sought that is not so intimately tied to the concern to represent potential states of full belief.

Even if the intended application is to this purpose, the application cannot be usefully implemented unless a characterization of logical consequence and logical truth is given for the language in question that is distinct from the relation of comparative strength that holds between potential states of full belief and the notions of logical consequence and logical truth constructible from that relation.

In addition, although all logical truths in the given language represent the weakest state of full belief, in applications one might want to restrict attention to a set of potential states where the weakest member is represented by mathematical truths and, perhaps, other sentences that, for one reason or another, we think should be judged true no matter what the potential state

84

of belief might be. If we are to avoid committing hostages to controversies concerning the set of sentences we wish to include in this category, the claim that only logically true sentences represent the weakest potential state will have to be left open and this will add another incentive to provide an independent characterization of logical truth.

A common way for one to proceed is, having specified the syntax and vocabulary for the language $\underline{L}$, to introduce a notion of an interpretation of the vocabulary in $\underline{L}$ in some (conceivably structured) set of objects or domain that yields unique truth value assignments "true" or "false" for all the sentences in $\underline{L}$.[1]

A sentence h is satisfiable in the domain if and only if there is some interpretation of $\underline{L}$ in the domain for which h comes out true. It is valid in the domain if it comes out true under all interpretations in the domain and is a logical truth of $\underline{L}$ if and only if it is valid under all interpretations in every domain of the specified kind. $\underline{H} \vdash h$ if and only all interpretations in every domain assign "true" to h if "true" is assigned to every element of $\underline{H}$.

A feature of this kind of semantical approach is that if a sentence is valid in a domain, its negation is not satisfiable in that domain. That is to say *positive* validity in a domain (truth under all interpretations in the domain) is equivalent to *negative* validity in the domain (negation not satisfiable in the domain). The distinction is not of great importance according to such semantical approaches because of the equivalence of the two kinds of validity; but it will prove useful in subsequent discussion.

For sophisticated spaces of potential states of full belief, there may be potential states that cannot be represented by single sentences. We can do better by representing potential states of full belief by theories or corpora that are sets of sentences in regimented $\underline{L}$ that are closed under logical consequence.[2] This representation has normative import. The inquiring agent is committed to judging true all potential belief states at least as strong as the potential state in which he or she is situated. The representation of the agent's belief state represents this commitment at least partially by stipulating that the agent is committed to judging true all potential states that can be represented by sentences in the corpus representing his or her belief state.

Not even the cleverest logician or mathematician is, of course, capable of fully implementing such a commitment and the requirement of rationality does not, strictly speaking, demand it. Rationality does insist on implementation of the requirement insofar as the agent is able to do so and, when the agent's capacity fails to measure up as it inevitably does, to undertake measures to improve computational capacity, emotional stability, and memory so as to enhance the ability to implement the commitments. (See Levi, 1991, ch. 2, for more on this notion of commitment.)

When deductive logic is applied in this fashion, it serves to characterize the commitment to closure under logical consequence that rationality demands of rational agents. It imposes constraints on the rational coherence of commitments to full belief and, as such, it has a normative force. The rationality requirement imposed by deductive closure regulates only attitudinal states that, like states of full belief, are truth value bearing.

As we have seen, many attitudinal states are neither true nor false. Yet, attempts are made to construct, among other things, logics of preference, of probability judgment, and of conditionals.

J. M. Keynes (1921) sought to construct an account of probability logic that somehow would mimic deductive logic as Bertrand Russell understood it so that probability logic would be a logic of truth. F. P. Ramsey (1990) resisted the idea. He thought that probability logic – which Keynes construed as regulating our credal probability judgments – could not be a logic of truth because credal probability judgments are neither true nor false. Yet Ramsey did think one could have a logic of probability that did impose constraints on the rational coherence of credal probability judgments and, in this sense, on the commitments undertaken by rational agents when making credal probability judgments.

The contrast between Keynes and Ramsey is an interesting one. Keynes supposed that the calculus of probabilities characterizes a relation between propositions or sentences that is a generalization of the relation of logical consequence or entailment allowing for degrees of entailment. Just as deductive logic may be applied to identify the conditions of rational coherence for belief states, Keynes thought that the principles of probability logic could be used to identify the conditions of rational coherence for credal states. The rationale for applying the principles of deductive logic as conditions of rational coherence for states of full belief is that the logical consequences of true assumptions are true. Hence, an agent who fully believes a set of assumptions to be true should judge true all their consequences. In the same vein, the probabilistic consequences of true assumptions are probably true to the appropriate degrees and should be so judged if one is sure of the premises.

Ramsey denied that probability judgments are truth value bearing and rejected the notion of a probabilized consequence relation between premises and conclusion. Like Keynes, however, he did think one could use the calculus of probabilities to characterize conditions of rational coherence for credal states. The rationale he invoked did not appeal to a probabilized consequence relation that was alleged to be truth preserving in probability but to principles of practical decision making. The "valid" judgments of probability logic could not be characterized in terms of a semantics according to which valid probability judgments preserve truth between evidence and credal probability judgments.

Those who interpret conditional statements as truth value bearing and conditional judgments likewise will find the possible-worlds interpretations of authors like Stalnaker and Lewis attractive for just that reason. It will then be possible to construct a logic of conditionals, instances of whose schemata are valid under all interpretations in all domains for suitable characterizations of a domain and an interpretation in a domain and where positive and negative validity are equivalent.

Those who, like myself, withhold truth-value-bearing status to conditional sentences and to the conditional judgments of serious possibility they express cannot understand schemata of conditional logic in this way. Principles of conditional logic will have to be understood in the first instance as prescriptions specifying those constraints on the acceptability of conditionals and their negations that ought to be satisfied no matter what the inquirer's state of full belief and no matter what his or her demands for information (as represented by an M-function) might be. This will correspond to the way principles of deductive logic, as applied to states of full belief, enforce a closure principle on what is judged true no matter what the inquirer's state of full belief might be. And, since the strongest potential state of full belief is to be avoided when feasible, this application of deductive logic also marks the distinction between the inconsistent corpus representing this state and the other consistent potential corpora. Thus, conditional logic is, like deductive logic, a "logic of consistency" in Ramsey's sense, regulating the coherence of judgments of conditional serious possibility just as deductive logic regulates the coherence of full beliefs. The difference between them is that deductive logic is, in the intended application, a "logic of truth" as well. A logic of Ramsey test conditionals is, like the principles regulating probability judgment, a logic of consistency without being a logic of truth. This is the main philosophical thesis elaborated upon in this chapter.

A notion of valid schema for conditional logic suitable for Ramsey test conditionals will first be proposed. The ramifications of this notion of validity for determining what are principles of conditional logic will be elaborated.

As usual, Gärdenfors (1988) has addressed these issues. However, his ambition is to show that the logic of consistency for Ramsey test conditionals is formally identical with the logic of truth for conditionals understood after the fashion of Lewis and Stalnaker. I shall try to show that his view is mistaken on this score and that transference of the main themes of Ramsey's view about probability logic to conditional logic is more than an idle philosophical reflection. The principles of conditional logic will perforce be different from a logic of truth-valued conditionals in ways I shall try to explain.[3]

4.2   Validity in the Logic of Conditionals

Prevailing sentiment has it that if there is a logic of conditionals, it is a logic of truth. The most familiar kind of approach provides a semantics for conditionals utilizing notions of comparative similarity between possible worlds or equivalent ideas such as those discussed by Lewis (1973). Approaches of this kind constitute the basis for developing imaging tests for the acceptability of conditionals expressing suppositional reasoning discussed in the previous chapter.

According to such approaches, the notion of a valid schema in the logic of conditionals is understood along lines analogous to the understanding of valid principles of deductive logic as principles true under all interpretations. Validity cannot be understood this way if one favors a view of conditionals according to which they are neither true nor false under an interpretation but are, instead, acceptable or unacceptable at a belief state. The first task to be addressed is to sketch the salient differences between validity in conditional logic as a logic of consistency when such a logic is equivalent to a logic of truth and when it is not.

Consider a nonmodal language $\underline{L} = \underline{L}^1$ and let $\underline{ML}$ be a language containing $\underline{L}$, sentences of the form "h > f" where h and f are in $\underline{ML}$ and truth functional combinations of these. $\underline{L}^2$ is the subset of $\underline{ML}$ containing $\underline{L}$, sentences of the form "h > f" where h and f are in $\underline{L}$ and truth functional compounds of these. $\underline{L}^i$ is the subset of $\underline{ML}$ containing $\underline{L}^{i-1}$, sentences of the form "h > f" where h and f are in $\underline{L}^{i-1}$ and truth functional compounds of these.

A potential corpus in $\underline{L}$ is any deductively closed theory in $\underline{L}$.

A *belief revision system* $\langle K, * \rangle$ is an ordered pair consisting of (1) a set K of potential states of full belief closed under expansion and (2) a revision transformation mapping consistent potential states of full belief to potential states of full belief (Gärdenfors, 1978; 1988, sec. 7.2). The potential states of full belief are representable by potential corpora in $\underline{ML}$. These too must be closed under expansion. The expansion of $\underline{MK}$ by adding h is, as before, the smallest deductively closed potential corpus containing the logical consequences of $\underline{MK}$ and h.

The revision operator is customarily constrained by revision postulates of some kind. These postulates include some Ramsey test criteria specifying necessary and sufficient conditions for the acceptability of sentences of the type "h > g" (positive Ramsey tests) and "~(h > g)" (negative Ramsey tests) relative to $\underline{MK}$. If acceptability conditions are specified that violate the condition of adequacy for the Ramsey test but nonetheless stipulate that h > g is acceptable if and only if g is in $\underline{K}_h^*$ for some revision operator, I shall include them in a more general category of *revision-based tests*. I assume revision-based criteria constrain the domain of potential corpora in the following way:

(i) A potential corpus in $\underline{L}^1 = \underline{L}$ is any deductively closed set of sentences in $\underline{L}$.

(ii) A potential corpus in $\underline{L}^{i+1}$ is any deductively closed set $\underline{K}^{i+1}$ in $\underline{L}^{i+1}$ such that

  (a) $\underline{K}^{i+1}$ contains all sentences in $\underline{L}^{i+1}$ that must be accepted according to the Ramsey test or other revision-based criteria relative to the potential corpus $\underline{K}^i$ in $\underline{L}^i$ that is the intersection of $\underline{K}^{i+1}$ and $\underline{L}^i$.

  (b) If $\underline{K}^{i+1}$ is consistent, it contains no sentence in $\underline{L}^{i+1}$ whose acceptance is precluded relative to potential corpus $\underline{K}^i$ by the Ramsey test or other revision-based criteria.

(iii) A potential corpus in $\underline{ML}$ is the union of series of potential corpora inductively generated from a potential corpus in $\underline{K}$ via (i) and (ii).

This characterization of potential corpora in $\underline{ML}$ emphasizes the point that the Ramsey test or other revision-based criteria not only constrain the revision operator but also the space of potential corpora ingredient in a belief revision system.

Satisfiability and validity cannot, in general, be defined for the language $\underline{ML}$ and its several sublanguages in the usual truth semantics fashion. However, we can think of a sentence as satisfiable if and only if it is a member of some consistent potential corpus and use this as a basis for generating conceptions of validity:

A sentence ϕ in $\underline{L}^i$ is *satisfiable under an interpretation* $\langle K,* \rangle$ if and only if ϕ is in some consistent potential $\underline{K}^i$ in $\underline{L}^i$.

ϕ is *positively valid* in $\underline{L}^i$ under an interpretation iff it is in every consistent potential $\underline{K}^i$.

ϕ *is negatively valid* in $\underline{L}^i$ under an interpretation iff $\sim\phi$ is not in any consistent potential $\underline{K}^i$.

ϕ is (*positively, negatively*) *valid in a specified set of interpretations* in $\underline{L}^i$ iff it is (positively, negatively) valid under all interpretations in the set.

The first part of the discussion will focus on the logic of noniterated conditionals and, hence, will explore validity for sentences in $\underline{L}^2$. Some attention, however, will be directed to validity for sentences in $\underline{L}^3$ allowing for one iteration.

Three kinds of belief revision systems (brs's) will be explored:

(1) *Belief revision systems according to Gärdenfors* (brsg's) use AGM revision functions from consistent potential corpora in the nonmodal $\underline{L}$ to potential corpora in $\underline{L}$ constrained in addition, by RTGL without a reduction postulate so that the acceptability of iterated conditionals is determined according to the criteria outlined in section 3.7. AGM revisions (for potential corpora in $\underline{L}$) satisfy $(K*1) - (K*8)$ as spelled out in section 2.3. The contents of a potential corpus in $\underline{L}^i$ are uniquely determined by the corpus in $\underline{L}$ contained in it. Consequently, a brsg for potential corpora in $\underline{L}^i$ is uniquely determined by the brsg for potential corpora in $\underline{L}$. These are two alternative representations of the same belief revision system. I prefer working with the belief revision systems as represented in $\underline{L}$.

(2) *Belief revision systems according to Ramsey* (brsr's) use Ramsey revision functions from consistent potential corpora in the nonmodal $\underline{L}$ to potential corpora in $\underline{L}$ constrained in addition by RTR without a reduction postulate so that the acceptability of iterated conditionals is determined by the criteria outlined in section 3.7. Ramsey revisions satisfy $(K*1) - (K*3)$, $(K*r4)$, $(K*5) - (K*7)$, and $(K*r8)$ as explained in section 2.3. As before, the contents of a potential corpus in $\underline{L}^i$ are uniquely determined by the contents of its nonmodal part in $\underline{L}$ via RTR so that an alternative representation of a brsr is available.

(3) *Weak belief revision systems according to Gärdenfors* (brsgw's) use a different kind of brs. Because of his commitment to the reduction condition, he endorses a revision operator defined for potential corpora in $\underline{ML}$ satisfying $(K*1) - (K*3)$, $(K*4w)$ (called "weak preservation"), and $(K*6) - (K*7)$. He replaces $(K*8)$ by $(K*L)$, which runs as follows (1988, p. 151):

$(K*L)$ If $\sim(h > \sim g)$ is in $\underline{K}$, $(K_h^*)_g^+ \subseteq K_{h \wedge g}^*$.

Gärdenfors also uses $(K*5)$ but, as I shall argue, he should modify $(K*5)$ in ways that shall be explained subsequently.

Finally, Gärdenfors uses only the positive version of a revision-based criterion for the acceptability of conditionals where the revision operator satisfies the requirements mentioned previously.

If a belief revision system of any one of these varieties is represented by a set of potential corpora in $\underline{L}^i$ for $i > 1$ or in $\underline{ML}$, the associated transformation could not satisfy $(K*r4)$. brsg's and brsr's satisfy $(K*r4)$ for transformations of consistent potential corpora in the nonmodal $\underline{L}$. Consider, by way of contrast, a belief revision system associated with the Gärdenfors's revision-based test. Let the nonmodal part of $\underline{K}^i$ be $\underline{K}^1$ and consider the revision $(\underline{K}^i)_h^{*r}$ when neither h nor $\sim$h (both sentences in $\underline{L}$) are in $\underline{K}^i$. In general, the nonmodal part of the revision will not be the expansion of $\underline{K}^1$ by adding h. In this sense, the condition of adequacy secured by (K*r4) when applied to consistent potential corpora in $\underline{L}$ is violated. I shall call Gärdenfors's test a positive *quasi Ramsey test* (QRT) for the acceptability of conditionals. He deploys this test rather than positive RTGL because (as noted in section 3.6), positive RTGL together with the reduction postulate leads to contradiction if the belief revision system is nontrivial.

I shall call the belief revision systems on which positive quasi Ramsey tests are based *weak brsg's* or brsgw's. They embrace all the conditions imposed on brsg's, including (K*4w), except those Gärdenfors thinks should be jettisoned to avoid inconsistencies in nontrivial belief revision systems.

Unless an explicit disclaimer is made, I shall suppose that brsg's, brsr's and brsgw's are all weakly nontrivial in the sense of section 3.6. Indeed, I shall adopt the stronger yet still mild assumption that for any extralogical h in nonmodal $\underline{L}^1$, there is some consistent potential corpus in K that contains neither h nor ~h.

Using brsgw's, the quasi Ramsey test determines a corpus **QRT($\underline{K}$)** in $\underline{L}^2$ as a function of a potential corpus $\underline{K}$ in $\underline{L}^1$. $\underline{K}$ must be the intersection of the corpus $\underline{MK}$ with $\underline{L}^1$. **QRT($\underline{K}$)** must be contained in the intersection $\underline{K}^2$ of $\underline{MK}$ with $\underline{L}^2$. But it need not be identical with it. $\underline{K}^2$ contains all and only those sentences in $\underline{L}^1$ that are in $\underline{MK}$. But since only a positive quasi Ramsey test is used, **QRT($\underline{K}$)** may contain neither h > g nor ~(h > g) when neither h ⊃ g nor its negation are in $\underline{K}$. Relative to such a $\underline{K}$, one cannot add h > g to **QRT($\underline{K}$)** without altering $\underline{K}$ and respecting the positive quasi Ramsey test. But one can add ~(h > g) to **QRT($\underline{K}$)** without altering $\underline{K}$. The upshot is that the potential corpus $\underline{K}$ in the nonmodal language $\underline{L}$ does not uniquely determine $\underline{K}^2$ (and, more generally, $\underline{K}^i$).

If one uses brsg's, RTGL determines a corpus **RTGL($\underline{K}$)** in $\underline{L}^2$ as a function of a potential corpus $\underline{K}$ in $\underline{L}^1$ belonging to K. However, because all revisions are operations on potential corpora in $\underline{L}^1$ and both positive and negative RTGL are used, the only corpora to consider in $\underline{L}^2$ are those determined by the function **RTGL**. That is to say, the only potential corpus in $\underline{L}^2$ containing $\underline{K}^1$ in $\underline{L}^1$ is **RTGL($\underline{K}^1$)**.

If one uses brsr's, RTR determines a corpus **RTR($\underline{K}$)** in $\underline{L}^2$ as a function of potential corpora in K, which is, in the same spirit, the only potential corpus in $\underline{L}^2$ containing $\underline{K}$.

We can, if we wish, introduce different notations >*w, >*, and >*r to mark the difference between the three kinds of conditionals: QRT conditionals, RTGL conditionals, and RTR conditionals. We could construct theories in a suitably enlarged $\underline{L}^2$ that contain all three kinds of conditionals and identify valid formulas generated from using truth functional compounds of both kinds. For ease of exposition, I shall resist the urge to do this and shall instead use >, understanding it either according to QRT, RTGL, or RTR depending upon whether one is interested in the logic of conditionals derived from one kind of revision-based test or the other.

According to the approach I have proposed, once we have fixed on using * or *r, a belief revision system can be represented by the pair $\langle K, M \rangle$ where M is the informational value determining probability defined over K.[4] However, the same $\langle K, M \rangle$ will determine both a brsg $\langle K, * \rangle$ and a brsr $\langle K, *r \rangle$. Hence, unless we have stated in advance whether we are focusing on brsg's or brsr's, we cannot unambiguously use the representation $\langle K, M \rangle$. However, when we want to be neutral with respect to whether we are using brsg's or brsr's while supposing that the belief revision systems under consideration are of one type or the other, this representation is useful. In the same spirit, **RT($\underline{K}$)** will be used when we want to be neutral as to whether RTGL or RTR is being deployed.

Belief revision systems of the type brsgw are not characterizable as ordered pairs $\langle K, M \rangle$ in the manner just indicated. The pairs $\langle K, * \rangle$ must be used instead. Such belief revision systems are required to be closed under expansions, and the revision operator is characterized by the postulates for revision it satisfies.

None of the kinds of brs's under consideration specify truth conditions for conditionals. Instead, satisfiability of a formula in $\underline{L}^2$ under an interpretation in a domain amounts to the presence of the formula as a member of a consistent potential corpus in a brs (i.e., the surrogate for the domain).

If the set of positively and negatively valid formulas under an interpretation coincided, we should expect that (i) every negatively valid formula of conditional logic expressible in $\underline{L}^2$ should be in every consistent potential corpus in $\underline{L}^2$ and, hence, be positively valid and (ii) for any sentence h such that neither h nor ~h is positively valid, h should belong to some potential corpus in $\underline{L}^2$ and ~h to others.

Our epistemic criteria for positive and for negative validity cannot meet these demands.

Consider, for example, the sentence h $\supset$ (T > h) where neither h nor ~h contains a conditional or modal connective and neither sentence is a logical truth in $\underline{L}^1$. T is a logical truth in $\underline{L}$. Relative to any potential corpus containing h and to any potential corpus containing ~h, h $\supset$ (T > h) is acceptable. For if h is in the corpus, T > h is acceptable. Hence, h $\supset$ (T > h) is acceptable. If ~h is in the corpus, h $\supset$ (T > h) must also be acceptable. This is common ground for QRT, RTGL, and RTR.

When neither h nor ~h is in the corpus, ~(T > h) will be acceptable if a negative as well as a positive Ramsey test is endorsed as in RTGL and RTR. h $\supset$ (T > h) cannot be acceptable relative to such corpora; for ~h is by hypothesis not in the corpus. The negation of $h \supset (T > h)$ cannot be a member of such a corpus either; for h is not a member of the corpus.

The upshot is that if one uses brsg's or brsr's, $h \supset (T > h)$ is negatively but not positively valid. Neither this sentence nor its negation is positively valid. But the negation is not acceptable relative to any potential corpus in $\underline{K}^2$.

Suppose, however, that only a positive Ramsey or quasi Ramsey test is used. When neither h nor ~h is in K, neither $T > h$ nor its negation can be shown to be acceptable in virtue of Ramsey test considerations. And there is no way to guarantee the presence of $h \supset (T > h)$ with the positive Ramsey test.

The negation of $h \supset (T > h)$ is prevented from being acceptable relative to any consistent potential corpus. Clearly $h \supset (T > h)$ is negatively valid. But, as it stands, it is not positively valid. Thus far, QRT resembles RTGL and RTR. There is, however, an important difference.

If we wish to eliminate negatively but not positively valid formulas by some additional postulate, we may seek to do so by denying the status of a potential corpus to those corpora where the negatively valid formula is not present. In the case of brsg's and brsr's, corpora that do not contain $h \supset (T > h)$ could be eliminated by removing potential corpora that contain neither h nor ~h where h is in L. That is to say, one would be required to be completely opinionated concerning all nonmodal sentences and the brs would be weakly trivial.

QRT does not demand such drastic measures to achieve the desired result. To insure the validity of $h \supset (T > h)$, the use of the negative Ramsey test must be *prohibited*. By doing this, we may allow corpora containing neither h nor ~h to be in brsgw's. Potential corpora containing neither h nor ~h exist in brsgw's that contain $h \supset (T > h)$ as a member and there are others that do not do so. It is open to us, therefore, to explore adding another postulate to brsgw's that rules out the corpora that do not contain $h \supset (T > h)$ without forbidding suspension of judgment between a nonmodal proposition and its negation.

Once this move has been made, the logic of truth and the logic of consistency have been made to coincide for conditional logics.

I shall return to this point later. For the present, let us focus on the three belief revision systems without such amendments. Whether brsgw's, brsg's, or brsg's are used, $h \supset (T > h)$ is a negatively valid sentence in "flat" conditional logic (the logic for noniterated conditionals expressible in $\underline{L}^2$) but fails to be positively valid on our epistemic reading of validity.

This formula is an immediate consequence of the following formula that Gärdenfors takes to be a valid formula of conditional logic.

(A6) $h \wedge g \supset h > g$.

One need only substitute T for h and h for g to obtain h ⊃ (T > h). Clearly it cannot be positively valid. Yet, it is negatively valid for brsg's. It is not even negatively valid for brsr's as we shall see later.

In his own official characterization of validity (1988, p. 148, Def Val), Gärdenfors adopts negative validity as his standard and uses the quasi Ramsey test relying on brsgw's. He claims that (A6) corresponds to (K*4w) and offers a proof of the correspondence that, if cogent, would demonstrate the *positive* validity of (A6) (1988, p. 239, Lemma 7.4) based on the assumption that postulate (K*4w) holds. Aside from the reduction condition, (K*1), (K*2), and a positive pseudo–Ramsey test of acceptability of conditionals generated by the belief revision system, no other postulate regulating revision is allegedly used.

Whether (A6) is positively valid or merely negatively valid is a matter of some importance; for one would expect that positive validity should be the standard to adopt in constructing a logic of consistency that is not a logic of truth. We should, so it seems, want a valid sentence to be acceptable relative to every potential state of full belief for every K̲ in K in an interpretation. If all the axioms of the conditional logic favored by Gärdenfors (they shall be explicitly stated in section 4.3) are positively valid, then the derivation rules proposed by Gärdenfors for conditional logic can be shown to guarantee that the theorems are also positively valid. But not even the rule of detachment can guarantee that conclusions derived with its aid from negatively valid assumptions are negatively valid. If (A6) or any other postulates in Gärdenfors's system are negatively but not positively valid, it cannot be shown that the theorems are all even negatively valid.

Gärdenfors's declaration that (A6) is a valid formula of conditional logic is correct for negative validity in the restricted setting of brsg's when reduction does not hold and for brsgw's when reduction does hold; but that would not appear to be the sort of validity we should be focusing on.

So, if Gärdenfors's proof actually showed (A6) to be positively valid for belief revision systems where * satisfies (K*4w) and reduction holds, Gärdenfors's proposed modifications of AGM revision to make it compatible with reduction would furnish a belief revisional alternative to the possible-worlds semantics as a means for characterizing valid schemata of conditional logic.

(A6) is a (both positively and negatively) valid formula according to Lewis's semantics for the logic VC (1973, ch. 6) which is favored by him for conditionals under the "subjunctive" interpretation. Gärdenfors has had the ambition to obtain a formally identical logic of conditionals to VC. If (A6) is not positively valid in the epistemic sense, the merit of his claim is very much in doubt. So it will be instructive to consider Gärdenfors's proof.

If K consists of potential corpora in $\underline{L}^1$ as I have been supposing in the discussion of brsg's and brsr's, h ∧ g ⊃ h > g cannot be in $\underline{K}$ for any $\underline{K}$, so that h > g cannot be in $\underline{K}^+_{h \wedge g}$. However, it might be in $RT(\underline{K})$ for every potential corpus $\underline{K}$.

Gärdenfors, however, adopts the full reduction condition. For the purpose of examining his proof, I shall do the same. Because, as long as h and g are in $\underline{L}^1$, the proof involves only sentences in $\underline{L}^2$, and we can focus on theories in that language. Gärdenfors, of course, restricts himself to the positive QRT. I shall, for the sake of the argument, follow him in this. Finally, Gärdenfors does not assume the full (K*4). He gives up use of (K*r4) or preservation while retaining (K*4w). His argument for the positive validity of (A6) proceeds by indirect proof.

Assume that there is some consistent $\underline{K}^2$ in the set of potential corpora in $\underline{L}^2$ and a pair of sentences h and g (which we can assume without loss of relevant generality to be nonmodal sentences in $\underline{L}^1$) such that h ∧ g ⊃ h > g is not in $\underline{K}^2$. Gärdenfors then claims to show that the assumption that such a consistent $\underline{K}^2$ exists leads to contradiction.

If existence of such a corpus were inconsistent, it would show by indirect proof that for every pair h and g and every potential corpus in $\underline{L}^2$, h ∧ g ⊃ h > g is in the corpus. The positive validity of (A6) for the given brsgw would be demonstrated.

Consider then $\underline{K}^{2\prime}$ which is the expansion of $\underline{K}^2$ by adding (h ∧ g) ∧ ~(h > g). Because the set of potential states of full belief is closed under expansions, $\underline{K}^{2\prime}$ is a potential corpus in $\underline{L}^2$ if $\underline{K}^2$ is.

Gärdenfors also claims that $\underline{K}^{2\prime}$ is consistent. He offers no justification for this step and we shall need to scrutinize it more closely in a moment. Suppose, for the moment, that the step is legitimate.

(K*4w) implies that $\underline{K}^{2\prime}$ is contained in and, hence, identical with both the AGM revision (and the quasi AGM revision) $\underline{K}^{2\prime}_h$ of $\underline{K}^{2\prime}$ by adding h. g, therefore, is in $\underline{K}^{2\prime}_h$ and, by positive RTGL, h > g is in $\underline{K}^{2\prime}$. But ~(h > g) is in $\underline{K}^{2\prime}$. So we have a contradiction.

There is, however, a flaw in Gärdenfors's argument. Gärdenfors appears to think that closure under expansion insures the consistency of $\underline{K}^{2\prime}$. But that is not so. To make Gärdenfors's argument work we need to assume, in addition to closure under expansion, consistent expandability. That is to say, we need to require that when a corpus contains neither h nor ~h, the expansion of the corpus by adding h (= the smallest potential corpus containing the logical consequences of the initial corpus and h) not only exists, as closure under expansion requires, but that it be consistent. No such assumption is included in Gärdenfors's list of constraints on revisions.

Far from proving positive validity of (A6) from his assumptions, Gärdenfors has demonstrated that the inconsistency of $\underline{K}^{2\prime}$ is derivable from his postulates and other explicitly stated assumptions such as closure under expansion together with the assumption that h ∧ g ⊃ h > g is not in $\underline{K}$. The inconsistency of $\underline{K}^{2\prime}$ shows that either consistent expandability fails or that (A6) is positively valid. Gärdenfors has not demonstrated that (A6) is positively valid.

As far as the argument goes, two types of brsgw's can be identified.

Type 1 brsgw's recognize the deductively closed set $\underline{K}^2$ hypothesized not to contain an instance of (A6) as a potential corpus. The inconsistency of $\underline{K}^{2\prime}$ demonstrated by Gärdenfors implies the failure of consistent expandability and not the positive validity of (A6).[5]

For type 1 brsgw's, the operation of the positive quasi Ramsey tests requires that consistent expandability fail in some cases.[6]

Type 2 brsgw's do not countenance $\underline{K}^2$ as a potential corpus. Relative to such brsgw's, (A6) is positively valid. Such brsgw's can be obtained by deleting potential corpora from type 1 brsgw's that fail to contain instances of (A6). Gärdenfors has taken for granted that all brsgw's are type 2. Is there any route alternative to the one he deploys to secure this result?

In order to approach this question, let us consider a new kind of belief revision system brsgw² where K is nontrivial, the revision operator satisfies (K*1)–(K*3), (K*4w), (K*6)–(K*7), (K*L), the reduction condition, and the quasi Ramsey test condition. All brsgw's are brsgw²'s but not conversely because (K*5) is not required.

Return to $\underline{K}^2$ as defined before and consider the *revision* $\underline{K}^{2*}_{(h \wedge g) \wedge \sim(h > g)}$ adding the counterinstance to (A6). If $\underline{K}^2$ is a potential corpus, this revision must also be one; for revision functions are required to yield potential corpora as values if they operate on consistent potential corpora. This revision must contain h > g by QRT, (K*4w), and (K*2) and must contain ~(h > g) by (K*2). So the revision is an inconsistent potential corpus.

Notice that the revision is a subset of the corresponding expansion $\underline{K}^{2\prime}$ by (K*3) but the expansion has already been shown to be inconsistent. It is in virtue of this that consistent expandability fails if $\underline{K}^2$ is a potential corpus.

Neither (K*6), (K*7), nor (K*L) does anything to preclude $\underline{K}^2$ from being a potential corpus in a brsgw². Hence, brsgw²'s can be type 1 as well as type 2.

Suppose we now add the constraint (K*5). This reduces the class of brsgw²'s to brsgw's. (K*3) guarantees that the *revision* $\underline{K}^{2*}_{(h \wedge g) \wedge \sim(h > g)}$ of $\underline{K}^2$ by adding (h ∧ g) ∧ ~(h > g) is a subset of the expansion $K^{2+}_{(h \wedge g) \wedge \sim(h > g)}$ of $\underline{K}$ adding (h ∧ g) ∧ ~(h > g). (K*5) insures that this revision is consistent provided $\underline{K}^2$ and (h ∧ g) ∧ ~(h > g) are consistent. We assume $\underline{K}^2$ to be consistent. (h ∧ g) ∧ ~(h > g) is also logically consistent. So (K*5) does apply.

But this contradicts the assumption that the revision that is derivable from QRT, (K*2), and (K*4w) is inconsistent. There can be no type 1 brsgw's. It does seem that we can prove Gärdenfor's claim that (A6) is positively valid by another route than the one he actually used.

This argument does not work. For even if one restricts attention to type 2 brsg's, $(h \wedge g) \wedge {\sim}(h > g)$ remains logically consistent. For any consistent corpus $\underline{K}^2$ containing $(h \wedge g) \supset (h > g)$, (K*5) requires that $K^{2*}_{(h \wedge g) \wedge {\sim}(h > g)}$ be consistent. But as long as (K*4w), the quasi Ramsey test and the reduction condition are in place, this revision cannot be consistent whether or not there is a consistent corpus in $\underline{L}^2$ that does not contain $(h \wedge g) \supset (h > g)$.

The reduction condition, the quasi Ramsey test, (K*1)–(K*3), (K*4w), and (K*5) are inconsistent. The set of brsgw's is empty.

In the previous chapters, I argued that abandoning (K*r4) is tantamount to abandoning the Ramsey test and that it would be preferable to abandon the reduction condition. The inconsistency result just announced indicates that in the face of the quasi Ramsey test, the reduction condition, (K*4w), and (K*5) appear to be in tension. If the ambition is to derive the axioms of VC from some form of quasi Ramsey test, (K*4w) must be retained. As Gärdenfors observes (1988, p. 150), (K*5) is stronger than the axiom (A7) of VC whose validity it is intended to secure:

(A7) $({\sim}h > h) \supset (g > h)$.

H. Arló-Costa has suggested replacing (K*5) by a weaker requirement necessary and sufficient given the quasi Ramsey test to establish the validity of (A7). Depending on whether the ambition is to insure positive or negative validity of (A7), the function of (K*5) can be performed by one of the following pair of conditions without depriving brsgw's of their status as brsgw²'s:[7]

PosVal:

Let $\langle K, * \rangle$ be a brsgw². If h is positively consistent (i.e., ~h is not positively valid) relative to $\langle K, * \rangle$, $\underline{K}^*_h$ is a (logically) consistent potential corpus in $\langle K, * \rangle$ for any (logically) consistent potential corpus $\underline{K}$ in $\langle K, * \rangle$.

PosVal secures the positive validity of (A7) in brsgw²'s.[8]

NegVal:

If h is negatively consistent (~h is not negatively valid) relative to $\langle K, * \rangle$, $K^*_{{\sim}h}$ is a (logically) consistent corpus relative to $\langle K, * \rangle$ for any (logically) consistent potential $\underline{K}$ relative to $\langle K, * \rangle$.

NegVal secures the negative validity of (A7) but not its positive validity.[9]

Return then to the status of (A6). In virtue of (K*4w), a corpus containing the negation of an instance of (A6) must be inconsistent. Hence, revision of a consistent corpus $\underline{K}$ by the negation of (A6) must contain (A6). This result and PosVal imply that (A6) is positively valid (and, hence, negatively valid). Type 1 brsgw^2's are ruled out. NegVal does not rule out type 1 brsgw^{2}s. It implies that (A6) is negatively valid.

Which of the two surrogates for (K*5) should be adopted?

NegVal guarantees that brsgw^2's contain consistent potential corpora that are revisions of consistent potential corpora by negatively consistent sentences. Such negatively consistent sentences are, of course, positively consistent as well. The critical question is whether there are any consistent potential corpora containing positively consistent sentences that are not negatively consistent. NegVal does not rule out such consistent potential corpora. Hence, there can be negatively valid sentences relative to a brsgw^2 that are not positively valid if NegVal is adopted.

PosVal states that it is sufficient for $\underline{K}^*_h$ to be a consistent potential corpus that h be positively consistent whether or not it is negatively consistent. But this means that if there are positively consistent sentences that are not negatively consistent, these can be members of consistent potential corpora even though their negations are negatively valid. This is a contradiction. So PosVal rules out of existence negatively but not positively valid sentences by eliminating consistent potential corpora that might exemplify them from K. Moreover, it does so without ruling out potential corpora containing neither h nor ~h for nonmodal extralogical sentence h in $\underline{L}$.

Thus, endorsing NegVal avoids conflating positive with negative validity and, to this extent, avoids identifying the logic of consistency for conditionals with a logic of truth. Yet, with the reduction postulate in place, NegVal calls for the violation of the condition of adequacy for Ramsey tests that presuppose (K*r4).

Adoption of PosVal entails an even more radical departure from the Ramsey test point of view. To adopt PosVal is to cross the Rubicon from an "epistemic" understanding of conditionals to an "ontic" one. brsgw^2's that satisfy PosVal are just such brsgw^2's for which positive and negative validity coincide. In particular, if we think of maximally consistent potential corpora in a brsgw^2 as possible worlds, PosVal insures that there is no sentence h such that for some possible world (maximally consistent potential corpus) neither it nor its negation is true at that world (is a member of that corpus).

Once the logic of consistency and the logic of truth have been reunited in this way, it will be reasonable to search for a semantics assigning truth conditions to conditionals that characterizes the appropriate notion of validity. Since axioms for VC will be rendered positively valid, the semantics could be of the sort offered by Lewis for that system. Consequently, the revision transformation should become an imaging transformation.

Both versions of Gärdenfors's quasi Ramsey test clearly violate the condition of adequacy for Ramsey tests. Neither is, therefore, suited to interpret conditionals understood as expressions of suppositional reasoning for the sake of the argument. Gärdenfors was driven to adopt a version of the quasi Ramsey test because of his endorsement of the reduction condition (condition O) and abandonment of (K*r4). Abandonment of (K*r4) for the sake of the reduction condition is already a serious break with the program of characterizing conditionals as expressions of suppositional reasoning even if only NegVal is adopted. Endorsing PosVal would be to enter still deeper into the realist camp.

Traversing the path on which Gärdenfors embarked in 1978 when he undertook to dispense with truth theoretic semantics for conditionals and yet sought to recover the axioms of VC with the resources of his account of belief revision requires (1) abandoning the condition of adequacy for Ramsey test conditionals embodied in (K*r4w) for the sake of the reduction condition and (2) returning to a conception of validity that equates positive and negative validity. The identification of positive and negative validity itself surrenders to a view of the logic of conditionals that conflates, in Ramsey's words, "the logic of consistency" with a "logic of truth." It is far from obvious that Gärdenfors intends to take the second step; but to the extent that he does not, he has failed to obtain the positive validity of the axioms of VC or to avoid a contradiction in the postulates for revision he did adopt.

Presystematic reflection should deter anyone from endorsing the validity in a conditional logic for Ramsey test conditionals of $(h \wedge g) \supset (h > g)$. To do so requires the inquirer to say that either it is false that h or that it is false that g or that h > g is acceptable relative to his or her current belief state. But if the agent's state of full belief $\underline{K}$ leaves him or her in doubt as to whether h and g are jointly true or one of them is false, the agent must also be in doubt concerning the truth of $h \supset g$. According to an epistemic reading of conditionals, h > g cannot be acceptable relative to $\underline{K}$. The agent can, in that belief state, acknowledge the serious possibility that both h and g are true even though he or she does not currently judge them both to be true. But the agent cannot acknowledge that h > g is acceptable relative to his or her current state of full belief if unbeknownst to the agent h and g are both true. That would make the acceptability of h > g from the agent's point of view depend in part on the unknown truth values of h and g and not merely on what the agent *judges* to be true. This runs counter to both the spirit and the letter of Ramsey test approaches to conditionals as encoded in both RTGL and RTR.

The same point may be brought out more emphatically in the special case of (A6) claiming the validity of $h \supset (T > h)$. This says that the truth of h, whether fully believed by the agent or not, is sufficient to secure the acceptability of T > h relative to the agent's state of full belief. Since the acceptability of T > h relative to $\underline{K}$ is equivalent to the membership of h in K, we have the result that if h is true, h is fully believed to be true.

Only an agent who was convinced that he or she fully believed all truths could accept this. On the position I have adopted, a rational agent is committed to judging true or fully believing all consequences of his or her current state of full belief. But only a pretentious or preposterous agent would claim that he or she is currently committed to judging true all truths. If (A6) were indeed valid for Ramsey test conditionals, however, rational agents would be so committed. Fortunately that is not the case.

Arló-Costa has observed that imaging revision operations based on a comparative similarity relation over the possible worlds meeting semantic conditions securing the validity of VC should satisfy postulates on imaging revision strong enough to guarantee coincidence of positive and negative epistemic validity of the axioms of VC. Arló-Costa (1990) suggested one set of postulates. (K*1)–(K*3), (K*4w), (K*6)–(K*7), (K*L), and PosVal together with the reduction condition and QRT is another set. (K*L) can be replaced by postulates akin to (U6)–(U8) of Katsuno and Mendelzon (1992).[10]

The critical target of this section is not imaging or the semantic interpretation of conditionals that underpins it. I registered my worries about imaging in Chapter 3. My complaint was that imaging is a formalism in search of an application. Ramsey test conditionals appear relevant to practical deliberation and scientific inquiry and to the practice of suppositional reasoning important to such activities. *Pace* Collins, imaging test conditionals do not express suppositional reasoning of the relevant kind.

Whatever the merits of imaging might be, however, one cannot gain the results of imaging without incurring the costs. Gärdenfors has sought to show, to the contrary, that one can.

Many will no doubt continue to think that (A6) is sufficiently compelling to justify the conviction that a kernel of truth is to be found in it. Here is that kernel:

(A6t) $(T > (h \vee g)) \supset (h > g)$.

(A6t) is positively valid for the conditional logic operative when acceptability is according to RTGL, the reduction condition is abandoned, and validity is positive validity for brsg's.

(A6t) corresponds to the claim that if h ∧ g is in $\underline{K}$, h > g is acceptable relative to $\underline{K}$. (A6t) is itself acceptable relative to every $\underline{K}$ in a brsg in virtue of (K*4w). (A6t) represents the kernel of truth in (A6).[11]

The main point belabored in this section is that the concept of a valid formula of conditional logic for Ramsey test conditionals should not be confused with validity according to a truth theoretic semantics that equates positive and negative validity. The epistemic character of Ramsey test conditionals argues against such an approach.

The next question that needs to be addressed is what is left of Gärdenfors's axiomatization of VC when these conclusions about validity are endorsed.

4.3 Logic of Noniterated Ramsey Test Conditionals

To begin, the principles of conditional logic will be restricted to sentences in $\underline{L}^2$ where all sentences of the type h > g are constructed from sentences h and g in the nonmodal $\underline{L}^1$ and where a sentence in $\underline{L}^2$ is acceptable relative to $\underline{K} = \underline{K}^1$ representing in $\underline{L}^1$ a belief state if and only if it is a member of **RT**($\underline{K}$). Thus, the conditional logic does not concern iterated conditionals and, in this sense, is "flat."

Because no use is made of the reduction condition, there is no obstacle to using either AGM revision * or Ramsey revision *r as the revision operator in a belief revisions system. There is no need to resort to positive quasi Ramsey tests. Use can be made of both positive and negative components of RTGL or of RTR. Consequently, validity can be characterized either using brsg's or brsr's.

Whether we characterize validity using brsg's or brsr's, the revision functions will satisfy postulate (K*1), which states that the revision of a corpus by adding h is also a corpus. Gärdenfors (1988, p. 149) points out that as long as brs's satisfy this requirement, the following schemata are (positively) valid:

(A1) All truth functional tautologies.

(A2) $(h > g) \wedge (h > f) \supset (h > g \wedge f)$.

(A3) h > T.

Moreover, any sentence derivable from (A1) – (A3) via the following derivation rules is also positively valid. (The proof for h and g in $\underline{L}^1$ mimics the one given in Gärdenfors, 1978.)

(DRI) Modus ponens

(DR2) If $g \supset f$ is a theorem, then $(h > g) \supset (h > f)$ is also a theorem.

The axioms and the derivation rules just given constitute the system Gärdenfors calls **CM**.

We cannot claim, of course, that all positively valid formulas relative to the set of brsg's or the set of brsr's are derivable from **CM**. However, if we were to focus on the *minimally constrained class of brs's* where K is closed under expansions and revisions of consistent elements of K and it is assumed that the revision operator satisfies the Ramsey test and (K*1), we could construct a *minimalist* conception of positive and of negative validity. If a formula in $\underline{L}^2$ is positively valid in the minimalist sense, it is derivable in **CM**.

(K*2) states that h is a member of $\underline{K}_h^*$. This holds for revisions in both brsg's and brsr's.

From this we obtain the following axiom scheme:

(A4) h > h.

We also obtain the following derived rule of inference:

(DR3) If h ⊃ g is a theorem, so is h > g.

In the face of (A4) and (DR3), (A3) becomes a theorem.

(K*3) or inclusion yields the following schema for positively valid sentences. It licenses modus ponens for the logic of noniterated conditionals.

(A5) (h > g) ⊃ (h ⊃ g).

As already stated, (A6) is introduced as a postulate for conditional logic by Gärdenfors. We list it here for the record.

(A6) (h ∧ g) ⊃ (h > g).

The arguments against the positive validity of (A6) discussed in section 4.2 apply even when the brs's are restricted to brsg's and (K*4w) is in force. However, the following is positively valid for brsg's:

(A6t) (T > h ∧ g) ⊃ (h > g).

However, (A6t) fails to be either positively or negatively valid when RTR is the Ramsey test deployed. In that case, (K*4w) is not satisfied by the revision function *r though (K*r4) is.

Of course, if recovery is required as a condition on admissible contractions, RTR becomes equivalent to RTGL, (K*4w) is satisfied, and (A6t) is positively valid.

Since the reduction condition is not in force, we do not have to follow Gärdenfors's practice of eschewing (K*r4) in deriving his conditional logic. There is no obstacle to imposing (K*r4) on brsg's and brsr's. (K*r4) sustains the positive validity of the following schema:

(A6r) ~(T > h) ∧ ~(T > ~h) ⊃ [(h ⊃ g) ⊃ (h > g)].

The positive validity of (A6r) holds both for RTGL conditionals and RTR conditionals. According to Lewis's semantics, either h or ~h must be true at any specific possible world and, hence, the antecedent of the conditional in (A6r) must be false in all possible worlds. So (A6r) is valid in

Lewis's framework as well as this one. However, T > h is equivalent to h in the Lewis semantics. Hence, (A6r) is equivalent to the following:

(A6l) h ∧ ~h ⊃ h > g.

(A6l) is positively valid according to both Ramsey and imaging tests. It is a truth functional tautology in $\underline{L}^2$ and, hence, is derivable from (A1). As just noted, (A6r) is obtainable from (A6l) by replacing h by T > h and ~h by T > ~h in (A6l) if one uses imaging. One cannot derive (A6r) from (A6l) in this way for Ramsey test conditionals.

~[~(T > h) ∧ ~(T > ~h)] is not positively valid – that is, acceptable relative to all $\underline{K}$. Hence, (A6r) is not derivable from (A1) and substitution. If one does follow Gärdenfors and gives up (K*r4) or preservation as a constraint on belief revision systems, (A6r) would not be positively valid. But (A6l) would be.

(K*5) states that, for consistent $\underline{K}$, $\underline{K}_h^*$ is consistent if h is. When the reduction condition is in force, (K*5) conflicts with (K*4w). There is no conflict when the reduction condition is abandoned. So it may be imposed on brsg's. Of course, (K*4w) does not apply to brsr's. So there is no conflict in that case.

Consequently, search for a replacement for (K*5) is unnecessary. Indeed, when reduction is abandoned, PosVal and NegVal are equivalent to (K*5). The reason is that the only sentences to which these principles can apply are sentences in $\underline{L}^1$. The only positively or negatively valid sentences involved are logical truths in $\underline{L}^1$.

(K*5) applies only to consistent $\underline{K}$. Since $\underline{K}_h^*$ is consistent if and only if not both g and ~g are in it, (K*5) implies that h is consistent if and only if the joint acceptability of h > g and h > ~ g relative to consistent $\underline{K}$ is precluded by the Ramsey test. But if h is *inconsistent*, ~h is in every potential corpus and, in particular, in the revision $\underline{K}_f^*$ for every f. So f > ~h should be acceptable relative to $\underline{K}$.

Thus, we have the following schema:

(A7g) [(h > g) ∧ (h > ~ g)] ⊃ (f > ~h).

Substituting ~h for g in (A7g) and appealing to (A4), we obtain the following:

(A7) (h > ~h) ⊃ (f > ~h).

(A7g) is positively valid both for brsg's and brsr's, as is, of course, (A7). It is necessary and sufficient for the positive validity of (A7g) that the revision operator satisfy (K*5) for consistent $\underline{K}$.

Gärdenfors denies that the validity of (A7) is sufficient for claiming that the revision function satisfies (K*5). But the positive validity of (A7) is sufficient. To be sure, if we were dealing with a set of brs's relative to which (A7) is negatively but not positively valid, the revision function would not satisfy (K*5). For reasons already explained, I am requiring positive validity.

(K*7) and (K*8) imply that revision satisfies the following condition:

(K*⟨7,8⟩) If g is in $\underline{K}_h^*$ and h in $\underline{K}_g^*$, $\underline{K}_h^* = \underline{K}_g^*$.

(K*⟨7,8⟩) insures the positive validity of the following schema for brsg's:[12]

(A8) $(h > g) \vee (g > h) \supset ((h > f) \supset (g > f))$.

When RTR and brsr's are deployed, (K*r8) replaces (K*8). (K*<7,8>) cannot be endorsed as well without restoring (K*8).[13] Hence, (A8) cannot be positively valid for brsr's. However, the following is positively valid for brsr's:

(A8r) $[(h \vee g > f) \wedge (h > g) \wedge (g > h)] \supset [(h > f) \supset (g > f)]$.

(K*7) implies that the following is positively valid for both brsg's and brsr's:[14]

(A9) $(h > f) \wedge (g > f) \supset (h \vee g > f)$.

(K*8) applies to AGM revisions and brsg's. It states that if ~g is not in $\underline{K}_h^*$, then $(\underline{K}_h^*) \subseteq K*h \wedge g$. This constraint secures the positive validity of the following for brsg's when the positive and negative Ramsey tests are enforced.[15]

(A10) $(h > f) \wedge \sim(h > \sim g) \supset (h \wedge g > f)$.

(K*8) is not satisfied by Ramsey revisions and, hence, (A10) is not positively valid for brsr's. However, there is a weaker version of (A10) corresponding to (K*r8) that is positively valid.

(A10r) $(h > f) \wedge \sim(h > g) \wedge \sim(h > \sim g) \supset (h \wedge g > f)$.

Nine of the ten axioms for VC given by Gärdenfors are both positively and negatively valid for all brsg's provided only the "flat" portion of VC containing no iterated conditionals is considered, both the positive and negative Ramsey tests are used, and no reduction postulate is deployed. (A6), however, is negatively valid without being positively valid. This axiom should be excluded to be replaced by (A6t).

We have, therefore, shown that Gärdenfors's ambition to derive VC with the aid of a Ramsey test interpretation of conditionals fails even when the Ramsey test is RTGL and attention is restricted to the logic of noniterated conditionals.

Still Gärdenfors comes very close to succeeding. (A6t) may seem a close approximation to (A6) when interpreted within the Ramsey test framework.

There is, however, reason to question (A6t) for RTR conditionals. (A6t) depends for its positive validity on (K*4w). (K*4w) is violated by brsr's appropriate to RTR unless the recovery condition for contractions is endorsed.

All is not lost. When neither h nor ~h is in $\underline{K}$, the preservation condition is satisfied. This is the substance of (K*r4) imposed on revision according to Ramsey. If recovery is rejected and (K*r4) endorsed, (A6r) remains positively valid.

For brsr's not only is (K*4) replaced by (K*r4), (K*8) is replaced by (K*r8). And this undermines both the positive and negative validity of (A8) and (A10). (A8) can be weakened to (A8r) and (A10) to (A10r).

These results apply to the logic for noniterated Ramsey test conditionals. In section 4.2, we noted that when the logic of iterated conditionals is explored on the assumption that the reduction condition is satisfied, nothing short of imaging will yield the positive validity of axioms for VC.

As was emphasized in Chapter 3, however, tests for the acceptability of iterated conditionals can be developed without adopting the reduction condition. In the next section, the logic of iterated conditionals along these lines will be examined.

4.4 The Demise of Modus Ponens

In the past (Levi, 1988), I have registered skepticism concerning the viability of a logic of conditionals covering iterated conditionals of all varieties entertainable. Thanks to the prodding of S. O. Hansson (1992), I now think I overstated the situation. It would have been better to say that a logic of iterated conditionals that is based on either RTGL or RTR tests for acceptability of noniterated conditionals and their negations without the reduction postulate, along the lines of the proposals in section 3.7, is weaker than the corresponding logic for noniterated conditionals. I have not explored the question for conditionals with arbitrary iterations. It will suffice for my purposes to consider situations where there are at most two iterations – that is to say, for conditionals whose acceptability relative to $\underline{K} = \underline{K}^1$ in $\underline{L}^1$ implies membership in a corpus in $\underline{L}^3$.

In particular, I shall show that (A5), which is positively valid in the noniterative version of the logic of Ramsey test conditionals, fails to be positively valid if just one iteration is allowed. This is true both for brsg's and brsr's.

In addition, (A5) fails to be negatively valid for brsr's, although it is negatively valid for brsg's.

The failure of negative validity has already been supported by Van McGee (1985) through counterexamples. In effect, once iterated conditionals are considered within the framework of section 3.7 using RTR, modus ponens ceases to be valid for conditionals.

Three putative counterinstances to modus ponens in the iterative setting may help motivate the following discussion. I do not use the counterinstances to refute modus ponens. Modus ponens remains negatively valid for brsg's so that those who insist on RTGL may wish to retain it.

The three alleged counterinstances serve then to illustrate verdicts that flow from adopting the approach to conditionals proposed in section 3.7 when RTR is used. To the extent that the examples are compelling as counterinstances, they offer support for endorsing RTR rather than RTGL.

Example 1: (McGee, 1985): Just before the 1992 election for president of the United States, an American of reasonably sound mind might have endorsed (a) and (b) as given here but not (c) in violation of modus ponens for iterated conditionals:

(a) If a Republican is not elected, then if Clinton is not elected, Perot will be.
(b) A Republican will not be elected.
(c) If Clinton is not elected, Perot will be.

Example 2: Consider a spinner and dial divided into three equal parts 1, 2, and 3. It is known that the spinner was started and landed in part 1. It follows that it landed in an odd-numbered part. Clearly also, if the winner had landed in an odd-numbered part then, if it had not landed in part 1, it would have landed in part 3.

Example 3: Consider the same spinner as before. It remains known that it landed in an odd-numbered part. Suppose it is known that it did not land in part 1. It is still acceptable to judge that if the spinner had landed in an odd-numbered part, then if it had not landed in part 1, it would have landed in part 3.

In both examples 2 and 3 we would refuse to countenance the claim that had it not landed on part 1, it would have landed on part 3.

These examples can be understood as showing that if h is in $\underline{K}$ and $h > (g > f)$ is acceptable relative to $\underline{K}$ (where h, g, and f are in $\underline{L}^1$), $g > f$ may not be acceptable relative to $\underline{K}$. This cannot be so, however, if schema (A5) is negatively valid for sentences in $\underline{L}^3$. Under the circumstances described in examples, (A5) cannot be negatively valid either.

According to the approach to iterated conditionals developed here, examples 1–3 are counterinstances to modus ponens only when RTR is used and the brs's are brsr's. When RTGL is adopted and brsg's are used to characterize positive and negative validity, examples 1–3 are not counterinstances to modus ponens. To the extent that these examples seem presystematically compelling as counterinstances, they add support to the claim that RTR is to be preferred to RTGL as a criterion for the acceptability of informational-value-based conditionals.

Prior to explaining why (A5) fails to be negatively valid for brsr's but is negatively valid for brsg's, I will show that whether brsr's or brsg's are used to characterize validity, (A5) is not positively valid.

Once more an example will be useful in order to illustrate the point.

Example 4: Consider "If this match were struck (h), then if the surface were dry (g), the match would light (f)." That might be acceptable at $\underline{K}$. Suppose the agent does not know whether or not the match is struck but knows that the surface is not dry. g > f is acceptable at $\underline{K}^*_h$. But it is not acceptable at $\underline{K}$ even though h $\supset$ (g $\supset$ f) is in $\underline{K}$. Hence, (A5) is *not* positively valid. In ignorance as to whether the match was struck, the agent would reject the claim "If the surface were dry, the match would light."

This example falls short of being a counterinstance to (A5). The reason is that the agent is supposed to be in suspense as to whether the match is struck or not. Whether RTGL or RTR is used, (A5) is not acceptable relative to K. But its negation is not acceptable either. So there a counterinstance has not been offered.

McGee (1989) rightly points out that his counterinstances to modus ponens do not undermine modus ponens for noniterated conditionals. This point has already been sustained in the previous section. However, McGee's theoretical account of the acceptability of conditionals is based on the work of E. Adams (1975), whose acceptability conditions depend on probabilistic considerations. Nothing in Ramsey's brief partial formulation of a criterion for the acceptability of conditionals nor in completions of that partial criterion such as RTGL or RTR relies on such considerations. It is desirable, therefore, to show that McGee's insights are sufficiently robust to remain in force when probability is not an issue. (Probability will prove relevant in the discussion of "inductively extended" Ramsey tests in later chapters. But this is not a present concern.) The criteria for acceptability of iterated conditionals allow for exploration of the status of (A5) when it is construed as covering iterated conditionals without appealing to probabilities at all.

If (A5) were positively valid for iterated as well as noniterated conditionals, the following would be positively valid as well in $\underline{L}^3$.

(A5it) h > (g > f) $\supset$ [h $\supset$ (g > f)].

To satisfy the negation of (A5it) in a brs, there must be a consistent K relative to which h > (g > f) is acceptable, and g > f is not acceptable while h is in $\underline{K}$. If there is no such $\underline{K}$, (A5it) is negatively valid.

Because both positive and negative parts of the Ramsey test of choice are being deployed, relative to any given K, either h > (g > f) or its negation is acceptable. Hence, to satisfy (A5it) in a brs, there must be a consistent $\underline{K}$ relative to which either the negation of h > (g > f) is acceptable or h ⊃ (g ⊃ f) is in $\underline{K}$. To be positively valid in a brs, (A5it) must be acceptable relative to all $\underline{K}$ in the brs. Hence, for every $\underline{K}$, either ~[h > (g > f)] is acceptable at $\underline{K}$ (i.e., ~(g > f) is acceptable at $\underline{K}_h^*$) or h ⊃ (g ⊃ f) is in $\underline{K}$.

If h > (g > f) is not acceptable relative to $\underline{K}$, its negation is and the acceptability of (A5it) is trivial – likewise if ~h is in $\underline{K}$.

If h > (g > f) is acceptable relative to $\underline{K}$, (A5it) is acceptable relative to $\underline{K}$ if h ⊃ (g > f) is. Suppose that neither h nor ~h is in $\underline{K}$. In that event, h ⊃ (g > f) is acceptable relative to $\underline{K}$ if and only if g > f is acceptable relative to $\underline{K}$.

Thus, to construct a $\underline{K}$ that fails to satisfy (A5it) and, hence, refutes its positive validity, one must construct a $\underline{K}$ meeting the following conditions: (a) h > (g > f) is acceptable relative to $\underline{K}$, (b) neither h nor ~h is in K, and (c) g > f is not acceptable relative to $\underline{K}$ (so that ~(g > f) is). Under conditions (a) and (b), condition (c) can obtain only if ~g is in $\underline{K}$.

Example 4 may serve as an illustration of the general construction just offered. This example is not a counterinstance to (A5it). It does not establish the failure of (A5it) to be negatively valid. But it does show that it is not positively valid. And the failure of positive validity obtains whether or not RTGL or RTR conditionals are under consideration.

Is (A5it) negatively valid? It is for RTGL conditionals but not for RTR conditionals. That will become apparent shortly.

The fact that (A5) fails to be positively valid for RTGL conditionals implies a weakening of conditional logic for noniterated conditionals when a very modest dose of iteration is allowed. (A5it) is a substitution instance of (A5) so that (A5) fails to be positively valid. Nonetheless, there is a shadow version of (A5it) analogous to the shadow version (A6t) of (A6) that avoids the problem posed by cases like example 4.

(A5itt) h > (g > f) ⊃ [(T > h) ⊃ (g > f)].

In the case where neither h nor ~h is in $\underline{K}$, ~(T > h) is acceptable relative to $\underline{K}$ so that (A5itt) becomes acceptable as well provided RTGL is used. (This will be explained subsequently.)

Is (A5itt) positively valid? As will be explained shortly, it is when RTGL is used but not when RTR is adopted. Moreover, the positive validity for brsg's of (A5itt) does not reinstate the positive validity of (A5). However, it does secure the positive validity of the following:

(A5t) (h > g) ⊃ (T > h ⊃ g).

This result does not entirely settle the matter. McGee's counterexamples to modus ponens (one of which is example 1) consider predicaments seen to be cases where h > (g > f) is acceptable relative to <u>K</u>, h is in <u>K</u> so that T > h is acceptable relative to <u>K</u>, but g > f is not. Under these circumstances, (A5itt) is neither negatively nor positively valid. (A5it) is neither negatively nor positively valid. And the same holds for (A5).

The question to be addressed is whether the account of conditionals under inspection allows for such counterinstances. Are examples 1–3 genuine counterinstances to modus ponens?

I shall show that if the belief revision systems are brsg's, so that the conditionals are RTGL conditionals, the answer is negative. On the other hand, when the conditionals are RTR conditionals and recovery fails, examples 1–3 are counterinstances to (A5itt).

The salient difference between RTGL conditionals and RTR conditionals is that RTR conditionals are sensitive to whether the recovery condition on contractions holds or not, whereas RTGL conditionals are not. This means that revision according to Ramsey satisfies weaker postulates than AGM revision unless recovery obtains. However, the recovery postulate does make a difference (as I shall show) as to whether McGee's examples can undermine the negative validity of (A5itt).

With recovery, (A5itt) is both positively and negatively valid whether we consider RTGL conditionals or RTR conditionals. One cannot construct a McGee counterexample consonant with the recovery condition.

Without recovery such counterexamples are constructible for RTR. (A5itt) fails to be negatively (and, hence, positively) valid for RTR conditionals.

Let h > (g > f) be acceptable relative to <u>K</u> and h be in <u>K</u> (so that T > h is acceptable relative to <u>K</u>). If these conditions hold, <u>K</u> must contain the following:

(i) h ⊃ (g ⊃ f).
(ii) h.

As a logical consequence of (i) and (ii), K contains the following:

(iii) g ⊃ f.

In order to show that (A5itt) is not negatively valid and, hence, not positively valid, we need to identify conditions under which <u>K</u> contains (i)–(iii), h > (g > f) is acceptable at <u>K</u>, while ~(g > f) is acceptable as well. Example 4 does not allow <u>K</u> to contain h or ~h. Although it illustrates cases where instances of (A5it) are not acceptable at some consistent <u>K</u>, it does not illustrate cases where instances of (A5itt) are not acceptable. To do that, we need a counterinstance to (A5itt). h must be in $\underline{K}$. So we shall assume that (i)–(iii) are in $\underline{K}$ and h > (g > f) is acceptable at $\underline{K}$. Both the positive and negative validity of (A5itt) depend on whether there are any counterinstances where ~(g > f) is also acceptable at $\underline{K}$.

If neither g nor ~g is in $\underline{K}$, both RTGL and RTR demand that g > f be acceptable at $\underline{K}$ in virtue of the presence of (iii) in $\underline{K}$. No counterinstance to (A5itt) can occur. So attention must be restricted to cases where g is in $\underline{K}$ and to cases where ~g is in $\underline{K}$.

Our examination of these two cases will show that

(a) If RTGL is used, in both cases g > f must be acceptable at $\underline{K}$. No counterinstance to (A5itt) occurs.

(b) The same holds if RTR with recovery is used.

(c) If RTR without recovery is used, counterinstances to (A5itt) are constructible where g > f is not acceptable at $\underline{K}$ for both cases.

If recovery fails and RTR is used, the acceptability of h > (g > f) at $\underline{K}$ depends on whether g > f is acceptable at K^{*r}_h, which may be a proper subset of $\underline{K}$. Suppose that it is.

Suppose further that neither g nor ~g is found in the Ramsey contraction $\underline{K}^{-r}_h$ of $\underline{K}$ removing h and neither is added when h is restored to form K^{*r}_h. Even though either g or ~g must be in $\underline{K}$, neither of them need be in the Ramsey revision when recovery fails. To secure the acceptability of h > (g > f) at $\underline{K}$, (i) must have been retained in the Ramsey contraction. Hence, (iii) must be in the Ramsey revision. The absence of both g and ~g from K^{*r}_h and the presence of (iii) suffices to establish the acceptability of g > f at K^{*r}_h and with it the acceptability of h > (g > f) at $\underline{K}$. But the acceptability of g > f at K^{*r}_h does not imply its acceptability at $\underline{K}$.

Consider then the acceptability of g > f at $\underline{K}$. There are two cases to consider:

Case 1: g is in $\underline{K}$. To avoid the acceptability of g > f at $\underline{K}$, (iii) must be removed in the Ramsey contraction removing g. Once (iii) is removed, f will be removed as well. Hence, either (i) or (ii) must be removed. Suppose (i) is retained and (ii) – that is, h – is removed. Moreover, when g is returned on expansion to form the Ramsey revision, recovery fails by not restoring h – that is, (ii). So (iii) is not returned. g > f is not acceptable at $\underline{K}$. Yet h > (g > f) is acceptable by following the recipe given before. Give up h and, in so doing, remove g and f. Restore h. (iii) is in the resulting Ramsey revision by h and g > f is acceptable at K^{*r}_h even though ~(g > f) is acceptable at $\underline{K}$.

Thus, we have a counterinstance to (A5itt) in case 1. *The counterinstance could not have been constructed using RTR unless recovery is allowed to fail.*

Case 2: ~g is in $\underline{K}$. Contract removing ~g from $\underline{K}$. Suppose (iii) is withdrawn. So is f if it is in $\underline{K}$ at all. Suppose that h is withdrawn as well. When the contraction is expanded by adding g, (iii) is not returned to $\underline{K}$ and g > f proves unacceptable at $\underline{K}$. Returning to the recipe for establishing the acceptability of h > (g > f), we require only that when h is withdrawn at the contraction step that ~g be withdrawn (and with it (iii)) while (i) is retained. Moreover, when h is returned, ~g is not returned. Once more we have a counterinstance of (A5itt), which could not have been constructed without abandoning recovery and using RTR.

We have exhausted the opportunities for formally constructing counterinstances to (A5itt) so that claims (a)–(c) are sustained.

Examples 1 and 2 discussed earlier in this section exemplify case 2.

Let h = A Republican is not elected (in the U.S. election of 1992), g = Clinton is not elected, and f = Perot is elected. A person might very well be convinced that (i) = "Either a Republican or Clinton or Perot is elected," (ii) = "A Republican is not elected," (iii) = "Either Clinton or Perot is elected," ~g and "Clinton is not a Republican." In removing h from $\underline{K}$, both (i) and "Clinton is not a Republican" would not be removed since they are well entrenched in the background information. Consequently, ~g = "Clinton is elected" must be given up. In giving this up, (iii) and "Either Clinton is not elected or Perot is not elected" will both be given up because, in the absence of (ii), neither would be more entrenched than the other. Precisely the same items will be removed and retained in contracting from $\underline{K}$ by removing ~g. (i) and "Clinton is not a Republican" will be retained. But removing ~g would also reopen the question of (ii) and so lead to its removal. For the same reasons as before, both (iii) and "Either Clinton is the winner or Perot is not" will also be removed.

$\underline{K}^{*r}_{h}$ contains h. It will also contain (iii). But it will not contain ~g thereby revealing a breakdown of recovery. So adding g is an expansion yielding f and securing the acceptability of g > f at $\underline{K}^{*r}_{h}$ and h > (g > f) at $\underline{K}$.

On the other hand, adding g to $\underline{K}^{-}_{~g} = \underline{K}^{-}_{h}$ does not return (ii) or (iii) so that $\underline{K}^{*r}_{g}$ ($= \underline{K}^{*}_{g}$) does not contain f and g > f is not acceptable at $\underline{K}$. This is not only a coherent story but one that should reflect a commonsense response to the situation. We have a case 2 failure of modus ponens.

Example 2 also illustrates case 2. It utilizes a statistical example where the desirability of abandoning recovery may seem more compelling.

Example 3 illustrates case 1. Suppose as in case 2 that it is known that the spinner landed in an odd-numbered part (h). Suppose also it is known that it did not land on part 1 (g). We still are prepared to accept that if the spinner had landed on an odd-numbered part, then if it had not landed on part 1, it would have landed on part 3 (f). But we would continue to refuse to accept the claim that had not it landed on part 1, it would have landed on part 3. It might have landed on part 2.

Thus, (A5itt) and (A5t) fail to be negatively valid for RTR conditionals. So they are not positively valid after all.

Devotees of modus ponens will conclude that RTR conditionals are unacceptable or that recovery must be secured at all costs.

Keep in mind, however, that appeal to recovery and RTGL does not establish the positive validity of (A5) and (A5it) but of (A5t) and (A5itt). Only a pale shadow of modus ponens survives.

Moreover, RTR is more sensitive than RTGL to the way judgments of serious possibility conditional on suppositions of the implementation of experiments are made where the outcome is stochastic or uncertain. (See the example of the foolish gamble in section 2.3.)

Finally, the counterinstances to modus ponens according to RTR do often seem convincing as counterinstances.[16]

These considerations do not constitute a demonstration but they do suggest that once one has endorsed a suppositional interpretation of conditionals and attends to the informational-value-based variety, RTR appears to be superior to its leading competitor.[17] If RTR is adopted on the basis of these considerations, (A5) is neither positively nor negatively valid.

This discussion has rested content with showing that a logic of conditionals along the lines I favor is going to be rendered even more impoverished than the logic of "flat" or noniterated conditionals. There will be, to be sure, some structure to such a logic. It is well worth exploring what the character of the logic of iterated conditionals might be. That, however, is a task I will leave to others.

4.5 Undermining and Undercutting

Consider the following two claims.

(i) If you had come to the party, you would have met Grannie. (h > g)
(ii) If you had come to the party with Gramps, you might not have met Grannie. $\sim$(h $\wedge$ f > g)

When both (i) and (ii) are acceptable relative to $\underline{K}$, f may be said to *undermine* g with respect to h. However, it does not necessarily *undercut* g with respect to h. f undercuts g with respect to h when (i) and (iii) are acceptable relative to $\underline{K}$.

(iii) If you had come to the party with Gramps, you would not have met Grannie. (h $\wedge$ f > $\sim$g)

If conditionals were truth functional material implications, there could be neither undermining nor undercutting. However, conditional logic allows for it.

To be sure, there are restrictions imposed on the extent to which undermining or undercutting may take place. Thus, (A10) states that f cannot undermine g with respect to h if ~(h > ~f) is acceptable relative to $\underline{K}$. (A10), however, is positively valid only if either RTGL is used or recovery holds for admissible contractions. If RTR is used without recovery, undermining can take place even though ~(h > ~g) is acceptable. Suppose X knows that coin a has been tossed and has landed heads. According to RTR, the conditional judgment "If Y bets that coin a will land heads, Y will win" is acceptable according to X. However, "If Y bets that a will land heads and a is tossed, then Y will win" is not acceptable. (A10r) indicates that not only should ~(h > ~f) be acceptable but so should ~(h > f).

These sufficient conditions for preventing f from undermining g with respect to h are also sufficient conditions for preventing f from undercutting g with respect to h. However, undercutting is prohibited in cases where undermining is possible.

(K*3) (inclusion) and the fact that when h ∧ f is consistent with consistent $\underline{K}$, $\underline{K}_h^+ \subseteq \underline{K}_{h \wedge f}^+$ where $\underline{K}_{h \wedge f}^+$ is consistent imply the following:

(No Undercutting Principle) If g is in $\underline{K}_h^*$ and ~f is not in $\underline{K}_h^+$, ~g is not in $\underline{K}_{h \wedge f}^*$.

Suppose that ~f is not in $\underline{K}_h^+$ and g is in $\underline{K}_h^*$. g is in consistent $\underline{K}_h^+$ by (K*3) and, hence, in consistent $\underline{K}_{h \wedge f}^+$. So ~g is not in $\underline{K}_{h \wedge f}^+$. By (K*3), ~g is not in $\underline{K}_{h \wedge f}^*$.

The argument from (K*3) to no undercutting continues to hold if we replace * with *r and RTGL with RTR. Hence, the following is a valid principle of conditional logic whether or not we deploy RTGL or RTR:

(NU) [(h > g) ∧ (h ∧ f > ~g))] ⊃ (h ⊃ ~f).

4.6 Reasoning from Suppositions

The Ramsey test approach toward the acceptability of conditionals interprets them as conditional judgments of serious possibility where the conditions are "suppositions" judged to be true for the sake of the argument. We are now in a position to say a little more about what it is to suppose that h is true for the sake of the argument.

Reasoning from a supposition is not merely a matter of considering the logical consequences of the supposition along with the propositions consistent with it. It is reasoning from the supposition that h is true along with a background of other assumptions. The problem is to identify what that background is.

In discussing Ramsey tests for the acceptability of conditionals, three kinds of background have been explored:

(1) When consensual backgrounds are the shared assumptions of the agent's belief state and some other target belief state, reasoning from supposition in such cases is to expand the consensus corpus and identify its contents.

(2) When RTGL is invoked, the background corpus for reasoning on the supposition that h is true is the admissible contraction $\underline{K}^{-}_{\sim h}$ of $\underline{K}$ by removing $\sim$h.

(3) When RTR is invoked, the background corpus for reasoning on the supposition that h is true is the Ramsey contraction $\underline{K}^{-r}_{h} = (\underline{K}^{-}_{\sim h})^{-}_{h}$.

No matter which of these interpretations of reasoning from suppositions is adopted, reasoning from suppositions adopted for the sake of the argument is deductive inference from a transformation of $\underline{K}$. That is to say, judgments of serious possibility conditional on h are judgments of serious possibility relative to a deductively closed theory.

Such inference has two salient characteristics for our purposes. (a) It is *explicative*. To recognize the logical consequences of what one fully believes or to identify what is seriously possible given one's full beliefs is to clarify or explicate doxastic commitments one already has. In itself, it involves no change in such commitments. (b) Deductive inference is *monotonic* in the sense that if g follows from a set $\underline{P}$ of premises logically, it follows from any set $\underline{P}^{*}$ that contains $\underline{P}$.

In this respect, the "logic" of suppositional reasoning is nothing more than monotonic deductive logic.

We have, however, left something out of account. Reasoning from suppositions involves deductive reasoning. But it is reasoning explicating the commitments represented by the expansion of a set of background assumptions by adding supposition h where the background assumptions are not items from the agent's current state of belief but are a selection from them. The items selected constitute a contraction of the current corpus of full belief determined either by an appeal to shared agreements with other views or by an assessment of the importance of the various items in the current doctrine to the satisfaction of the inquirer's demands for information. The inferential reasoning is deductive, explicative, and monotonic; but the network of background assumptions and suppositions constituting the premises of the reasoning are themselves the product of the stipulations and transformations based on those stipulations. In identifying Ramsey tests for conditionals and characterizing conditional logic, we were, in effect, studying the properties of the transformations deployed.

I have not objected to constructing a logic of conditionals based on the properties identified for such transformations; but it is misleading to call such a system a logic of *inference* from suppositions. Thus, in particular, we should not say that when g is in $\underline{K}^*_h$, g is legitimately inferred from h relative to $\underline{K}$. It is misleading to claim that g is, relative to $\underline{K}$, a special kind of consequence of h distinct from a monotonic logical consequence.

g is indeed inferred from h and an optimal contraction of $\underline{K}$ removing $\sim$h. This inference is purely deductive and monotonic. g is not, however, inferred from h and $\underline{K}$ itself except in the case where the recommended contraction is identical with $\underline{K}$ itself.

There is to be sure another kind of reasoning involved – to wit, the derivation of the contraction removing $\sim$h from K. But that reasoning is of a practical variety concerned with the best way to minimize loss of informational value given the stipulated demand to remove $\sim$h. To reach a conclusion here, one needs a characterization of all the available contraction strategies removing $\sim$h from $\underline{K}$ and an assessment of their informational values. Then the injunction to minimize loss of informational value and some method for breaking ties in optimality will yield a recommendation.

Practical reasoning is itself subject to regulation by principles of rationality and, hence, to logics of value judgment, credal probability judgment, and rational choice as well as full belief. And given the particular epistemic goal of minimizing loss of informational value in choosing among contractions removing h, one can derive conclusions about the kinds of contractions relative to which reasoning for the sake of the argument may rationally operate. However, the practical reasoning involved here ought not to be confused with the task of reasoning from h and the contraction strategy chosen. Suppositional reasoning is reasoning *from* suppositions and not *to* suppositions. Such reasoning is from suppositions in conjunction with a background derived by practical reasoning *from* the initial doctrine $\underline{K}$ and the goals as represented by the *M*-function to the admissible contraction. (This is so for Ramsey contractions as well as admissible ones.)

Observe that so construed, an account of suppositional reasoning cannot constitute a "logic" of belief change or "theory" change. There is no doubt that in belief change, one moves from one state of doxastic commitment to another. When an inquirer aims to obtain an answer to some as yet unsettled question, the inquirer identifies (via abduction) a set of potential answers to the question under study. These answers figure as cognitive options – potential expansion strategies. The inquirer may, in the course of deliberation, engage in suppositional reasoning by contemplating, let us say, the expansion of $\underline{K}$ by adding h. But he might also consider other options, such as expansion by adding f or by adding the disjunction of h and f.

To have a satisfactory account of rational belief change one needs an account of criteria for choosing among potential expansion strategies for the purpose of promoting cognitive goals. One needs to justify adding h to the corpus. In supposing that h for the sake of the argument, no such justification is needed.

A similar observation applies to contraction. In contraction, one needs to show that there is good reason to give up information, to argue that it is better to give up one proposition rather than another and that, given that a proposition h is to be given up, such and such a strategy is the best for doing this. In suppositional reasoning, only the third of these tasks needs to be addressed. One does not need to justify the decision to remove something. One is doing that for the sake of the argument. Nor does one need to consider whether to remove h or to remove g. That too is built into the supposition stipulated.

Thus, the "logic of theory change" so elegantly presented in Alchourrón, Gärdenfors, and Makinson (1985) is only misleadingly so called; for it provides no account of how to justify adding one theory rather than another to the evolving doctrine, no account of criteria for parameter estimation in statistics, no account of when one should and when one should not give up settled information, and the like. Without criteria for legitimate expansions of belief states, there is no logic of belief change. There is at best a logic of pseudobelief change – that is, change for the sake of the argument.

There is a similar complaint to be registered concerning contraction. There is nothing in the Alchourrón, Gärdenfors, and Makinson approach describing conditions under which contracting by removing h is to be recommended over contracting by removing h' or over not contracting at all. The only part of the story about contraction they tell concerns the following question: Given that one is going to contract by removing ~h, what criterion should be used to choose an admissible contraction? This question is certainly an important one for an account of rational belief change. But it deals with only that aspect of the problem that is relevant to supposition for the sake of the argument.

In the case of revision, Alchourrón, Gärdenfors, and Makinson are not only incomplete but misleading. For them, revision (i.e., AGM revision) is a basic mode of belief change. For someone concerned with rational belief change (and, I presume, someone interested in the "logic " of belief change is interested in rational belief change), no revision can be directly justified if respect for truth is to be observed. An inquirer who is certain that ~h is true and who is concerned to avoid error should never deliberately replace ~h with h as a revision requires. One may obtain the net effect of revision by first opening one's mind by contracting to a state of suspense if such a move is justified. From that vantage point one might find good reason to expand by adding h. All legitimate belief change should be decomposable into sequences of legitimate expansions and contractions (Levi, 1991).

Defenders of the importance of revision insist that sometimes inputs inconsistent with current doctrine are thrust upon us. To avoid inconsistency, we should engage in revision. As I have argued elsewhere (Levi, 1991; see also Hansson, 1991), sometimes the net effect is not retaining the input and giving up something else but giving up the input and retaining the background. And sometimes it is giving up both input and background. By insisting that revision is basic, Alchourrón, Gärdenfors, and Makinson are tacitly presupposing that when information inconsistent with the current corpus is legitimately added, the net effect should be to retain the input and remove some item from the background. That claim may sometimes hold. But frequently it will not. In an adequate account of rational belief change, we need to identify conditions under which one does retain the input and removes items from the background and conditions under which one does the opposite. Alchourrón, Gärdenfors, and Makinson preclude consideration of this important issue. In this case, their theory is not merely incomplete as an account of rational belief change, it is seriously misleading.

The idea of revision is, I submit, of marginal importance in an account of rational belief change or, if one likes, to a "logic" of belief change. It is rather more important in a discussion of suppositional reasoning. When h is inconsistent with $\underline{K}$, to suppose that h is true is to explore the logical implications of h together with the admissible contraction removing ~h from $\underline{K}$. That is to say, it is to explore, for the sake of the argument, the use of the (AGM or Ramsey) revision of $\underline{K}$ by adding h as a standard for serious possibility. There is no need to justify the contraction step or the subsequent expansion step yielding the revision. By supposing that h is true, one is proposing to explore the consequences of h and the best contraction of $\underline{K}$ removing ~h.

Thus, the contribution of Alchourrón, Gärdenfors, and Makinson is best seen as a contribution to an account of reasoning for the sake of the argument and not as an account of the logic of belief change.

To say this is not to denigrate the contribution of Alchourrón, Gärdenfors, and Makinson to discussions of rational belief change. A firm grasp on reasoning for the sake of the argument is of paramount importance to an account of rational belief change. The large point I am pressing here in elaborating upon earlier work (Levi, 1991) is that it is not the whole story. To suppose otherwise can be extremely misleading.

Ramsey test criteria for the acceptability of conditionals construe conditionals as conditional judgments of serious possibility. The condition is supposed true for the sake of the argument. The logical consequences of the supposition and a contraction of the current corpus are explored. Alternatively, a revision of the current corpus by the supposition is used, for the sake of the argument, to make judgments of serious possibility. As we have seen, the Alchourrón, Gärdenfors, Makinson theory of supposition can be used in a fundamental way to give an account of Ramsey test conditionals. To be sure, some rather substantial amendments need to be made. Consensus Ramsey tests replace Gärdenfors contraction with consensus contraction. This kind of contraction is not informational value based and, hence, fails to deploy the Alchourrón, Gärdenfors, Makinson theory in any interesting way. RTR also deviates from that theory by introducing Ramsey revisions as of fundamental importance in suppositional reasoning and by rejecting the recovery condition on contractions. But Ramsey revisions, like AGM revisions, are constructed with the aid of the notion of an informational-value-based contraction.

I have introduced another important deviation from the Alchourrón, Gärdenfors, Makinson theory here. Because contraction removing some sentence h from K can be implemented in so many different ways, doing so legitimately should be understood as choosing an option (contraction strategy removing h) that promotes some kind of cognitive goal. Previously (in Levi, 1980, and in earlier publications), I suggested that the goal should be to minimize loss of informational value. As Gärdenfors (1988) rightly points out, I did not explain how to implement this idea in detail. Alchourrón, Gärdenfors, and Makinson do offer criteria for identifying recommended contractions removing h from <u>K</u>; but these proposals fall far short of being satisfactory: (a) They rule out certain contraction strategies (meets of saturatable contractions that are not maxichoice) without even saying so. (b) Even if we rest content with the contraction strategies they consider (meets of subsets of maxichoice contractions), they rely exclusively on an evaluation of maxichoice contractions and do not derive an evaluation of meets of subsets of maxichoice contractions that explains how choosing anything other than a maxichoice contraction could be optimal. That is to say, they fail to give a decision theoretic rationalization of their criteria.

Gärdenfors (1988) has subsequently furnished us with an account of epistemic entrenchment, which can be used to derive an evaluation of meets of subsets of maxichoice contractions from evaluations of maxichoice contractions. However, the methods of evaluation can be extended to meets of subsets of saturatable contractions and the initial evaluation of saturatable contractions can be derived from an informational value determining probability function M as explained previously (Levi, 1991). Once this is done, it becomes clear that a non–ad hoc rationalization of the restriction of contractions to meets of subsets of maxichoice contractions cannot be given. The recovery postulate becomes utterly untenable.

If AGM revision were the only transformation of interest, the failure of recovery would not matter, as Makinson pointed out (1987). In an account of rational belief change, AGM revision is of no special interest at all (except for the sociological fact that many philosophers and logicians have been interested in it). In an account of reasoning from supposition, I have suggested that Ramsey revision is better both from a philosophical point of view and with respect to presystematic judgment. And Ramsey revision is sensitive to the failure of recovery.

Thus, even as an account of suppositional reasoning, the approach of Alchourrón, Gärdenfors, and Makinson needs some substantial modification. The modifications, however, do not change what seem to me the core ideas of the approach.

With this understood, we should be in a position to conclude that the Alchourrón, Gärdenfors, Makinson theory is identical with the Ramsey test theory of conditionals. If we set rhetorical differences to one side, both theories are theories of suppositional reasoning and, indeed, are the same theory of such reasoning.

As we have seen, however, Gärdenfors does not understand matters in this way. Because of his commitment to the Reduction thesis and the idea that the Ramsey test should be applied to iterated conditionals, he has felt obliged to surrender (K*r4) or the preservation condition as a constraint on the sorts of revisions operative in Ramsey tests and, hence, has driven a wedge between the theory of conditionals and suppositional reasoning. I have argued that Gärdenfors's loyalty to the reduction condition that has led to this curious turn of affairs is unwarranted. Once it is overcome, conditional reasoning and the (improved) theory of Alchourrón, Gärdenfors and Makinson are seen as different expressions of the same approach to suppositional reasoning.

5 Nonmonotonicity in Belief Change and Suppositional Reasoning

5.1 Inference and Transformation of Belief States

In recent years, logicians and computer scientists have produced a substantial literature on what is sometimes called "nonmonotonic reasoning" or "nonmonotonic logic." Identifying the domain under investigation by students of nonmonotonic reasoning is not easy. In particular, it is not always clear whether the topic is reasoning involved in genuine belief change or in suppositional thinking. The same ambiguity in the domain of intended applications found in studies of belief revision appears here. For reasons similar to those pertaining to theories of belief revision, a charitable construal of nonmonotonic reasoning should equate it with some form of suppositional reasoning.

Gärdenfors and Makinson (1991) have already proposed exploring the similarities between theories of belief revision and nonmonotonic reasoning. In particular, they have suggested that the claim that h nonmonotonically implies f relative to $\underline{K}$ (h $|\!\sim$ f @ $\underline{K}$) should be interpreted as holding if and only if f is in $\underline{K}^*_h$. Because theories of nonmonotonic reasoning have tended to eschew iteration, the test for nonmonotonic implication resembles the positive RTGL version of the Ramsey test for the acceptability of conditionals. When denials of nonmonotonic implication are allowed, the corresponding test is negative RTGL.

Unlike Gärdenfors and Makinson, I do not require that the revision operator be AGM revision. In the previous discussion, I have argued in favor of the importance of Ramsey revision. In the discussion that follows, I shall urge the importance of considering still a larger class of *expansive transformations* (see section 5.3 for a characterization of this class).

Students of nonmonotonic reasoning have tended to focus on features of suppositional reasoning acknowledged to be formally similar to what was called undermining and undercutting in section 4.5. To use the trite and true example of Tweety, the supposition that Tweety is a bird nonmonotonically implies that Tweety is able to fly. But adding the information (consistent with Tweety's being a bird) that Tweety is a penguin not only undermines the reasoning but undercuts it.

The examples in section 4.5 may be adjusted to illustrate the same point. Supposing that you come to the party tomorrow nonmonotonically implies that you will meet Grannie. But supposing that you come to that party with Gramps does not do so. Indeed, it nonmonotonically implies that you will not meet Grannie.

In section 4.6, I contended that suppositional reasoning of the sort explicated using AGM or Ramsey revision does not characterize, in general, *inference* from suppositions. Insofar as nonmonotonic reasoning is restricted to this kind of suppositional reasoning, it cannot be considered a form of inference either.

To be sure, such reasoning involves the use of inference. But the inference is explicative, monotonic, deductive inference from a contraction of the corpus $\underline{K}$ designed to yield a corpus to which the supposition can be consistently added in the case of AGM revision and to yield a corpus relative to which the supposed premises are left in suspense. The contraction of $\underline{K}$ is a transformation of $\underline{K}$ but it is not itself a step in an argument sustaining an inference from the suppositional premises and $\underline{K}$ to a conclusion.

Nonmonotonicity is generated by belief revisional reasoning when such reasoning yields undermining of reasoning from h to g by f relative to $\underline{K}$ as described in section 4.5.

K*8 and the Gärdenfors–Makinson strategy for equating nonmonotonic implication with AGM revision * guarantee that inconsistency of f with $\underline{K}_h^*$ is a necessary condition for undermining. To satisfy this condition, h $\wedge$ f is inconsistent with $\underline{K}$ whether or not h is consistent with $\underline{K}$. Hence, revision by adding h $\wedge$ f to $\underline{K}$ requires giving up sentences in $\underline{K}$. The presence of a case of nonmonotonicity calls for the contravention of belief.

When Ramsey revision is used, undermining presupposes that f is inconsistent with $\underline{K}_h^{*r}$ or entailed by it. When f is inconsistent, the situation is the same as with AGM revision. If f is entailed by $\underline{K}_h^{*r}$, then, to obtain undermining, $\underline{K}_{h\wedge f}^{*r}$ must be a proper subset of $\underline{K}_{h\wedge f}^{+}$. If $\underline{K}_{h\wedge f}^{*r} = \underline{K}_{h\wedge f}^{+}$ (as when recovery is satisfied), $\underline{K}_h^{*r} = \underline{K}_h^{+}$ as well. In that case, adding f cannot undermine reasoning from h to g. Such reasoning is straightforward monotonic *inference*.

In sum, when attention is focused on AGM or Ramsey revision in explicating nonmonotonic reasoning, the only cases where nonmonotonicity can arise occur when reasoning from h to g is undermined by f because revising $\underline{K}$ by adding h $\wedge$ f calls for removing assumptions from $\underline{K}$. In all of these cases, the reasoning requires a nondegenerate contraction step in at least one of the revision transformations being compared. For this reason, nonmonotonic reasoning of this kind cannot be inference to the conclusion g from the suppostion h $\wedge$ f relative to the corpus $\underline{K}$.

Makinson and Gärdenfors (1991) are sensitive to the point I am making and seek to respond to it by making a distinction rehearsed in the following two paragraphs:

There are two ways in which theory change can be said to be nonmonotonic. Firstly, if one sees the revision of a theory A by a proposition x as an *operation on A modulo x*, it is much "worse" than merely nonmonotonic. It does indeed fail monotony: we may have $A \subseteq B$ but $A^*x \nsubseteq B^*x$. . . . But, and this is our main point, on all variants the operation fails an even more fundamental condition, of reflexivity or inclusion. This is the condition that if A is any set of propositions then each element of A may be inferred from A, i.e., that $A \mid\sim x$ for all $x \in$ A where $\mid\sim$ is the nonmonotonic inference relation; or in other notation, that $A \subseteq C(A)$ where $C(A)$ is $\{x: A\mid\sim x\}$. Relations $\mid\sim$ . . . failing this condition may well represent processes of reasoning, but they hardly deserve to be called inferences monotonic or otherwise. And such is the case of revision. We may very well have $A \nsubseteq A^*x$; indeed, this will happen *whenever* the proposition x is consistent but is inconsistent with A. (1991, p. 188)

Makinson and Gärdenfors are clearly concerned with explaining how AGM revision may be regarded as licensing certain kinds of nonmonotonic inferences. The proposal considered in the paragraph just cited points out that if the set $\underline{A}$ is regarded as a set of premises in the inferences licensed by the revision $\underline{A}_x^*$, inferences from premises to themselves will not always be legitimated because of the belief contravening character of AGM revision.

Makinson and Gärdenfors continue as follows:

On the other hand, if we change our *gestalt* and see the revision of a theory A by a proposition x as an *operation on x modulo A*, we get a quite different sense in which theory revision is nonmonotonic. The revision of A by x always contains x. . . . In other words, the revision of a theory by a proposition x can be seen as an inference – but as *inference from the proposition x* rather than from the theory A. This inference operation will moreover be nonmonotonic, in that we may have $x \vdash y$ (where $\vdash$ is classical consequence), but not $A^*x \vdash A^*y$. . . . With this *gestalt*, we are in a position to understand the link between nonmonotonic logic and the logic of theory change. The key idea is this:
(1) See the revision of a theory A by a proposition x, forming a theory A^*x, as a form of nonmonotonic inference *from* x, (Makinson and Gärdenfors, 1991, p. 188)

What sort of *inference* is licensed by an operation on h modulo $\underline{K}$? That is to say, what are the premises of the inference and what is the conclusion? Gärdenfors and Makinson clearly recognize in the first citation that $\underline{K}$ cannot be included in the premises. They suggest that the premise should be h instead. But h cannot in general be the sole premise.

Construing revision of $\underline{K}$ by adding h as nonmonotonic inference from h is just as questionable as construing $\underline{K}_h^*$ as inference from $\underline{K}$ to the revision.

Formally, revision is a transformation of $\underline{K}$. It represents a change in belief state. The change may be "for the sake of the argument" where no attempt at justification is made but the change is merely stipulated or it may be one for which a justification is offered. But *what* is justified in the latter case is the implementation of the change represented by the transformation. So neither the change nor the transformation representing it are the inferences justifying the change.

The premises of the inference are h combined with the set of premises in $\underline{K}^-_{\sim h}$. But the inferences from this set of premises are all *monotonic* and deductive.

The only inference involved in AGM or Ramsey revision concerns what may be inferred from h together with a "background theory" that is a transformation (to wit, a contraction) of the initial corpus $\underline{K}$. The inference itself is deductive inference from these premises. One can represent the revision as the set of logical consequences of h and the admissible or Ramsey contraction removing ~h.

There is a third alternative. We can consider the inference to be from the supposition h and the initial uncontracted background theory $\underline{K}$. But precisely for the reasons Makinson and Gärdenfors cite in the first paragraph, this cannot be an inference; for the revision of $\underline{K}$ by h need not contain all elements of $\underline{K}$.

Rather than speaking of inference, it would be preferable to speak of nonmonotonic transformations of belief states in suppositional reasoning or of nonmonotonic reasoning that is not inference.

Are all examples of nonmonotonic reasoning of this kind?

Many students of nonmonotonic reasoning appear to be ambivalent about this matter. Thus, R. C. Moore (1983) writes:

If we know that Tweety is a bird, in the absence of evidence to the contrary, we will normally assume that Tweety can fly. If, however, we later learn that Tweety is a penguin, we will withdraw our prior assumption. If we try to model this in a formal system, we seem to have a situation where a theorem P is derivable from a set of axioms S, but is not derivable from some set S′ that is a superset of S.

It is clear that when Moore writes that the claim that Tweety can fly will be withdrawn when the information that Tweety is a penguin is added to the belief state, he is thinking of a sort of transformation that looks very much like a revision. But Moore also seems to suggest that if we initially believe or suppose that Tweety is a bird, we then "assume" (does this mean "conclude"?) that Tweety can fly and presumably also that Tweety is not a penguin "in the absence of evidence to the contrary." This suggests that the agent begins with a corpus lacking information about Tweety's species or aviational capabilities. The agent then expands by adding "Tweety is a bird." Having done this he then goes further and adds "Tweety can fly" and presumably also that Tweety is not a penguin.

The addition of "Tweety is a bird" to $\underline{K}$ does not logically imply that Tweety can or cannot fly. Nothing entails that Tweety is or is not a penguin. Yet, nothing in K is given up in order to "leap" to the conclusion that Tweety can fly and is not a penguin. The objection lodged against considering reasoning based on AGM or Ramsey revision as inference from the supposition and $\underline{K}$ does not apply. But the inference is not deductive. The inference is *ampliative* rather than explicative or deductive. I shall often call it inductive inference from the initial corpus and the assumption that Tweety is a bird.

One can analyze this process in one of two ways: (a) From initial $\underline{K}$ and b = Tweety is a bird, it is inferred inductively that $\sim$p = Tweety is not a penguin. $\underline{K}$, b, and $\sim$p are then taken to imply logically that f = Tweety can fly. (b) From $\underline{K}$ and b, it is inferred inductively that f and from that it is inferred deductively that $\sim$p.

These are two alternative ways to represent the inference from b to f. Whichever one uses, the reasoning is by broad consensus considered non-monotonic. It also avoids the strictures against considering it an inference to which Makinson and Gärdenfors point. The inference from b and $\underline{K}$ to f is, in both cases, a genuine non-belief-contravening and ampliative inference.

If the nonmonotonic reasoning involved here is a genuinely ampliative inference from b and $\underline{K}$ to f, how may it be undercut?

Can some new premise such as p be added to $\underline{K}$ and b that does not warrant the inference that f (undermining) and, indeed, warrants the inference that $\sim$f (undercutting)? The interesting point is that we can do this without giving up any item in $\underline{K}$ or any item in $\underline{K}_b^+$. Just add p (which entails b) to $\underline{K}$. This yields $\sim$f.

This process does not contravene the contents of $\underline{K}$ or even the expansion of this by adding b. It only contravenes the results $\underline{I}$ of induction from $\underline{K}_b^+$. $\underline{I}$ contains $\sim$p so that adding p to it calls for a revision. But adding p to $\underline{I}$ is not critical to determining whether the inductive inference from b and $\underline{K}$ to f is nonmonotonic. To show that the inference is undercut by the addition of p, p must be added to the set of premises consisting of $\underline{K}$ and b and not to the set of premises consisting of members of $\underline{I}$.

It appears that the Tweety example is a case of genuine nonmonotonic inference. Neither of the reasonings compared to show the nonmonotonicity requires nondegenerate contraction of $\underline{K}$. Moreover, the inference is ampliative or inductive.

Moore (and not Moore alone) does not appear to see matters quite in this fashion. Instead of comparing the results of reasoning from $\underline{K}$ and b with the results of reasoning from $\underline{K}$ and p, in the passage cited, Moore compares the results of adding b to $\underline{K}$ and drawing the inductive conclusions constituting $\underline{I}$ with the results of adding p to $\underline{I}$. This latter move calls for belief contravention of the inductively extended theory. $\underline{I}$ must be contracted by removing f. A deductive and monotonic inference is then drawn from this contraction and p to ~f. The net effect is a revision of $\underline{I}$. This transformation is nonmonotonic but no inference from p and $\underline{I}$ to ~f is involved.

Thus, even though this and many other examples of nonmonotonic reasoning discussed in the literature can be properly identified as instances of nonmonotonic inference of the ampliative or inductive type, there is a widespread tendency illustrated by what Moore says to argue for the nonmonotonic character by appealing to a belief revision of a corpus distinct from $\underline{K}$ – to wit, the corpus $\underline{I}$. But nonmonotonicity, according to Makinson and Gärdenfors must guarantee that the reasonings from b and from p should be relative to the same corpus. On this point, I agree with them.

I am claiming that all nonmonotonic reasonings fall into two categories (or sequences of these):

(A) those that are ampliative inferences from the initial corpus $\underline{K}$ and the suppositional premises; and

(B) those that are noninferential transformations of the initial corpus by the suppositional premises involving a contraction of that corpus and a monotonic explicative deductive inference from that contraction and the suppositional premises.

The informal discussions of Moore and other students of nonmonotonic reasoning suggest that the intended applications they have in mind are focused primarily on ampliative type (A) inference (although other types may not be excluded). Nonmonotonic reasoning is commonly held to involve leaping to conclusions.

Nonetheless, when it comes to exhibiting the structure of such nonmonotonicity, Moore and others engage in an effort to make such reasoning look as much like belief-contravening, noninferential transformations of type (B) as possible. Indeed, many authors attempt to suppress the ampliative and inductive aspects of the matter because they mean to resist the intimation that probabilistic or statistical considerations might be relevant. Others, like Gabbay (1985) and Makinson (1989) have been interested in understanding nonmonotonic reasoning as a special case and monotonic reasoning as another of a more general type of "cumulative" reasoning. Makinson and Gärdenfors, who are open and sensitive to questions about statistical and inductive reasoning, explore the extent to which they can assimilate nonmonotonic reasoning to the model based on AGM revision.

Nonmonotonic *inference* cannot be so assimilated. I do not mean this to be a mere logomachy concerning the correct usage of "inference." The claim would be uninteresting if ampliative inference were never legitimate. I claim that it often is. Both noninferential and inferential nonmonotonic reasonings are important.

Makinson and Gärdenfors (1991) represent "g is in $\underline{K}_h^*$" as "h $|\sim$ g @ $\underline{K}$" and use this representation to derive postulates about the nonmonotonic inferential relation "$|\sim$ @ $\underline{K}$" from postulates about revision. Such a relation licenses inferences from suppositions and contractions of $\underline{K}$ prepared to accept such suppositions. However, since the contraction of $\underline{K}$ deployed in drawing inferences from h may be different from the contraction of $\underline{K}$ used in drawing inferences from h and f, the set of inferences from suppositions does not constitute a set of inferences from a fixed set of background assumptions. g may be legitimately inferred from h and the contraction or Ramsey contraction of $\underline{K}$ but not from the contraction or Ramsey contraction together with h and f. The relation $|\sim$ @ $\underline{K}$ is nonmonotonic but it does not license *inference*.

Like conditional judgment, the relation captures the kind of nonmonotonicity involved in belief contravening suppositional reasoning. The initial, informal definition of nonmonotonic reasoning offered undermining as a critical mark of the presence of nonmonotonicity. As we have seen, undermining is a phenomenon that is allowed by conditional logic interpreted as a logic of belief-contravening suppositional reasoning.

For precisely this reason, the relation $|\sim$ @ $\underline{K}$ as explored by Gärdenfors and Makinson is a useful one. Strictly speaking, however, it is not an inferential relation. The satisfaction of such relations does not license inferences exclusively but belief-contravening suppositional reasoning as well. Such relations are *pseudoinferential relations*.

Pseudoinferential relations like conditionals may be characterized by reference to transformations of belief states. The link thereby established between conditionals and pseudoinferences affords a helpful vantage point for fresh consideration of the logic of Ramsey test conditionals and of suppositional reasoning for the sake of the argument. The critical comments just addressed to Makinson and Gärdenfors concerning their efforts to build bridges between their proposals concerning belief change and nonmonotonic reasoning are not intended to question the importance of their contribution but the misleading conflation of inference and reasoning.

Are there genuinely nonmonotonic inferences? We do not need to look to the much abused Tweety to find them. They include statistical reasonings of both qualitative and quantitative varieties, inductive inferences, theory choice, and predictive inference. We rule out the prospect logically consis-

tent with what we believe that the contents of a kettle of water standing on
a surface at room temperature will freeze while the kettle turns red hot. We
conclude that a large number of tosses of a fair coin will yield heads approximately 50 percent of the time. No such inferences are involved in either
AGM or Ramsey revisions. If, as I think, the intended applications of nonmonotonic reasoning proposed in the computer science literature do require
consideration of nonmonotonic *inference* and not merely nonmonotonic
transformations of the belief-contravening variety, conflating the reasonings licensed by nonmonotonic pseudoimplication relations with nonmonotonic inferences underrates the importance of taking inductive inference
into account in offering accounts of suppositional reasoning and of rational belief change.

Revision (of one or the other variety) is the only type of nonmonotonic
transformation of $\underline{K}$ that has been considered thus far. Expansion is the
only type of monotonic transformation. If we rest content with these kinds
of transformations, we are not going to come to grips with the subject of
nonmonotonic, ampliative inference. Our net must be cast wider.

The next step in this discussion will be to propose a characterization of a
larger class of transformations within the framework of which a useful distinction between monotonic and nonmonotonic transformations may be
made. Once this framework has been constructed, we shall be in a position
to see whether the account of conditionals and suppositional reasoning can
be extended in an interesting and important way to cover other types of
transformations than those we have been considering and to determine
whether, in that framework, we can find a useful role for something that
properly may be considered nonmonotonic inference.

5.2 Expansive Transformations

The contrast between monotonic and nonmonotonic transformations is a
distinction between transformations of a corpus $\underline{K}$ both of which successfully add some sentence h to $\underline{K}$. That is to say, the transformation function
is a function $T(\underline{K}, h)$ of two arguments – a potential corpus (I shall restrict
consideration to consistent corpora) and consistent sentence h in $\underline{L}$ where
the value of the function is a corpus containing h. The expansion function
and Gärdenfors, Ramsey, and consensus revision functions are *additive* in
this sense.

Contraction removing h is not an additive transformation. We could
consider extending the distinction between the monotonic and nonmonotonic to cover contractions that would end up in the second category. But the
contrast between monotonic and nonmonotonic transformations of interest is a distinction between additive transformations.

For our purposes, however, attention will be focused on a smaller class of additive transformations – those I shall call *expansive*. Our first order of business, therefore, is to give a more formal characterization of expansive transformations.

> $T(\underline{K},h)$ is an *expansive transformation in the narrow sense* if and only if it is of one of the following types:
> (1) An expansion of $\underline{K}$ adding h.
> (2) An expansion adding h of an admissible contraction of $\underline{K}$ removing ~h if present.
> (3) An expansion adding h to a Ramsey contraction of $\underline{K}$ removing h if present and ~h if present.

Expansive transformations in the narrow sense include expansions, AGM revisions, and Ramsey revisions. Consensus revisions are not included. Nor are imaging revisions. They do belong to the larger class of additive transformations and, if it were desirable to do so, we could enlarge the scope of the discussion to include them and other sorts of additive transformations. But imaging transformations do not represent conclusions of suppositional reasoning in a sense satisfying the conditions of adequacy for Ramsey test conditionals. Strictly speaking, neither do consensus revisions. Our focus here is on such suppositional reasoning. Consequently, I am beginning with the three kinds of transformations just listed and generating a larger range of expansive transformations from them.

> $E(\underline{K})$ is a *generalized expansion* of $\underline{K}$ if and only if $E(\underline{K})$ is a function from potential corpora representing belief sets to potential corpora representing belief sets, $\underline{K} \subseteq E(\underline{K})$ and $E(\underline{K})$ is consistent if $\underline{K}$ is.

In the following discussion, E is always a generalized expansion function.

> $T(E, \underline{K}, h)$ is an *extended expansion* of $\underline{K}$ *by adding h relative* to E if and only if $T(E, \underline{K}, h) = E(\underline{K}^+_h)$.

> $T(E, \underline{K}, h)$ is an *extended AGM revision of $\underline{K}$ by adding h relative* to E if and only if $T(E, \underline{K}, h) = E(\underline{K}^*_h)$.

> $T(E, \underline{K}, h)$ is an *extended Ramsey revision of $\underline{K}$ by adding h relative to* E if and only if $T(E, \underline{K}, h) = E(\underline{K}^{*r}_h)$.

> $T(K, h)$ is an *expansive* transformation of $\underline{K}$ by adding h if and only if there is some generalized expansion E-function of $\underline{K}$ such that $T(\underline{K}, h) = T(E, \underline{K}, h)$ where $T(E, \underline{K}, h)$ is either an extended expansion, AGM revision, or Ramsey revision of $\underline{K}$ by adding h relative to $\underline{K}$.

Expansions, AGM revisions, and Ramsey revisions are all expansive transformations in this sense. But there can be others. These additional expansive transformations are constructed out of expansions, AGM revisions, and Ramsey revisions with the aid of the generalized expansion function $\underline{E}$.

Expansive transformations satisfy analogues of (K*1), (K*2), (K*r4), and (K*6) but, as we shall see later, need not in general satisfy (K*3), (K*5), (K*7), (K*8), or even (K*r8). Extended revisions, whether AGM or Ramsey, do satisfy (K*5) (where $\underline{K}$ is consistent). And where $E(\underline{K}) = \underline{K}$ for all $\underline{K}$, the expansive transformations are either expansions, AGM revisions, or Ramsey revisions.

For present purposes, the important feature that distinguishes expansive transformations from other additive transformations is that an expansive transformation licenses all the conclusions allowed by unextended suppositional reasoning and, in addition, those conclusions warranted, on this basis, by the generalized expansion in question.

The generalized expansion function E is a function of potential corpora alone. Formally, it can be defined in terms of the extended expansion transformation derived from it. $E(\underline{K}) = E(\underline{K}_T^+)$ where T is a logical truth. So we could have begun with extended expansion functions rather than generalized expansion functions.

No matter how one begins, generalized expansions do not invoke a specific "input" sentence that is stipulated true by supposition for the sake of the argument. However, in the cases of interest in this discussion, the generalized expansion function does yield as output a corpus that is the expansion of the argument $\underline{K}$ by adding some sentence h. But h is not stipulated to be true for the sake of the argument. It is added to $\underline{K}$ as a conclusion of an argument from $\underline{K}$. In this case, the h to be added is determined by $\underline{K}$ according to the given generalized expansion transformation.

This kind of function can be used to represent formally a rule for drawing inferences from a potential corpus to a hypothesis not in the corpus. The inference purports to justify adding the hypothesis to the corpus on the basis of the assumption that the items in corpus are true. Once the hypothesis is added, the inquirer is committed to adding via deduction all of the logical consequences of the initial corpus and the hypothesis added by the nondeductive inference from the initial corpus.

A proper account of rational belief change should address inferential expansions of this sort instead of focusing attention exclusively on expansion by stipulated supposition. However, even if one is concerned with suppositional reasoning, one might be interested in generalized expansions and the nondeductive inferences ingredient in them.[1]

Suppose, for example, $\underline{K}$ contains the information that coin a is an unbiased or fair coin so that there is an equal chance of 0.5 that a will land heads on a toss and land tails on a toss and that the chance distribution for relative frequencies on n tosses is binomial. With this background information about objective physical probability or chance, we may invite the inquirer to suppose that the coin will be tossed 1,000 times.

The inquirer might understand the invitation to be to expand $\underline{K}$ by adding to it the information h that the coin is tossed 1,000 times and to identify the contents of $\underline{K}$ – that is, the deductive consequences of $\underline{K}$. The resulting expansive transformation of $\underline{K}$ is monotonic. And the only kind of inference involved is monotonic deductive reasoning. Moreover, both RTGL and RTR deliver the following conditional judgment:

(a) If coin a is tossed 1,000 times, it may fail to land heads approximately 50 percent of the time. ~(h > g)

However, many people would not make the conditional judgement (a). They would instead endorse the following:

(b) If coin a is tossed 1,000 times, it will land heads approximately 50 percent of the time. (h > g).

This verdict is unacceptable according to RTGL and RTR. Presystematic judgment, however, seems ambivalent. There is a pull in favor of obeying RTGL or RTR and endorsing (a). But there seems to be a rival tendency to endorse (b).

In my opinion, there is an equivocation in the demands of suppositional reasoning that emerges in the tension concerning the acceptability of (a) and (b). On the one hand, the invitation to suppose that h is true – that the coin is tossed 1,000 times – is an invitation to consider the logical consequences of this supposition and the initial $\underline{K}$ (that contains neither h nor ~h). On the other hand, the invitation to suppose that h is true is to consider not only such logical consequences but also the additional conclusions that are justified on the basis of $\underline{K}_h^+$ via some sort of nondeductive and, in this case, statistical or inductive inference. In effect, one is invited to consider the contents of the corpus obtained by a generalized expansion of $\underline{K}_h^+$. But this, by definition, is an extended expansion of $\underline{K}$ by adding h.

If the supposition h here had been incompatible with $\underline{K}$ (it was known that coin a was not tossed 1,000 times), a similar equivocation emerges between considering an AGM or Ramsey revision by adding h or extended versions of such revisions.

I think there is, indeed, an ambiguity in English in the way we understand sentences like (a) and (b). However, there seem to be resources available for disambiguating cases where (b) is acceptable from cases where it is not. When (b) is counted as acceptable it is sometimes rephrased as follows:

(b') If a is tossed 1,000 times, it will probably (in all likelihood, in all probability) land heads approximately 50 percent of the time.

In such cases, the qualifier "probably" does not convey the judgment that conditional on a's being tossed 1,000 times, the probability is high that a will land heads 500 times. This reading forces natural language into the procrustean bed of probabilism that refuses to countenance full belief and allows only probabilistic degrees of belief. The qualifier "probably" as I understand it corresponds to the use of the term "probable inference" to mean inference where the premises do not deductively imply the conclusion. When such inference is legitimate, it still warrants coming to believe fully the conclusion on the basis of full belief in the truth of the premises. Thus, the inference to the conclusion that coin a lands heads approximately 50 percent of the time from the premise that it is tossed a thousand times is not a deductive and explicative inference but is, instead, ampliative and, perhaps, inductive. To see that this is so, consider the following:

(c) If die b is tossed once, it will probably not land showing one spot up.

This is not acceptable on the basis of any sensible extended expansion rule capturing a legitimate inference except in unusual circumstances.[2] Given the background information that the die is fair, supposing that it is tossed once does not warrant either deductively or inductively any conclusion concerning the precise outcome of the toss. The inquirer should in typical circumstances suspend judgment.

There is, to be sure controversy about this matter; but that controversy arises from the mistaken view that an inductively generalized expansion recommends adding new information if its probability (in a sense conforming at least qualitatively to the requirements of the calculus of probability) is sufficiently high. That view confronts the following dilemma:[3] Either transformations of this sort are not generalized expansions at all because they do not transform deductively closed theories into deductively closed theories or they transform such theories into inconsistent ones even in cases where the new information added by supposition is consistent with the background.

States of full belief are linguistically representable by theories closed under deduction. The first alternative precludes regarding the transformation as representing a legitimate transformation from one state of full belief to another. The second counsels induction into inconsistency. Neither alternative is palatable.

Efforts to remain faithful to high-probability criteria in the face of this dilemma are reminiscent of efforts to finesse the problem of Evil. What needs to be explained is why serious authors continue to trisect angles with straight edge and compass. I conjecture that it arises by misidentification of credal probability with another index of inductive support that, if it reaches a certain threshold, warrants adding the information in question into the corpus. I shall say some more about this point later on in sections 6.7 and 6.8.

The simple statistical examples introduced suggest the plausibility of the following claims:

(1) There are expansive transformations that are based on generalized expansion functions finding applications in both genuine belief change and in suppositional and conditional reasoning.

(2) Generalized expansions involve inference from an initial corpus taken as premises to a conclusion that is then added to the initial corpus together with the deductive consequences. Thus, such expansion involves a kind of ampliative inference distinct from nonampliative and monotonic deductive inference.

In the light of all this, we may then ask whether, among the generalized expansions of relevance to inquiry, there are any nonmonotonic ones and, if so, are the inference patterns associated with them nonmonotonic? In order to address these questions we need first to introduce the distinctions between monotonic and nonmonotonic expansive transformations and between monotonic and nonmonotonic inferences more formally.

5.3   Monotonic and Nonmonotonic Expansive Transformations

If $T(\underline{K}, h)$ is an expansive transformation (relative to generalized expansion function E), $T(\underline{K}, h)$ is *monotonic* if and only if for every pair of consistent corpora $\underline{K}$ and $\underline{H}$ such that $\underline{K} \subseteq \underline{H}$, $T(\underline{K}, h) \subseteq T(\underline{H}, h)$.

If $T(\underline{K}, h)$ is expansive, it is *finitely monotonic* if and only if for every pair of consistent corpora $\underline{K}$ and $\underline{H}$ for which there is a sentence d such that $\underline{H} = \underline{K}_d^+$, $T(\underline{K}, h) \subseteq T(\underline{H}, h)$.

If $T(\underline{K}, h)$ is expansive, it is *conjunctively monotonic* if and only if for every consistent $\underline{K}$ and consistent h ∧ f, $T(\underline{K}, h) \subseteq T(\underline{K}, h \wedge f)$.

Thesis 1:

All expansive transformations that are monotonic are conjunctively monotonic.

Thesis 2:

All expansive transformations that are conjunctively monotonic are finitely monotonic.

An expansive transformation $T(\underline{K}, h)$ has the form $E[F(\underline{K}, h)]$ where F is either an expansion, AGM revision, or Ramsey revision. Theses 1 and 2 will be shown to be true for each of these three cases.

These theses fail for transformation functions that are not expansive. For example, imaging transformations are not expansive. They are monotonic; but they are not conjunctively monotonic.

Theses 1 and 2 will be proved for extended expansions, extended AGM revisions, and extended Ramsey revisions in that order.

Extended Expansions

The generalized expansion function E is monotonic if and only if whenever $\underline{K} \subseteq \underline{H}$, $E(\underline{K}) \subseteq E(\underline{H})$. It is finitely monotonic if and only if $E(\underline{K}) \subseteq E(\underline{K}_d^+)$ for every $\underline{K}$ and d.

The expansion transformation $\underline{K}_h^+$ is both monotonic and conjunctively monotonic. As a consequence, if E is monotonic, then the associated extended expansion function $E(\underline{K}_h^+)$ must be both monotonic and conjunctively monotonic. If E is finitely monotonic but not monotonic, the associated expansion strategy will be finitely monotonic and conjunctively monotonic. If E fails to be finitely monotonic, the associated expansion strategy will fail to be either finitely monotonic or conjunctively monotonic.

So monotonicity is sufficient for the conjunctive monotonicity of extended expansion functions but not necessary. However, finite monotonicity is necessary and sufficient for conjunctive monotonicity. Theses 1 and 2 are satisfied.

As we shall see later, there are important extended expansion functions that fail to be monotonic in the finite sense and, hence, conjunctively monotonic. They are based on generalized expansion functions that fail to be finitely monotonic.

Extended AGM Revisions

Extended revisions occur both when belief contravention is and is not required:

Case 1: No belief contravention:

When ~h and ~(h ∧ f) are not in $\underline{K}$, the AGM revisions of $\underline{K}$ by adding h and by adding h ∧ f are both expansions. Hence, whether in such cases, finite monotonicity and conjunctive monotonicity are satisfied depends upon whether or not E is monotonic, finitely monotonic, or neither.

Case 2: Belief contravention required:

Let h and, hence, h ∧ f be inconsistent with some corpus $\underline{H}$. Such a situation can arise in one of two ways:

(α) There is a consistent corpus $\underline{K}$ containing neither h nor ~h but which contains the following sentence:

(b) = h ⊃ ~f. [Thus, ~(b) = h ∧ f.]

In this case, let $\underline{H} = \underline{K}_f^+$. $\underline{H}$ contains the set {(b), f, ~h}.

(β) There is a consistent corpus $\underline{J}$ containing neither h nor ~h but which contains f. In this case, let $\underline{H} = \underline{J}_b^+$. $\underline{H}$ is the same set as before.

$\underline{K} \subseteq \underline{H}$.

$\underline{J} \subseteq \underline{H}$.

Hence, any monotonic transformation T of potential corpora by adding sentences (i.e., which satisfy (K*1) and (K*2)) must be such that the following obtain:

(1) $T(\underline{K}, h) \subseteq T(\underline{H}, h)$.

(2) $T(\underline{J}, h) \subseteq T(\underline{H}, h)$.

Conjunctive monotonicity requires that the following be satisfied:

(3) $T(\underline{H}, h) \subseteq T(\underline{H}, \sim(b))$.

(4) $T(\underline{H}, h) \subseteq T(H, h \wedge (b))$.

Neither h nor ~h is in $\underline{K}$. Hence, the extended AGM revision must be identical with $E(\underline{K}_h^+)$. Hence, it contains the elements of the following set:

(A) = {h, f, ~(b)}.

Similarly, $T(\underline{J}, h)$ contains elements of the following set:

(B) = {h, ~f, (b)}.

Consider now $T(\underline{H}, h)$. As we have seen, $\underline{H}$ contains the following set:

(C) = {f, ~h, (b)}.

Since the T-function is an extended AGM revision, adding h to $\underline{H}$ calls for the removal of ~h and either f or (b). The result will yield a corpus containing either the set (A) or the set (B). If the set (A) results, condition (1) is satisfied but (2) is violated. If (B) obtains, condition (1) is violated but (2) is satisfied. Either way, the demands of consistency have forced a violation of monotonicity.

Consider now $T(\underline{H}, \sim(b))$. It must contain set (A). $T(\underline{H}, h \wedge (b))$ must contain set (B). Hence, if $T(\underline{H}, h)$ contains (A), (3) is satisfied but (4) is violated. If, on the other hand, it contains (B), (3) is violated and (4) satisfied. Either way, conjunctive monotonicity fails.

Hence, if the transformation function is an extended AGM revision, the transformation is neither conjunctively monotonic nor monotonic in belief-contravening cases where h and h $\wedge$ f are inconsistent with H. Moreover, in the proofs, the failures of monotonicity are all failures of finite monotonicity. Theses 1 and 2 are satisfied.

Extended Ramsey revisions

Such revisions behave exactly like extended AGM revisions except when the input sentence is in the corpus.

Consider then a situation where $\underline{K}$ contains the consequences of $g \wedge f$ but does not contain either h or ~h. To fix ideas, suppose that g = fair coin a is tossed 1,000 times and f = fair coin a lands heads exactly 500 times. h = Clinton will be reelected in 1996. Let $\underline{H} = \underline{K}_h^+ = \underline{K}_h^{*r}$. It contains the logical consequences of the set {h, g, f} and contains $\underline{K}$. Hence, $T(K, h \wedge g) = E(\underline{K}_{h \wedge g}^{*r})$ should be a subset of $T(\underline{H}, h \wedge g) = E(\underline{H}_{h \wedge g}^{*r})$ if the T-function is monotonic.

The condition fails. $\underline{K}_{h \wedge g}^{*r}$ contains h, g, and f since neither $h \wedge g$ nor its negation is in $\underline{K}$. $h \wedge g$ is, however, in $\underline{H}$. Hence, $\underline{H}_{h \wedge g}^{*r}$ will be the expansion by $h \wedge g$ of the contraction removing $h \wedge g$ from $\underline{H}$. This result will contain h and g but not f.

The same conclusion emerges if we keep h and g but replace f by f' for f' asserting any one of the other 1,000 relative frequencies that might result from the 1,000 tosses. Letting $\underline{H}'$ replace $\underline{H}$ for that case, $\underline{H}_{h \wedge g}^{*r} = \underline{H}'^{*r}_{h \wedge g}$. Hence, $E(\underline{H}_{h \wedge g}^{*r}) = E(\underline{H}'^{*r}_{h \wedge g})$. From this, it follows that the T-function must be nonmonotonic and that conjunctive monotonicity must also fail. Once more, finite monotonicity and conjunctive monotonicity stand or fall together as theses 1 and 2 require.

Conclusion

These considerations show that finite monotonicity is necessary and sufficient for conjunctive monotonicity in all types of expansive transformation functions. This proves the two theses.

The connection between monotonicity and conjunctive monotonicity in the domain of expansive transformations of belief states or potential corpora is partially a product of the definition of expansive functions. Keep in mind, however, that the phenomenon of nonmonotonicity has been examined in the context of suppositional reasoning extended by generalized expansions. Consideration is being given to methods for extending expansions, AGM revisions, and Ramsey revisions. These transformations meet the conditions of adequacy for suppositional reasonings expressed by Ramsey test conditionals. In effect, expansive transformations comprehend all cases where new information is successfully incorporated into the initial corpus when condition (K*r4) is satisfied.

The method by means of which expansive transformations are constructed here also indicates explicitly the different sources of failure of monotonicity. For extended expansions, nonmonotonicity and conjunctive nonmonotonicity are the result of the nonmonotonicity of the generalized expansion function E. For extended AGM revisions, such nonmonotonicity can result either from the nonmonotonicity of E or from belief contravention for the sake of avoiding inconsistency. In the case of extended Ramsey revisions, nonmonotonicity can also arise in belief-conforming cases because, in such revisions, beliefs in the corpus are called into question and then restored.

5.4   Nonmonotonic Inferences

Nonmonotonic inference has two paradigmatic features of inference worth emphasizing:

(1) It is reasoning *from* a set of truth-value-bearing premises to a truth-value-bearing conclusion where full belief that the conclusion is true is justified on the basis of full belief that the premises are true.

(2) The reasoning does not require ceasing to believe any assumptions already taken for granted.

My main concern in this chapter and those which follow is to elaborate on the distinction between nonmonotonic inference and forms of reasoning representable by nonmonotonic transformations of belief states where condition (1) may be satisfied but condition (2) is not. Even so, it is worth emphasizing that (1) is a central feature of inference as it is being examined here.

The main philosophical thesis being advanced in this chapter is that the only kinds of inference in a sense satisfying (1) and (2) that are nonmonotonic are statistical, inductive or, perhaps, more generally ampliative – that is, where the only premises are in $\underline{K}$ and the conclusion is some sentence h not logically implied by $\underline{K}$ together with the logical consequences of $\underline{K}$ and h.

There are no doubt more generous conceptions of inference. One can reason from probability assignments of .1 to h and .3 to g relative to $\underline{K}$ implying that h and g are not both true to a probability assignment to h $\vee$ g of .4. There are no truth-value-bearing premises and no truth-value-bearing conclusion. Condition (1) is not satisfied. In this sense, there is no inference. But obviously there is some sort of reasoning involved and there can be no objection to calling it "inference" as long as no confusion arises. I have been insisting on a narrower construal because in the discussion of nonmonotonic reasoning, there does appear to be a tendency to think of the inference to be analogous to a deductive inference. Deductive inference is taken to be the paradigmatic kind of monotonic inference. Systematic investigations of deviations from monotonicity in inference are undertaken as investigations of deviations from monotonicity in that paradigm. (See, e.g., Gabbay, 1985;

Makinson, 1989.) And this suggests that nonmonotonic inference is modeled not after the pattern of probabilistic reasoning, reasoning about preferences or values, and reasoning to a decision. It is supposed to be construed as reasoning from truth-value-bearing premises (perhaps for the sake of the argument) to a truth-value-bearing conclusion where full belief in the premises would justify full belief in the conclusion. This is what condition (1) requires.

It is in this setting that I think it worth emphasizing that if either AGM or Ramsey revision is a prototypical nonmonotonic transformation employed in suppositional reasoning, the only kind of inference involved is deductive and, hence, monotonic. Genuine nonmonotonic inference from premises to conclusion emerges only when we allow for extended expansions based on nonmonotonic generalized expansion functions. To the extent that such nonmonotonic inferences are either inductive or are patterned on inductive and statistical inference, the view that "logics" of nonmonotonic inference are nothing more than canons of good inductive inference will have been vindicated.

Such conclusions are perfectly compatible with the view that combining the background information with the supposition stipulated to be true that forms the basis for reasoning for the sake of the argument often yields a nonmonotonic transformation of the initial belief state that is neither inductive nor inductive like. And, as I have acknowledged, it may be useful to introduce a concept of suppositional reasoning in the sense that the set of claims in the expansive transformation of $\underline{K}$ by supposing that h is true is the set of conclusions suppositionally inferred from h and the admissible contraction removing ~h (or the admissible contraction removing h in the case of Ramsey contraction) from $\underline{K}$ and to equate the status of the pseudoinferential relation $h \mid\!\sim g @ \underline{K}$ with the status of the transformation defining it as monotonic or nonmonotonic.

This concession does not detract from the importance of the claim that the only sort of nonmonotonic inference in the strict sense is patterned on inductive or statistical inference.

In the previous section, expansive monotonic and expansive nonmonotonic transformations were formally defined. The next step is to characterize inferences and pseudoinferences in terms of these ideas and then identify nonmonotonic inferences and pseudoinferences.

Let $\underline{A}$ be a set of sentences in $\underline{L}$ and g a sentence in L. Let $\mid\!\approx$ be a relation between sets of sentences on the left and a single sentence on the right. $\vdash$ is the relation of deductive consequence and $Cn(A) = \{x: A \vdash x\} = K$.

Any relation $\mid\!\approx$ meeting the following condition is a *strictly inferential relation*:

$\underline{A}$, $h \mid\!\approx g @ E (= h \mid\!\approx g @ \underline{K}$ and $E)$ if and only if g is in $E(\underline{K}_h^+)$.

For some generalized expansions, the relation will be one that would be applied cogently in genuine inquiry. But reasonable or not, any generalized expansion function can be regarded as a candidate rule for deriving truth-value-bearing conclusions from specific truth-value-bearing premises inferentially. And inferences may be divided into two important categories:

If $E(\underline{K}) = \underline{K}$ for all $\underline{K}$, the strictly inferential relation is $\vdash$ – that is, logical consequence – and is *explicative*. I shall say also that the identity E-function is explicative as well.

If $\underline{K} \subset E(\underline{K})$, the inferential relation is *ampliative* in the sense that it licenses inferences from $\underline{K}$ to members of a nondegenerate expansion of $\underline{K}$. The E-function will be called ampliative as well.

Consider any expansive transformation $T(\underline{K}, h)$ where $\underline{K} = Cn(\underline{A})$ and define the corresponding *pseudoinferential relation* $\underline{A}, h \mathrel{|\!\sim} g @ E (= h \mathrel{|\!\sim} g @ \underline{K}$ and E) as follows:

$\underline{A}, h \mathrel{|\!\sim} g @ E$ if and only if g is in $T(E, \underline{K}, h)$ where, for some generalized expansion function E, $T(E, \underline{K}, h)$ is an expansive function of one the following three kinds:

    (a) $E(\underline{K}_h^+)$. (extended expansion)
    (b) $E(\underline{K}_h^*)$. (extended AGM revision)
    (c) $E(\underline{K}_h^{*r})$. (extended Ramsey revision)

The inferential relation characterized by a generalized expansion is both a strict inferential and pseudo inferential relation. When E is explicative, the pseudoinferential relation generated by extended AGM revision is precisely the relation defined by Makinson and Gärdenfors (1991).

The definition of a pseudoinferential relation can be recast as follows:

$\underline{K}, h \mathrel{|\!\sim} g @ E$ if and only if $h \mathrel{|\!\sim} g @ \underline{K}$ and E if and only if $(\underline{M} \cup \{h\}) \mathrel{|\!\approx} g @ E$ where $\underline{M}$ is derived from $\underline{K}$ by one of the following methods:

    (a) $\underline{M} = \underline{K}$. (extended expansion)
    (b) $\underline{M} = \underline{K}_{-h}^-$. (extended AGM revision)
    (c) $\underline{M} = \underline{K}_h^{-r}$. (extended Ramsey revision)

In the case (a) for extended expansion, it is obvious that $h \mathrel{|\!\sim} g @ \underline{K}$ and E if and only if $(\underline{K} \cup \{h\}) \mathrel{|\!\approx} g @ E$.

Any strict inferential relation will satisfy the following condition:

(I1) If $\underline{A} \vdash g$, $\underline{A} \mathrel{|\!\approx} g @ E$.

A strict inferential relation is *cumulatively transitive* (Makinson, 1989, p. 4) if and only if it satisfies the following condition as well:

(I2) If $\underline{A} \mathrel{|\!\approx} f @ E$ and $\underline{A} \cup \{f\} \mathrel{|\!\approx} g @ E$, then $\underline{A} \mathrel{|\!\approx} g @ E$.

A strict inferential relation $|\approx$ @ E is *monotonic* if and only if the following condition holds:

(I3) If $\underline{A} |\approx g$ @ E, then $(\underline{A} \cup \{h\}) |\approx g$ @ E.

Any pseudoinferential relation will satisfy the following condition:

(S1) If $h \vdash g$, then $h |\sim g$ @ $\underline{K}$ and E.

The pseudoinferential relation is *cumulatively transitive* if and only if it satisfies the following condition as well:

(S2) If $h |\sim f$ @ $\underline{K}$ and E and $h \wedge f |\sim g$ @ $\underline{K}$ and E, then $h |\sim g$ @ $\underline{K}$ and E.

A pseudoinferential relation is *monotonic* if and only if the following condition holds:

(S3) If $h |\sim g$ @ E and K, then $h \wedge f |\sim g$ @ E and $\underline{K}$.

Claim 1:

A strictly inferential relation relative to E is monotonic if and only if E is finitely monotonic.

Proof: By theses 1 and 2 of section 4.3, E is finitely monotonic if and only if it is conjunctively monotonic. This holds if and only if $E(\underline{K}_h^+) \subseteq E(\underline{K}^+h \wedge f)$. Hence, (I3) is satisfied and the strictly inferential relation is monotonic.

Claim 2:

A pseudoinferential relation relative to E and $\underline{K}$ is monotonic if and only if *both* E is monotonic and the transformation of $\underline{K}$ is an extended expansion (so that the pseudoinferential relation is also strict).

Proof: No extended (AGM or Ramsey) revision can be a finitely monotonic transformation function. Hence, no such function can be conjunctively monotonic. Hence, for such functions, the corresponding pseudoimplication relation must fail to satisfy (S3). So a pseudoimplication relation is monotonic and satisfies (S3) only if it corresponds to an extended expansion that is at least finitely monotonic. But extended expansions are finitely monotonic if they are determined by monotonic generalized expansions and only if the generalized expansions determining them are finitely monotonic.

Since we are not going to discuss finitely monotonic but nonmonotonic generalized expansion functions, in the subsequent discussion I shall simply speak of monotonic generalized expansion functions without the qualification.

Thus, there are two sources of nonmonotonicity in suppositional inference: (1) nonmonotonicity due to the use of an extended Gärdenfors or Ramsey revision, which involves contravening some of the assumptions in the initial corpus $\underline{K}$; (2) nonmonotonicity in the generalized expansion function, which will also manifest itself in the strict inferential relation defined in terms of the generalized expansion function. If there is no nonmonotonicity in E, then the only sort of inference in the strict sense involved in nonmonotonic suppositional reasoning is monotonic.

The identity E-function is, of course, monotonic. Moreover, it is the only such E-function. The only kinds of inferences it licenses are deductive, explicative, and monotonic.

Inferences licensed by other E-functions are ampliative. In section 5.6, it will be shown that all and only such ampliative E-functions are nonmonotonic generalized expansions. Hence, all and only ampliative inferences are nonmonotonic.

Popperians, among others, insist that all legitimate inference is explicative. No nonmonotonic generalized expansion functions are legitimate. The nonmonotonicity in suppositional inference is an artifact of belief contravention and not of any inference in the strict sense.

The question of the legitimacy of changes in belief states representable by nonmonotonic E-functions is equivalent to the question of the legitimacy of ampliative inferences.

If there are legitimate ampliative inferences, the nonmonotonicity in suppositional reasoning can be due to nonmonotonicity in the inference in the strict sense from the supposition h and the contraction of the initial corpus removing ~h (or, in the case of extended Ramsey revision, h) that constitute the premises of the inference. Belief contravention is no longer the sole source of nonmonotonicity in suppositional reasoning.

The elementary statistical examples cited previously argue for the existence of legitimate ampliative and, hence, nonmonotonic inferences in the strict sense. But ampliative inference is involved in choosing between theories on the basis of data and background information, in any sort of extrapolation from data to a universal generalization or claim, in deriving predictions concerning the outcomes of experiments on the basis of knowledge of chances or statistical probability, or in making predictions on the basis of records of past behavior. The term *inductive inference* has been used on one occasion or the other to cover some or all inferences of these types. I suggest that we may then regard all and only those inferences that are ampliative to be inductive. Consequently, we may speak of inductive expansions instead of generalized expansions and write $\underline{K}^i$ instead of $E(\underline{K})$. Extended expansions of the type $E(\underline{K}_h^+)$ become inductively extended expansions $\underline{K}_h^{+i} = (\underline{K}_h^+)^i$, and inductively extended revision of both the AGM and Ramsey varieties may be defined accordingly.

Inductive inferences have been acknowledged to be nonmonotonic for at least two centuries although not with that terminology. It is also widely acknowledged to be ampliative. If there were ampliative but monotonic transformations and inferences available, we might have had to preserve a distinction between generalized expansion as a genus and inductive expansion as a species. In the light of the arguments given in section 5.6, this proves unnecessary.[4]

I have been arguing that accounts of suppositional reasoning ought to take induction into account. This consideration raises the question as to whether Ramsey test conditionals of the sort we have discussed previously do justice to the full range of conditional appraisals of serious possibility that needs to be considered. The coin tossing examples of section 5.2 already suggest that neither RTGL nor RTR can do justice to all presystematic judgments of serious possibility. The Ramsey test criterion for the acceptability of conditionals, the principles of conditional logic, and the logic of nonmonotonic suppositional reasoning all need to be reconsidered. Section 5.5 states new inductively extended versions of RTGL and RTR for use in subsequent discussion.

5.5 Inductively Extended Ramsey Tests

Positive and negative informational-value-based Ramsey tests, whether they are RTGL or RTR, treat conditionals as judgments of serious possibility relative to transformations of the current corpus of full belief. The transformations are representable as expansions of contractions. In all cases the expansion to be made is obtained by adding a sentence (the one given in the "if" clause) to the contracted corpus. There is no question of justifying the addition of this sentence. We are supposing that the given sentence is true "for the sake of the argument."

Inductively extended Ramsey tests take this transformation ($\underline{K}_h^*$ or $\underline{K}_h^{*r}$) of the initial corpus $\underline{K}$ and transform it still further by generalized expansion.[5] However, the expansion is not obtained by stipulating that a given sentence be added. Instead, the following question is considered: If $\underline{K}_h^*$ ($\underline{K}_h^{r*}$) were the current corpus instead of $\underline{K}$, what set of sentences would be added justifiably by inductive inference? The inductively extended AGM revision $\underline{K}_h^{*i}$ (Ramsey revision $\underline{K}_h^{*ri}$) is the inductive expansion of the AGM (Ramsey) revision.

Formally, the inductive expansion operator is a generalized expansion operator E of the sort we have already considered. What this inductive expansion operator will be depends on the criteria for legitimate inductive expansion from a given corpus and not on a stipulation that a given sentence be added by expansion. The question of formulating criteria for inductive expansion shall be deferred for the present. I shall proceed as far as possible without injecting the controversial issues addressing this question inevitably faces. For the present, I shall suppose that we have a special sort of generalized expansion transformation $E(\underline{K}) = \underline{K}^i$ that is the deductive closure of $\underline{K}$ and some sentence whose addition to $\underline{K}$ is justified via inductive inference. $\underline{K}_h^{+i} = (\underline{K}_h^+)^i$ is the inductively extended expansion of $\underline{K}$ by adding h. $\underline{K}_h^{*i} = (\underline{K}_h^*)^i$ and $\underline{K}_h^{*ri} = (\underline{K}_h^{*r})^i$ are inductively extended Gärdenfors and Ramsey revisions respectively.

Inductively extended Ramsey tests may now be introduced formally:

The Inductively Extended Ramsey Test according to Ramsey [IERTR]:

Positive IERTR (PIERTR):

h I > g is acceptable relative to consistent $\underline{K}$ iff g is a member of $\underline{K}_h^{*ri}$.

Negative IERTR (NIERTR):

~(h I > g) is acceptable relative to consistent $\underline{K}$ iff g is not a member of $\underline{K}_h^{*ri}$.

The Inductively Extended Ramsey Test according to Gärdenfors and Levi [IERTGL]:

Positive IERTGL:

h I > g is acceptable relative to consistent $\underline{K}$ iff g is a member of $\underline{K}_h^{*i}$.

Negative IERTGL:

~(h I > g) is acceptable relative to consistent $\underline{K}$ iff g is not a member of $\underline{K}_h^{*i}$.

As already stated, the discussion of criteria for legitimate inductive expansion will be postponed for the present. However, some of the properties of inductive expansion and inductively extended expansion can be identified in a fairly noncontroversial manner without reference to such criteria. I have in mind the *ampliativity* of induction and its *nonmonotonicity*. We shall discuss these properties before turning to more controversial matters.

5.6 Ampliativity

Inductive inference takes as its premises the information contained in the agent's corpus $\underline{K}$.[6] The conclusion of the inference is the addition of some sentence g to that corpus together with the logical consequences of $\underline{K}$ and g. g is not a logical consequence of $\underline{K}$. Otherwise the inference is deductive and, hence, cannot be considered inductive unless, for ease of systematic discussion, it is desirable to think of it as degenerate induction. In this sense, an inductive inference licenses a change in belief state representable by the shift from $\underline{K}$ to the expansion $\underline{K}_g^+$. But the sentence g is not added to $\underline{K}$ by stipulation for the sake of the argument. Nor is it added as the result of following some reliable program for using inputs from an external source in "routine expansion" (Levi, 1980, 1991). The expansion is justified on the basis of information in $\underline{K}$ as the best answer from among those recognized as potential answers to a question under investigation. And under these conditions, the inference from $\underline{K}$ to g is legitimate.

Thus, the *inductive inference* from premises $\underline{K}$ to the conclusion g warrants the *inductive expansion* realized by a shift from the initial belief state $\underline{K}$ to $\underline{K}_g^+$. This inductive expansion should not, however, be confused with the expansion $\underline{K}_g^+$ of $\underline{K}$ by adding g. The latter expansion does not indicate whether the addition of g is implemented by stipulation for the sake of the argument, via routine expansion or inductive expansion. When we want to represent the transformation as an inductive expansion justified by inductive inference, I shall use the expression $\underline{K}^i$. This representation does not state what is added to $\underline{K}$ but only indicates that what is added is what is warranted by inductive inference. When we want fuller information, we can say that $\underline{K}^i = \underline{K}_g^+$.

Inductive inference is inference in the strict sense, just as deductive inference is. However, it is *ampliative* rather than *explicative*. Deduction from information in $\underline{K}$ elaborates on information to which the agent who endorses $\underline{K}$ as his or her belief state is committed. If the deduction leads the agent to recognize aspects of the commitment that he or she had not previously recognized, the change in beliefs that takes place, important though it may be, is not a change in the agent's commitments but only in the agent's ability to fulfill his or her commitments. It is in this sense that deduction is explicative. Inductive inference is ampliative in that it brings about a change in doxastic commitment and does so by adding new information not logically implied by the initial belief state to the belief state rather than removing it. That is why it is associated with inductive expansion and why it is ampliative.

Examples of inductive inference abound. Inferring that the water in the kettle will not freeze from its being placed over a flame on the stove and the statistical claim that water almost never freezes upon the kettle's being placed over a flame on the stove are one kind of case. Inferring (immediately before election day) that a plurality of Americans would vote for Clinton on the basis of a sampling of voters' opinions is another. And adopting the theory of gravitation contained in General Relativity rather than Newton's theory is still a third.[7]

The ampliativity of inductive expansion amounts to the claim that such expansion can be and typically is nondegenerate. $\underline{K} \subseteq \underline{K}^i$ where, in general, $\subseteq$ is replaceable by $\subset$.

Sometimes we are concerned with the result of first adding some claim h to $\underline{K}$ by expansion and then drawing an inductive inference from the result. The ampliative feature of inductive expansion may be captured in this case with the aid of the notion of *an inductively extended expansion* $\underline{K}_h^{+i}$ of $\underline{K}$ by adding h. Here we are not claiming that h is added by induction. h may be added by routine expansion, by a previous inductive expansion, or as a supposition. The point is that once $\underline{K}$ is expanded (in the standard sense) by adding h, it is expanded further via induction. We may then formulate the ampliativity condition as follows:

Ampliativity:

(a) $\underline{K}_h^+ \subseteq \underline{K}_h^{+i}$.

(b) For some $\underline{K}$ and h, $\underline{K}_h^+ \subset \underline{K}_h^{+i}$.

Whether one is an inductivist or an antiinductivist, the ampliativity of inductively extended expansion ought to be noncontroversial. Even antiinductivists might explore, for the sake of the argument, the properties of legitimate inductive inferences (expansions) were there (counter to fact, on their view) such inferences. Consequently, it is useful to consider the conclusions that can be reached about inductive expansion appealing to just this feature of such inferences. It turns out that ampliativity has the following important implication for inductively extended revision both in the Gärdenfors and in the Ramsey sense and, hence, for conditional logics based on IERTGL and IERTR.

Ampliativity Thesis:

Ampliative inference is nonmonotonic.

There are two kinds of ampliative *E*-functions: ampliative but constant and variably ampliative functions. Inferences licensed by both types of *E*-functions will be shown to be nonmonotonic, thereby establishing the ampliativity thesis.

A) Ampliative but Constant E-functions:

E is an ampliative but constant transformation if and only if for every consistent $\underline{K}$ and every e consistent with $E(\underline{K})$, $E(\underline{K}_e^+) = [E(\underline{K})]_e^+$.

Ampliative but constant *E*-functions are called such because if $E(\underline{K}) = \underline{K}_d^+$, and e is consistent with $E(\underline{K})$, $E(\underline{K}_e^+) = \underline{K}_{d \wedge e}^+$. For obvious reasons, we may also say that *E*-functions that are ampliative but constant have the *permutability* property.

B) Variably Ampliative E-functions:

E is variably ampliative if and only if it is ampliative but is not ampliative but constant.

Subthesis 1: All ampliative but constant E-functions are nonmonotonic.

Consider a background corpus $\underline{\mathbf{B}}$ that may be stronger than the minimal corpus of logical, mathematical and other allegedly incorrigible truths but is, nonetheless, consistent. Let $\underline{E}$ be ampliative but constant and let $E(\underline{\mathbf{B}}) = \underline{\mathbf{B}}_d^+$. d is not entailed by $\underline{\mathbf{B}}$. Otherwise ampliativity fails. But then there is at least one c consistent with $\underline{\mathbf{B}}$ but inconsistent with $\underline{\mathbf{B}}_d^+$. $\underline{\mathbf{B}} \subseteq \underline{\mathbf{B}}_c^+$. Yet, $E(\underline{\mathbf{B}})$ is not contained in $E(\underline{\mathbf{B}}_c^+)$. Not only has monotonicity failed but so has finite monotonicity. This result will hold for arbitrary consistent $\underline{\mathbf{B}}$ including the weakest corpus in the set of all potential corpora. This suffices to prove the thesis.

Subthesis 2: All variably ampliative E-functions are nonmonotonic.

As before let $E(\underline{\mathbf{B}}) = \underline{\mathbf{B}}_d^+$. Let e be consistent with $\underline{\mathbf{B}}$. There are two cases to consider:

Case 1: e is consistent with $E(\underline{\mathbf{B}})$. Suppose that $E(\underline{\mathbf{B}}_e^+)$ is not identical with $\underline{\mathbf{B}}_{d \wedge e}^+$. If $\underline{\mathbf{B}}_d^+ \subseteq E(\underline{\mathbf{B}}_e^+)$, nonmonotonicity has been exhibited. If $\underline{\mathbf{B}}_d^+ \subset E(\underline{\mathbf{B}}_e^+)$, $E(\underline{\mathbf{B}}_e^+)$ must equal $\underline{\mathbf{B}}_{d \wedge e \wedge e'}^+$ for some consistent e' not entailed by $\underline{\mathbf{B}}_{d \wedge e}^+$ but consistent with it. Form $\underline{\mathbf{B}}_{e \wedge \sim e'}^+$. This corpus is consistent. Compare $E(\underline{\mathbf{B}}_{e \wedge \sim e'}^+)$ with $E(\underline{\mathbf{B}}_e^+)$. The former does *not* contain the latter as it should if monotonicity were to be satisfied.

Case 2: e is inconsistent with $E(\underline{\mathbf{B}})$ (but consistent with $\underline{\mathbf{B}}$). $E(\underline{\mathbf{B}}_e^+)$ will not be contained in $E(\underline{\mathbf{B}})$. A failure of monotonicity in the E-function is inevitable whether E is variably ampliative or ampliative but constant. However, as long as condition (2) of the definition of ampliative but constant E-functions is satisfied, the failure of monotonicity can be attributed to the need to revise $E(\underline{\mathbf{B}})$ and nothing else. In any case, thesis 2 has been sustained.

As promised, the nonmonotonicity of all ampliative generalized expansions has been established. Given that nonampliative expansion is explicative, we have shown that ampliativity and nonmonotonicity of generalized expansion and, hence, the inferences licensed by such expansions coincide. This provides us with the excuse for equating inductive inference (pre-systematically acknowledged to be ampliative and nonmonotonic) with ampliative inference in general.[8]

The distinction between inductive expansion functions that are permutable (i.e., ampliative but constant) and those that are not (and, hence, are variably ampliative) will prove useful in subsequent explorations of the extent to which the formal properties of inductively extended AGM and Ramsey revision do and do not resemble the formal properties of AGM and Ramsey revision. This dichotomy may be further refined by introducing the following two partial permutability properties:

Importability of Inductive Expansion:

For every h, $E(\underline{K}_h^+) \subseteq (E[\underline{K}])_h^+$.

Exportability of Inductive Expansion:

For every h consistent with $E(\underline{K})$, $(E[\underline{K}])_h^+ \subseteq E(\underline{K}_h^+)$.

An ampliative E-function is permutable (and, hence, ampliative but constant) if and only if it possesses both properties for all hypotheses consistent with $E(\underline{K})$ and for all consistent $\underline{K}$. Variably ampliative E-functions might be importable but not exportable, exportable but not importable or *fully variable* – that is, neither exportable nor importable. Does the mere ampliativity of an E-function rule out any of these possibilities?

Consider a hypothesis h consistent with $E(\underline{K}) = \underline{K}_c^+$. Such a hypothesis comes in two varieties:

(a) $h \vdash c$.
(b) $h = c \vee c'$ where c' is inconsistent with $E(\underline{K})$.

$[E(\underline{K})]_h^+$ is identical with $\underline{K}_h^+$ in case (a). However, $\underline{K}_h^+ \subseteq E(\underline{K}_h^+)$ by ampliativity. The upshot is that $[E(\underline{K})]_h^+ \subseteq E(\underline{K}_h^+)$. Hence, the demands of exportability are satisfied in case (a).

In case (b), $\underline{K}_c^+ = [E(\underline{K})]_h^+$. $E(\underline{K}_h^+)$ may or may not be contained in $\underline{K}_c^+$ and vice versa. The status of both exportability and importability is left open.

Thus, insofar as we appeal exclusively to the acknowledged ampliativity of inductive expansion, we can justify the conclusion that it is nonmonotonic; but we cannot decide the question of exportability or importability.

But there are some other conclusions that can be reached about induction without presupposing anything more than its ampliativity.

5.7   Ampliativity and Inclusion

The ampliativity of inductively extended expansion implies that both *i and *ri lack some important properties possessed by AGM and Ramsey revisions. In particular, it implies that *i and *ri fail to satisfy (K*3).

(K*3) $\underline{K}_h^* \subseteq \underline{K}_h^+$. (inclusion)

When *i (or *ri) is substituted for * (or *r), (K*3) becomes the following:

(K*i3) $\underline{K}_h^{*i} \subseteq \underline{K}_h^+$. (strict inductive inclusion)

Clearly this must fail. When neither h nor ~h is in $\underline{K}$, $\underline{K}_h^* = \underline{K}_h^+$ by the definition of AGM revision. (The same holds for Ramsey revision.) $\underline{K}_h^{+i} = \underline{K}_h^{*i}$ by definition of *i and the fact that neither h nor ~h is not in $\underline{K}$. Ampliativity implies that $\underline{K}_h^+ \subseteq \underline{K}_h^{+i}$. Assuming that ampliativity is nontrivial, there will be some occasions where $\underline{K}_h^+ \subset \underline{K}_h^{+i}$. In those cases, $\underline{K}_h^+ \subset \underline{K}_h^{*i}$ in violation of (K*i3). A similar argument holds when *ri is substituted.

Inclusion (i.e., (K*3)) is a consequence of (K*7) as long as revision is AGM or Ramsey.

(K*7) $\underline{K}_{h \& g}^* \subseteq (\underline{K}_h^*)_g^+$.

The reason is that both AGM and Ramsey revisions by a logical truth T are the identity transformation. Substitute a logical truth $\underline{T}$ for h and h for g and (K*7) yields (K*3). The two postulates hold for all consistent $\underline{K}$ in K when the revision operator is AGM or Ramsey.

When *i (or *ri) is substituted for * (or *r), (K*7) becomes the following:

(K*i7) $\underline{K}_{h \& g}^{*i} \subseteq (\underline{K}_h^{*i})_g^+$.

Neither inductively extended AGM nor inductively extended Ramsey revision by T is the identity transformation of $\underline{K}$. (K*i3) is not derivable from (K*i7). Substituting T for h and h for g, we get the following:

(K*i3m) $\underline{K}_h^{*i} \subseteq (\underline{K}^i)_h^+$. (weak inductive inclusion)

We are entitled to derive (K*i3) from (K*i7) in this way because we assume only that AGM (Ramsey) revision by T and expansion by T are identity transformations. (K*i7) does not imply (K*i3) but only (K*i3m). Neither (K*i3m) nor (K*i7) are refuted merely by the failure of inclusion.

(K*i3m) and, hence, (K*i7), as principles regulating inductively extended AGM *i, imply that the inductive expansion transformation i is importable in the sense of section 5.6.

Conversely, given the importability of inductive expansion together with the fact that both * and *r satisfy (K*7) and (K*r8), we get (K*i7) and, hence, (K*i3m).

From (K*7) and (K*r8), we obtain the following: If neither f nor ~f belongs to $\underline{K}_h^*$, $\underline{K}_{h \wedge f}^* = (\underline{K}_h^*)_f^+$. Consequently, if neither f nor ~f belongs to $\underline{K}_h^*$, $(\underline{K}_{h \wedge f}^*)^i = [(\underline{K}_h^*)_f^+]^i$ – that, is $\underline{K}_{h \wedge f}^{*i} = (\underline{K}_h^*)_f^{+i}$.

Importability entitles us to claim that the following holds: $(\underline{K}_h^*)_f^{+i} \subseteq (\underline{K}_h^{*i})_f^+$.

Combining this with the previous result yields (K*i7) for cases where neither f nor ~f is in $\underline{K}_h^*$. (K*i7) holds trivially when ~f is in $\underline{K}_h^*$.

When f is in that corpus, it continues to hold trivially for inductively extended AGM revision *i. It does not hold for inductively extended Ramsey revision.

These results may be summed up as follows:

Main Inclusion Thesis:

    (1) Neither inductively extended AGM revision *i nor inductively extended Ramsey revision *ri satisfies strict inductive inclusion (K*i3).

(2) Importability of inductive expansion is both necessary and sufficient for satisfaction of (K*i7) and (K*i3m) by inductively extended AGM revision *i.

(3) Importability of inductive expansion is necessary for satisfaction of (K*i7) and (K*i3m) by *ri but it is not sufficient unless f is not in $\underline{K}^{*r}_h$. (K*i7) is necessary and sufficient for (K*i3m) when *i is being used. Necessity holds for *ri but sufficiency presupposes that f is not in $\underline{K}^{*r}_h$.

The permutability of inductive expansion is sufficient but not necessary for its importability. Hence, the sufficiency results of the main inclusion thesis apply to permutable E-functions.

Thus, adopting an ampliative but constant inductive expansion function is sufficient but not necessary to secure (K*i7) and (K*i3m).

As shall be argued in section 5.8, adopting an inductively extended version of (K*8) presupposes that the inductive expansion transformation is exportable.

Thus, if inductively extended versions of (K*7) and (K*8) are required to be satisfied jointly, inductive expansion must be ampliative but constant. Conversely, if inductive expansion is ampliative but constant, the inductively extended versions of (K*7) and (K*8) will be satisfied.

Chapter 6 elaborates an account of principles of inductive expansion. According to those principles, in some contexts, inductive expansion will fail to be ampliative but constant (permutable). In some of these contexts, exportability will fail. In others, importability will fail. And in still others both exportability and importability will fail. Neither (K*i7), (K*i8), nor (K*ir8) should count as a universally mandatory constraint on inductively extended revision.

Nonetheless, there may be some interesting class of contexts where inductive expansion rules must be ampliative but constant so that both (Ki*7) and (Ki*8) are jointly satisfied. Chapter 6 will identify such sets of conditions.

Gärdenfors (1988, pp. 150, and 239, lemma 7.3) shows that (K*3) is satisfied by a belief revision system <K, *> if and only if the following schema of conditional logic is positively valid in <K,*>:

(A5) (h > g) ⊃ (h ⊃ g).

To claim that (A5) is positively valid in <K,*> is to say that no matter what consistent $\underline{K}$ in the language $\underline{L}$ representing a potential belief state in K we consider, all instantiations of (A5) in $\underline{L}^2$ will be in the corpus in $\underline{L}^2$ determined by $\underline{K}$ according to the Ramsey test. Consider then $\underline{K}^{*}_h$ where this is the AGM revision of $\underline{K}$ by adding h. "h > g" is in the corpus **RT**($\underline{K}$) in $\underline{L}^2$ determined by $\underline{K}$ and the Ramsey test if and only if g is in $\underline{K}^{*}_h$. Otherwise ~(h > g) is in **RT**($\underline{K}$). To secure the validity of (A5) we need to consider the case where g is in $\underline{K}_h^*$. When * is AGM revision, (K*3) holds so that g must be in $\underline{K}_h^+$. And this implies that "h ⊃ g" and, hence, (A5) are also in $\mathbf{RT}(\underline{K})$. The same argument goes through for Ramsey revision *r. On the other hand, if we replace * by *i or *r by *ri, (K*i3) cannot be secured. In the case where "h I> g" is in the inductively extended $\mathbf{RT}(\underline{K})$, "h ⊃ g" need not be.

Thus, failure of (K*i3) for *i and also *ri means that neither IERTGL nor IERTR will secure the validity of (A5) when ">" is replaced by "I>." That is to say, the following will be neither positively nor negatively valid:

(A5: version 1) (h I> g) ⊃ (h ⊃ g).

It has already been established that (A5) or "modus ponens" fails as a principle of conditional logic for iterated conditionals when iterations are based on RTR. (A5: version 1) has now been shown to be invalid in the logic of inductively extended Ramsey test conditionals of both types even when no iteration is involved.

(K*3) and (A5) support via the translation manual given in section 5.4 the following principle for pseudo inferential relations:

If h |~ g then T |~ h ⊃ g.

This principle is sometimes called *weak conditionalization* (Gärdenfors and Makinson, 1993).

It is customary here to understand "|~ h ⊃ g" to be a convenient abbreviation of "T |~ h ⊃ g." According to the definition of pseudoinferential relations in section 5.4, "|~ h ⊃ g" is not otherwise well defined.

When we consider *i (or *ri) and wish to formulate the inductively extended inclusion principles in terms of an inductively extended pseudoimplication relation, it is useful to distinguish between an unextended pseudoinferential relation "|~ " derived from * or *r from a corresponding inductively extended pseudoinferential relation "||~ ". Applied to (K*i3), the result looks like this:

Inductively Extended Weak Conditionalization: Version 1:

If h ||~ g, then |~ h ⊃ g.

Version 1 of the inductively extended weak conditionalization fails according to our definitions precisely because inductive expansion violates the inductively extended inclusion principle (K*i3). Notice that the first pseudoimplication relation is inductively extended whereas the second is not. Saying that g is a member of $\underline{K}_h^{+i}$ is translated into h ||~ g. On the other hand, saying that g is a member of $\underline{K}_h^+$ is translated as |~ h ⊃ g.

Observe, however, that the following second version of the inductively extended weak conditionalization principle is a consequence of the first version and is, indeed, weaker than it is.

Inductively Extended Weak Conditionalization: Version 2:

If h ‖~ g, then T ‖~ h ⊃ g.

Our translation manual allows us to derive version 2 from (K*i3m).
The second version of modus ponens in conditional logic looks like this:

(A5: version 2) h I> g ⊃ T I> (h ⊃ g).

These second versions translate into the claim that if g is in $\underline{K}^{*i}_h$, h ⊃ g is in $\underline{K}^i$.

If we assume that $\underline{K}^{+i}_h \subseteq (\underline{K}^i)^+_h$, both second versions are validated. This assumption is warranted only if inductive expansion is importable. In general, the second versions can and do fail. In general, therefore, even second-version modus ponens and weak conditionalization fail for inductively extended suppositional reasoning.

Gärdenfors shows that (K*7) corresponds to a schema (A9) for conditional logic. (K*i7) corresponds to the following:

(A9 for inductively extended conditionals): (h I > g) & (f I > g) ⊃ (h ∨ f I > g)

Because (K*i7) holds for *i (*ri) if and only if inductive expansion is importable, this version of (A9) holds for inductively extended RTGL (RTR) if and only if inductive expansion is importable. Hence, the following principle of inductively extended suppositional reasoning depends for its acceptability on the importability of inductive expansion:

Inductively Extended Or:

If h ‖~ g and f ‖~ g, then h ∨ f ‖~ g.

The failure of (K*i7) when inductive expansion is not importable is also tantamount to the failure of the following:

Inductively Extended Conditionalization:

If h ∧ f ‖~ g, then h ‖~ f ⊃ g.

Inductively extended revisions generate failures of modus ponens in inductively extended conditional logic and of version 1 of inductively extended weak conditionalization as a principle of inductively extended nonmonotonic inference under all circumstances. Version 2 of inductively extended weak conditionalization and conditionalization survive only if inductive expansion is importable. Yet, I claim that inductive expansion is a form of inference.

None of these principles seems as constitutive of the notion of an inference as the demand that an inferential relation be cumulatively transitive and obey the *Cut* condition S2 of section 5.4.

Inductively Extended Cut:

   If h ‖~ f and h ∧ f ‖~ g, h ‖~ g.

Consider the special case where h is a logical truth T and suppose that f is in $\underline{\mathbf{K}}^i$. Cut tells us that $\underline{\mathbf{K}}_f^{+i}$ is contained in $\underline{\mathbf{K}}^i$, which is identical with $(\underline{\mathbf{K}}^i)_f^+$. If inductive expansion were not importable in the special case where f is in $\underline{\mathbf{K}}^i$, inductively extended Cut would be violated. Inductively extended Cut is weaker than inductively extended conditionalization so that importability is sufficient for its satisfaction but not necessary except in the special case.

Cumulative transitivity is considered by some very difficult to resist if one is going to consider generalizations of monotonic inference relations that remain inference relations. If the only genuine nonmonotonic strictly inferential relations are inductive and if, as I contend, fully variable ampliative inductive expansion rules should not be precluded, the validity of Cut is not refuted as long as importability remains enforced for the special case just described. Those who insist that genuine inference relations should satisfy Cut will be obliged to demand that any acceptable account of inductive expansion insures that the limited importability required for its satisfaction be guaranteed.

According to the proposals in Chapter 6, Cut also fails. However, when a technique I call *bookkeeping* (section 6.10) is used to modify the characterization of inductively extended expansion, Cut survives.

5.8   Nonmonotonicity and Restricted Weakening

The condition of restricted weakening (K*8) is satisfied by * and (K*r8) is satisfied by *r. How do these principles stand in relation to *i and *ri?

(K*8) reads as follows:

(K*8) If ~f is not in $\underline{\mathbf{K}}_h^*$, then $(\underline{\mathbf{K}}_h^*)_f^+ \subseteq \underline{\mathbf{K}}_{h\wedge f}^*$.

When ~f is not in $\underline{\mathbf{K}}_h^*$, ~f is not in the contraction of $\underline{\mathbf{K}}$ removing ~h. Hence, ~f is not in the contraction of $\underline{\mathbf{K}}$ removing ~(h ∧ f). Indeed, under the conditions specified, the two contractions must be the same corpus $\underline{\mathbf{H}}$. Hence, by the definition of AGM revision, $(\underline{\mathbf{K}}_h^*)_f^+ = (\underline{\mathbf{H}}_h^+)_f^+$. The monotonicity of expansion guarantees that this corpus is identical with $\underline{\mathbf{H}}_{h\wedge f}^+ = \underline{\mathbf{K}}_{h\wedge f}^*$ again by the definition of AGM revision. As a consequence, (K*8) holds.

Substituting *i for * in (K*8), we obtain:

(K*i8) If ~f is not in $\underline{\mathbf{K}}_h^{*i}$, $(\underline{\mathbf{K}}_h^{*i})_f^+ \subseteq \underline{\mathbf{K}}_{h\wedge f}^{*i}$.

(K*i8) implies the following version of (K*i4) or inductively extended preservation:

(K*i4m): If ~h is not in $\underline{\mathbf{K}}^i$, $(\underline{\mathbf{K}}^i)_h^+ \subseteq \underline{\mathbf{K}}_h^{*i}$.

If we had replaced * by *i directly in (K*4), however, we would have obtained the following:

(K*i4m): If ~h is not in $\underline{K}$, $\underline{K}_h^+ \subseteq \underline{K}_h^{*i}$.

Whereas ampliativity itself shows that (K*i3) fails, ampliativity shows that (K*i4) must hold. The reason is that $\underline{K}_h^{*i} = \underline{K}_h^{+i}$ and ampliativity of the inductive expansion function guarantees the result.

(K*i4m) is stronger than (K*i4). It implies the exportation of inductive expansion property. Since (K*i4m) is a consequence of (K*i8), exportability is a necessary condition for the satisfaction of (K*i8).

Appealing to (K*7) and (K*8) as well as the exportability of inductive expansion, we have enough to secure satisfaction of (K*i8) by *i. Given that ~f is not in $\underline{K}_h^{*i}$, it is not in $\underline{K}_h^*$. Hence, by (K*7) and (K*8), $(\underline{K}_h^*)_f^+$ and $\underline{K}_{h \wedge f}^*$ are identical. So are their inductive expansions. But $(\underline{K}_h^{*i})_f^+$ is a subset of the inductive expansion $(\underline{K}_h^*)_f^{+i}$ of the first of these corpora by exportation of inductive expansion. And this suffices to establish (K*i8).

Thus, (given K*7 and K*8) exportability of inductive expansion is necessary and sufficient for (K*i8) (and K*i4m).

We have already established in section 5.7 that the importability of inductive expansion is necessary and sufficient for securing (K*i7) in the case of AGM revision. Now we are in a position to claim as anticipated in that section that the permutability of inductive expansion condition is necessary and sufficient for the satisfaction of both (K*i7) and (K*i8). This permutation condition, however, holds if and only if inductive expansion is ampliative but constant.

K*i8 supports corresponding principles of conditional logic and of suppositional reasoning:

Corresponding to (K*i8), we have the following:

(A10i) [= (IERW)]:

[~(h I> ~f) ∧ (h I> g)] . h ∧ f I> g.

(IERWI)

If h $\Vdash$~ ~f and h $\Vert$~ g, then h ∧ f $\Vert$~ g.

Major Claim 1: (K*i8), (A10i) or (IERW), and (IERWI) do hold for inductively extended AGM revision if and only if inductively extended expansion is exportable.

(K*i7) and (K*i8) both hold if and only if inductively extended expansion is ampliative but constant (i.e., permutable).

(K*8) fails for *r but (K*r8) is satisfied. If ~f is not in $\underline{K}_h^{*r}$ but f is, the Ramsey contraction $\underline{F}$ removing both h and ~h from $\underline{K}$ (depending on which, if either, is present in $\underline{K}$) need not be the same as the Ramsey contraction $\underline{F}'$ removing both h ∧ f and its negation from $\underline{K}$. Hence, even though expansion is monotonic, (K*8) cannot be supported. The qualification introduced in (K*r8) addresses this issue. Taken together with the monotonicity of expansion, the argument parallels the one for (K*8) in the case of *.

When inductively extended Ramsey revision or *ri is considered, the principle corresponding to (K*r8) is the following:

(K*ri8): If neither f nor ~f is a member of $\underline{K}\,^{*\text{ri}}_{\,h}$, then $(\underline{K}\,^{*\text{ri}}_{\,h})^{+}_{f} \subseteq \underline{K}\,^{*\text{ri}}_{\,h\wedge f}$.

The consequence of this principle resembling (K*r4) runs as follows:

(K*ri4m): If neither h nor ~h is in $\underline{K}^{i}$, $(\underline{K}^{i})^{+}_{h} \subseteq \underline{K}\,^{*i}_{\,h}$.

On the other hand, the consequence of this principle that is obtained as a substitution instance in (K*r4) states:

(K*ri4): If neither h nor ~h is in $\underline{K}$, $\underline{K}^{+}_{h} \subseteq \underline{K}\,^{*i}_{\,h}$.

This latter principle is weaker than (K*i4) and, hence, at least as secure. Similarly (K*ri4m) is weaker than (K*i4m). (K*ri8) is weaker than (K*i8). Hence, although exportation of inductive expansion is sufficient to secure (K*ri8) because it secures (K*i8), it is not necessary. A weaker exportability condition stating that exportability is present if neither h nor ~h is in K is necessary. Hence, the conjunction of (K*i7) and (K*ri8) does not entail permutability of inductive expansion.

When (K*ri8) is endorsed, we have the following:

(A10ri) [= (IERWr]:

[~(h I> ~f) ∧ ~(h I> f) ∧ (h I> g)] ⊃ h ∧ f I> g.

(IERWIr)

If h ⊪̸ ~f and h ⊪̸ f and h ‖~ g, then h ∧ f ‖~ g.

Major Claim 2: (K*ri8), (A10ri), and (IERWIr) hold for inductively extended Ramsey revision if the inductively extended expansion is exportable and only if it is weakly exportable.

When a pseudoimplication relation satisfies (RWI), it is sometimes said to satisfy a requirement of *rational monotony* (e.g., Kraus, Lehmann, and Magidor, 1990). The following is a special implication of it:

Cautious Monotony:

If h |~ f and h |~ g, h ∧ f |~ g.

The inductively extended version of this principle runs as follows:

Inductively Extended Cautious Monotony:

If h $\|\!\sim$ f and h $\|\!\sim$ g, h $\wedge$ f $\|\!\sim$ g.

When *i is being used, inductively extended cautious monotony holds if and only if exportation of inductive expansion is satisfied. If *ri is used, cautious monotony fails even in the presence of exportability. This is not surprising. Cautious monotony fails for the unextended Ramsey revision *r as well as the inductively extended variant.

5.9   Monotonicity, Undermining, and Undercutting

A strictly inferential relation $\mid\!\approx$ @ E is monotonic if and only if the conclusions reached from a set of premises $\underline{A}$ cannot be undermined by the addition of new premises. That is the substance of condition (I3) of section 5.4. Claim 1 of section 5.4 states that a strictly inferential relation $\mid\!\approx$ @ E is monotonic in this sense if and only if the generalized expansion transformation E is finitely monotonic.

Claim 2 of section 5.4 states that a pseudoimplication relation $\mid\!\sim$ between sentences h and g @ $\underline{K}$ and E is monotonic if and only if E is monotonic and the transformation of $\underline{K}$ used in the definition of $\mid\!\sim$ is an extended expansion. Once more, monotonicity precludes undermining.

Failure of monotonicity for strict inference might come in two ways:

   (i) *Undermining without Undercutting*: g is in $E(\underline{K})$ but neither g nor ~g is in $E(\underline{K} \cup \{h\})$.
   (ii) *Undermining with Undercutting*: g is in $E(\underline{K})$ while ~g is in $E(\underline{K} \cup \{h\})$.

Failure of monotonicity for suppositional reasoning can come in two parallel ways. If we let $T(E,\underline{K},h)$ be any expansive transformation, the two ways are as follows:

   (si) *Undermining without Undercutting*: g is in $T(E,\underline{K},h)$ but neither g nor ~g is in $T(E,\underline{K},h \wedge f)$.
   (sii) *Undermining with Undercutting*: g is in $T(E,\underline{K},h)$ while ~g is in $T(E,\underline{K},h \wedge f)$.

AGM revision * and Ramsey revision *r define pseudoimplication relations $\mid\!\sim$ that are nonmonotonic in both senses. However, as section 4.4 points out, the strictly inferential relation associated with * and *r is $\mid\!-$ (i.e., logical consequence) and this is monotonic. There is neither undermining nor undercutting with respect to the strictly inferential relation involved.

Both types of nonmonotonicity in suppositional reasoning derive from belief contravention. *The novelty of inductively extended revision is that it involves nonmonotonic inference in the strict sense.* That is to say, it allows for inductive expansion or, what is virtually the same thing, inductively extended expansion. There is no belief contravention involved. The strict inference relation allowed by inductive expansion is nonmonotonic, with respect to allowing both undermining with undercutting and undermining without.

There is a principle imposing a restriction on undermining with undercutting that applies to both AGM and Ramsey revisions.

As we saw in section 3.5, (K*3) guarantees the applicability of the no undercutting principle:

No Undercutting:

If g is in $\underline{K}^*_h$ and ~f is not in $\underline{K}^+_h$, then ~g is not in $\underline{K}^*_{h \wedge f}$.

The derivation of the no undercutting principle runs as follows: Consider $\underline{K}^*_h$. By (K*3) it is a subset of $\underline{K}^+_h$. Similarly by (K*7) $\underline{K}^*_{h \wedge f}$ is a subset of $(\underline{K}^*_h)^+_f$. Suppose that the expansion of $\underline{K}$ by adding h contains g while the expansion of $\underline{K}$ by adding h $\wedge$ f contains ~g. $\underline{K}$ must contain h $\supset$ g and h $\wedge$ f $\supset$ ~g. It follows, therefore, $\underline{K}$ must also contain h $\supset$ ~f. Hence, ~f is in the expansion of $\underline{K}$ by adding h. But this suffices to establish the no undercutting principle for both * and *r.

Conversely, we can derive (K*3) from no undercutting and (K*2). Substitute ~g for f in the no undercutting principle and we obtain the result that if g is in $\underline{K}^*_h$ but not in $\underline{K}^+_h$, (K*2) is violated. Given (K*2), (K*3) is established.[9]

Let us reformulate the no undercutting principle for inductively extended revision. There are two versions to consider.

(No Undercutting, Version 1):

If g is in $\underline{K}^{*i}_h$ and ~f is not in $\underline{K}^+_h$, then ~g is not in $\underline{K}^{*i}_{h \wedge f}$.

(No Undercutting, Version 2):

If g is in $\underline{K}^{*i}_h$ and ~f is not in $(\underline{K}^i)^+_h$, then ~g is not in $\underline{K}^{*i}_{h \wedge f}$.

Throughout this discussion, *ri may be substituted for *i.

Version 1 of no undercutting stands or falls with (K*i3) given (K*2). And this, we have seen, must fail.

Version 2 of no undercutting stands or falls with (K*i3m) given (K*2). As we have seen, (K*i3m) can be secured only if inductive expansion is ampliative but constant or inductive expansion is importable.

Both RTGL and RTR insure the following principle of conditional logic:

(NU) [(h > g) ∧ (h ∧ f > ~g] ⊃ (h ⊃ ~f).

Corresponding to (NU), there is the following constraint on suppositional reasoning:

(NUI) If h |~ g and h ∧ f |~ ~g, then |~ h ⊃ ~f.

The ampliativity of inductive expansion implies the failure of (K*i3) for inductively extended revisions and with it the failure of version 1 of the no undercutting condition.

The following variant of (NU) for I> is invalid:

(NU for inductively extended conditionals, Version 1): [(h I> g) ∧ (h ∧ f I> ~g] ⊃ (h ⊃ ~f).

The following variant of (NU) for I> is invalid when the revision operator is *ri and IERTR is used or when the revision operator is *i and IERTGL is used unless inductive expansion is importable.

(NU for inductively extended conditionals, Version 2): [(h I> g) ∧ (h ∧ f I> ~g] ⊃ T I> (h ⊃ ~f).

The following two versions of conditions on suppositional reasoning corresponding to (NUI) are counted out in the same way:

(NUI for inductively extended suppositional reasoning, Version 1): If h ||~ g and h ∧ f ||~ ~g, then T |~ h ⊃ ~f.

(NUI for inductively extended suppositional reasoning, Version 2): If h ||~ g and h ∧ f ||~ ~g, then T ||~ h ⊃ ~f.

The second versions of (NU) and (NUI) are both acceptable only under the conditions relative to which no undercutting itself is acceptable.

Thus far, we have discussed the nonmonotonicity of inductive expansion and inductively extended expansion and revision assuming only the non-controversial ampliativity property of inductive expansion. Before considering the ramifications of further assumptions introduced Chapter 6, it may be helpful to look at how nonmonotonicity is displayed in some inductive expansions. In Chapter 6, attention will be devoted to criteria for legitimate inductive expansion or inductive inference relative to which analysis of examples can be conducted without relying on shaky intuitions. In this chapter, I have been trying to indicate how far one can bypass intuition in defending the nonmonotonicity of inductive expansion by relying on the fact that induction is ampliative. So appeal to examples will perforce rely more heavily on the weak reed of intuition. Even so, illustrations of undermining without and undermining with undercutting may illuminate the abstract argument.

> *Example 1:* Let $\underline{K}$ consist of the following information: Cabs in the city that have been involved in accidents are tagged as such with the so called Mark of Cain. Of the cabs in the city 25 percent are Red Cabs that bear the Mark of Cain; 20 percent are Green Cabs with the Mark of Cain; 35 percent are Red Cabs without the Mark of Cain; 20 percent are Green without the Mark of Cain. Cabs show up at the Central Square randomly so that chances of picking up a cab of one kind or the other agree with the percentages given.
> *Example 2:* $\underline{J}$ includes different information about the percentages: 5 percent are Red with the Mark of Cain, 15 percent are Green with the Mark of Cain, 55 percent are Red without the Mark of Cain, and 20 percent are Green without the Mark of Cain.

Let h be the information that Jones picks up a cab in Central Square and f the information that Jones picks up a cab in Central Square with the Mark of Cain. g is the claim that the cab Jones picks up is Red.

In example 1, some may regard it plausible to reject the alternatives that Jones's cab is Green but will leave open the question as to whether the cab in question has the Mark of Cain. So g is in $\underline{K}_h^{+i}$ but neither f nor ~f is. However, if attention is turned to $\underline{K}_{h \wedge f}^{+i}$, the probability of g is $5/9 = 0.54$ and the probability of ~g is 0.46. Rejecting ~g might be regarded as too rash. Thus, neither g nor ~g is in $\underline{K}_{h \wedge f}^{+i}$.

Nothing said thus far offers a principled account of why these attitudes make sense. But all that is needed here is acknowledgment that they are coherent attitudes to endorse. For those who make the acknowledgment, conformity with neither (K*i8) nor (K*ri8) can be mandatory. Undermining without undercutting is coherent under circumstances where these principles forbid it. Inductive expansion is not exportable.

Example 2 illustrates undermining with undercutting under those circumstances where the no undercutting condition precludes it. Relative to $\underline{J}^i$ no conclusion can be drawn deductively or inductively concerning Jones picking up a Red Cab or one with the Mark of Cain because neither the assumption h that Jones picks up a cab in Central Square nor its negation is in that corpus. $(\underline{J}^i)_h^+$ contains only the deductive consequences of $\underline{J}^i$ and h and these include neither f nor ~f. $\underline{J}_h^{+i}$ might contain g and ~f. Still $\underline{J}_{h \wedge f}^{+i}$ might well contain ~g.

Notice also that example 2 illustrates a situation where (K*i3m) fails. $\underline{K}_h^{+i}$ is not a subset of $(\underline{K}^i)_h^+$. In this case, importability of inductive expansion and (K*i7) fail as well.

The nonmonotonicity of i and of +i does not involve belief contravention. Instead, it exploits a property very familiar to students of inductive and statistical inference – that the legitimacy of an inductive inference depends on a total (relevant) information condition. Call it *inductive nonmonotonicity*.

It would be rash to consider these examples and others like them convincing refutations of (K*i7), (K*i8), and (K*ri8). Presystematic judgments concerning such matters are bound to conflict. Without a principled account of inductive expansion that indicates when inductive expansion rules are or are not legitimate, the status of these principles may remain controversial. This is the topic of Chapter 6.

5.10 Ampliativity and Nonmonotonicity

Inductive inference is the only genuine species of nonmonotonic inference. My defense of this claim does not depend solely on a definition of "inference" (even though I think my use of "inference" is fairly standard). It depends also on the claim that there is a significant difference between the extent of the breakdown of nonmonotonicity in suppositional belief contravening reasoning and the breakdown due to inductive expansion.

In this chapter, I have sought to explore the extent to which appealing to the ampliative character of induction can yield results pertaining to this issue. The claims made have been defended without appealing to details of an account of legitimate inductive inference.

Appealing to ampliativity alone is sufficient to establish the nonmonotonicity of inductive expansion without any belief contravention. It also suffices to reject as illegitimate such constraints on inductively extended expansion and revision as (K*i3) and corresponding principles on inductively extended conditionals and suppositional reasoning such as modus ponens for noniterated conditionals.

The failure of (K*i3) for *i (*ri) demonstrates that such revision is not an AGM (Ramsey) revision transformation. Even so some postulates for AGM revision remain applicable to inductively extended AGM revision. These are (K*1), (K*2), (K*i4), (K*5), and (K*6). Ramsey revision does not satisfy (K*i4) but does satisfy (K*ri4). (Establishing these claims will be left to the reader to consider.)

Given (K*i7) and its consequence (K*i3m), (K*i8) and its consequence (K*i4m) for AGM revisions and (K*ri8) and (K*ri4m) for Ramsey revisions, the inductively extended revision transformations that would emerge would very nearly, if not perfectly, fit the requirements for AGM revision and Ramsey revision.

To sustain (K*i3m), (K*i7), and associated principles of inductively extended conditional logic and suppositional reasoning, inductive expansion transformations must be importable. To sustain (K*i4m), and (K*i8) for *i or (K*ri4m) and (K*ri8) for *ri, inductive expansion transformations must be exportable. To sustain both, inductive expansion must be permutable and, hence, ampliative but constant. Such permutability is what is required to secure the near agreement between the structural properties of AGM and Ramsey revisions and their inductively extended counterparts.

Permutability, importability, and exportability are substantial constraints on inductive reasoning. I have invoked a few statistical examples to argue that they should not be imposed. In the absence, however, of a more specific account of inductive expansion appealing to more than its ampliativity, little headway can be made in reinforcing the view that none of the constraints listed here ought to be mandatory for legitimate inductive expansion. At best, it can be said that the reader should be willing to consider refusing to regard these principles as mandatory constraints on inductively extended revision even if he or she is willing to endorse the corresponding constraints on unextended revision.

The upshot is that ampliativity alone suffices to establish not only that inductive expansion is nonmonotonic but that, because of the failure of strict inclusion (K*i3), neither inductively extended AGM revision nor inductively extended Ramsey revision satisfies the formal requirements demanded of its unextended counterparts.

Moreover, it remains very much an open question whether some of the other important properties of unextended revisions can be transferred to their inductively extended cousins. The next chapter is designed to reinforce the suspicions that they cannot while identifying some very special conditions under which they can.

6　Inductive Expansion

6.1　Expansion and Context

In inductive expansion, the inquirer shifts from an initial belief state $\underline{K}$ to a belief state $\underline{K}_d^+ = \underline{K}^i$ by an inference *from* $\underline{K}$ (the total background information and evidence) *to* d. Once d is added to the corpus $\underline{K}$, deductive closure takes over and secures the expansion.

The expansion is uniquely determined, therefore, by the initial belief state $\underline{K}$ and the "conclusion" d. This circumstance reflects the fact that the conclusion of a deductive inference is uniquely determined by the premises and the principles of deductive logic.

Inductive inference from $\underline{K}$ to d, however, is not uniquely determined in the same way by $\underline{K}$ and a system of fixed principles for inductive reasoning. Other contextual parameters or features additional to the total knowledge or evidence $\underline{K}$ must be taken into account.

Many philosophers and methodologists deny this contention because they fear that it condones a form of subjectivity in inquiry that will give comfort to the excesses of cultural relativists and antirationalists.

Others, such as Popper, Carnap, and their antiinductivist followers, nurse and protect the objectivity of science by denying the legitimacy of inductive expansion in science. To counsel abstinence from induction in this manner is an admission of failure to bring the fixation of belief under critical control. Since inquirers will extrapolate from data and add theoretical conjectures to background information as resources in subsequent inquiry regardless of what antiinductivists say, the admission of failure concedes to the social relativists and antirationalists that induction is a topic for psychology, sociology and history to study and that there is no point in attempting to identify prescriptive standards for assessing its legitimacy.

I am an inductivist. Not only do inquirers engage in ampliative inductive inference but there are at least some standards for discriminating between legitimate and illegitimate inductive expansion. This chapter explores some proposals concerning what such standards might be.

According to the view of the matter I favor, an inquirer is assumed to be engaged in answering a given question or solving a certain problem where the answer sought is not implied by what is already taken for granted. To adopt a potential answer as the solution is to engage in inductive expansion. A rational inquirer should adopt that potential answer or inductive expansion strategy that is best given the demands of the question under examination among the expansion strategies that are available.

To do this, of course, the inquirer must identify the set of available answers. Otherwise no good sense can be made of the demand to identify a best answer among those available to the agent. Thus, in addition to determining the list of available answers to the question under investigation, the inquirer needs to make some sort of evaluation of how well such potential answers succeed in promoting the aims of the inquiry. And this, in turn, calls for clarity concerning the aims of the inquiry. It emerges, therefore, that, in addition to the initial state of full belief, the inquirer will need to consider several other contextual parameters. These will include the set of potential answers to the problem under consideration and those contextual features relevant to identifying the goals of the inquiry and how well various potential answers measure up to promoting these goals.

In this chapter, a sketch will be offered of criteria for legitimate inductive expansion and the relevant contextual parameters I have outlined elsewhere (Levi, 1967, 1980, 1984a). Five parameters shall be identified. The legitimacy of expansion strategies is determined relative to values of these parameters.

To regard a parameter as contextual does not mean that the choice of a value of such a parameter is immune from criticism or that its selection is or should be determined by moral, political, or cultural considerations that are not open to discussion. I have discussed the possibility of a critical review of the choice of these parameters elsewhere (Levi, 1980; 1984a, ch.7). Addressing those matters here would divert attention from the main line of this argument. For the purposes of this discussion, it will be sufficient to identify the relevant contextual parameters and indicate how they conspire to determine recommended expansion strategies.

6.2   Potential Answers

One of the most urgent tasks of serious inquiry is the construction of informative potential answers to the question under study. This is the task of *abduction*, as C. S. Peirce called it. In some contexts, the problem under consideration calls for an explanation of some phenomenon, and a good potential answer would be an explanation. In another context, the preoccupation is with prediction. In yet another, it may be with the estimation of some parameter. In still another, the aim is to construct a theoretical scheme systematizing some domain.

Sometimes carrying out the abductive task is very difficult and sometimes it is relatively easy. Constructing a wide ranging theory with comprehensive explanatory power (e.g., general relativity theory, quantum mechanics, or quantum electrodynamics) is time consuming, difficult, and often requires a collective effort. On the other hand, if one is asked to estimate the number of invited guests who will come to a party, the range of potential answers is typically trivially easy to determine. There is no recipe for inventing potential answers to a question any more than there is a recipe for constructing proofs. Both abduction and proof are creative activities calling for the exercise of imagination. The same thing is true of designing an experiment. However, in abduction, in proof construction and experimental design, there are standards that can be evoked to evaluate the inventions of the imagination.

Peirce thought that in this sense there could be a logic of abduction just as there is a logic of deduction. He even was prepared to consider the products of abduction to be the conclusions of inferences of a special kind. Thus, one might start with premises consisting of known data to be explained and infer as a "conclusion" a potential explanation of the phenomena for further investigation. In my judgment, this use of inference is as misleading as is the claim that supposition involves nonmonotonic inference. To repeat what has been said before, "inference" is from premises to a conclusion where, if the premises are judged true, the conclusion is judged true. The "conclusion" of an "abductive inference" is not, by Peirce's own admission, judged true. Rather it is recognized as a potential answer to the question under study.

With this much understood, we still need to consider what constraints, if any, to impose on the set of potential answers to a given question if they are to meet minimal standards of coherence. The following remarks are addressed to this issue and are based on proposals I have introduced elsewhere. (Levi, 1967, 1980, 1984a, 1986a, 1991).

Someone might think that the set of such "cognitive options" or potential answers should consist of all expansions of $\underline{K}$ that are potential corpora in the language $\underline{L}$. But this view of the matter neglects the difference between expansions that do not furnish information relevant to the demands for information occasioned by the question under study and those that do. Investigators interested in the participants in the bombing of the World Trade Center will not consider some speculation as to the location of the earth's magnetic poles in the twentieth century B.C. as a potential answer to their question. In general, the space of relevant potential expansion strategies (= potential answers) is more restrictive than the set of potential expansions in the agent's conceptual space.

Let U be a set of sentences in $\underline{L}$ exclusive and exhaustive relative to $\underline{K}$. A potential answer relative to $\underline{K}$ and U is obtained by taking some subset R of U, adding the negations of the elements of R to $\underline{K}$, and closing under the consequence relation. R is the *rejection set* for the potential expansion. Whether U is finite or infinite, I assume that all elements of the power set of U are eligible to be such rejection sets. If the complement of a given rejection set is representable in $\underline{L}$ by a sentence h, the expansion strategy is $\underline{K}_h^+$. If there is no such sentence, let the expansion be the consequences of adding the set of negations of elements of R to $\underline{K}$. For the purposes of this discussion, we will focus on cases where a sentence h is available. With this understood, U characterizes the potential expansion strategies. For some time (see Levi, 1967; 1984a, ch. 5), I have called U the inquirer's *ultimate partition*. I shall continue to do so even though the term *ultimate* involves an excessive hyperbole. Inquirers may alter their ultimate partitions and with good reason. A partition is ultimate only in the sense that in the context of the given inquiry, there is no consistent expansion strategy stronger than expanding by adding an element of U that adds information meeting the demands of the problem.

Illustration 1: Suppose that it is known that coin a is an unbiased coin tossed a thousand times. The decision maker is interested in the relative frequency of heads and, hence, selects U to be the 1,001 seriously possible hypotheses concerning the relative frequency of heads. The potential answers are represented by the 2^{1001} rejection subsets of U. The complement of a rejection subset can be represented as a finite disjunction d of elements of U and the expansion strategy represented as $\underline{K}_d^+$. Notice that U is itself a rejection subset. To adopt this rejection subset is, in effect, to expand into inconsistency with what is known. Presumably this is illegitimate; but it is good to keep this option in the set of potential expansions. If a proposed criterion of inductive expansion sometimes recommends such expansion, it is unacceptable. The recommendation may not be recognized if expansion into inconsistency is not included in the range of potential answers.

U could be refined by taking as elements each of the 2^{1000} possible sequences of heads and tails as elements of the ultimate partition. However, if the inquirer does not regard the information that is added by specifying in detail what the sequence is rather than just the relative frequency of heads and tails, that refinement might not be adopted.

Illustration 2: Take the same unbiased coin tossed a thousand times. Now form the U consisting of the two outcomes of the 100th toss. There are four potential answers including the inconsistent one.

Illustration 3: Suppose one is interested in reaching generalizations regarding the color of ravens. To simplify, one might be interested in whether ravens are black or nonblack. There are several U one can pick. Here are two: U_1 = (all ravens are black, some ravens are not black); U_2 = the $2^4 - 1$ ways in which the Carnapian Q-predicates "black raven," "nonblack raven," "black nonraven" and "nonblack nonraven" could be empty or nonempty (given that at least one is nonempty).

For a given language $\underline{L}$, one might adopt as one's ultimate partition the set of all maximally consistent extensions in $\underline{L}$ of the corpus $\underline{K}$ to be expanded. This approach provides a unique objective-sounding recommendation and removes one contextual relativity from our account of inductive expansion. The expansion strategies available to an inquirer would then be constrained by $\underline{K}$ and by $\underline{L}$.

The objectivity allegedly achieved is purchased at the cost of absurdity. But no matter whether the inquirer was interested in the mean temperature in January in Upper Slaughter or in expected increase in the yield of tomato crop from using a new mode of fertilization, as long as $\underline{L}$ and $\underline{K}$ are the same, the available expansion strategies are the same. The friends of hyperobjectivity ought to be laughed out of court for this one.

Hyperobjectivists might fend off derision by acknowledging that lack of imagination and intellectual capacity will prevent inquirers from adopting a space of maximally consistent expansions of $\underline{K}$ as the ultimate partition. Even so, they can insist that inquirers should seek ways and means to extend their imaginations and computational capacity just as they should when confronted with limitations preventing logical omniscience (see Levi, 1991, ch. 2).

Inquirers do come to problem solving with diverse capacities. But limitations of imagination and intellectual capacity do not tell the whole story.

Even if an inquirer is quite capable of entertaining a given expansion strategy, the inquirer may refuse to do so simply because the answer proffered is not germaine to the demands for information of the question under consideration. A prediction as to the outcome of the 1996 presidential election is not a relevant answer to a question concerning the population of Azerbaijan in 1980.

Relativity to demands of information seems inescapable in any event. Identifying a standard $\underline{L}$ relative to which one could invoke a standard ultimate partition of maximally consistent theories is a will-o'-the-wisp just like invoking a space of possible worlds. The difficulty is the same in both cases. Our conceptual schemes are always capable, in principle, of further refinement. If we choose to stop at one level of refinement as we must in order to conduct some specific investigation, we are relativizing our conclusions to that partition.

It may, perhaps, be objected that if inductive expansion is relative to an ultimate partition, different inquirers with different partitions but the same corpora can legitimately expand in different and, perhaps, conflicting ways – a result alleged to be offensive to scientific objectivity.

In my judgment, scientific objectivity of the sort that humanly matters is not offended. If different inquirers X and Y reach conflicting conclusions owing to the use of different ultimate partitions and there is good reason to seek to iron out the conflict, it is possible to recognize a partition that is to be recommended for the purpose of settling the issues between X and Y in a non-question-begging way (Levi,1984a, ch. 7). When such context relativity seems to threaten an untenable relativism, psychologism, or historicism, we can disarm the threat if we can argue that potential conflicts due to contextual differences can be addressed through other inquiries. As long as we can give guidance as to what to do pending settling of the conflict through inquiry and some indication of the proper conduct of such inquiry, we do not need a prepackaged standard for preventing conflict from arising in the first place in order to preserve scientific objectivity.

6.3   New Error-Free Information

Given a corpus $\underline{K}$ and an ultimate partition U relative to $\underline{K}$, a set of potential expansion strategies or answers to the question under consideration is determined. These expansion strategies are, in effect, the options available to the inquiring agent in a decision problem where the aim is to add new information to the current state of full belief while avoiding error.

The best expansion strategy among the available potential expansion strategies should, relative to this goal, be an option that maximizes the acquisition of new information with minimum risk of error. That is to say, when seeking to justify endorsing that expansion strategy, the inquirer should seek to show that this is so.

To be sure, inquirers can have all sorts of personal objectives in seeking to expand their belief states. They may be seeking an answer conforming to their ideological or theological commitments or one that will give them comfort or one that will enable them to gain a reputation or earn a fortune. But inquirers may and often do pursue cognitive goals as well. I suggest that in scientific inquiry, the proximate goal of efforts to expand the corpus should be such a cognitive goal – namely, to gain new error-free information. I do not know how to convince the skeptical that goals of this kind ought to be the aims of specific efforts at expansion and will not devote space to pretending to argue for my claim. The only argument I have is to be obtained by elaborating on this view of the aims of inquiry and exploring the ramifications of the approach to inference grounded on this view.

Although the proposals to be made are undeniably controversial, their main features closely resemble central features of several alternative approaches to inductive inference to a sufficient degree that at least some readers should find the conclusions to be drawn concerning inductively extended Ramsey tests compelling.

6.4   Freedom from Error

Given that the common feature of the goals of all inquiries seeking expansion of the evolving doctrine (removal of doubt) is to obtain new error-free information, we need to explore in some greater detail what this means.

When an inquirer is in a belief state represented by corpus $\underline{K}$, the inquirer is committed to taking for granted (to being certain) that all sentences in $\underline{K}$ are true – that is, his or her current corpus is error free.

This does not mean that the inquirer denies the *logical* possibility that sentences in $\underline{K}$ are false. However, the inquirer completely discounts such possibilities in his or her deliberations. He or she has eliminated them as *serious* possibilities. The inquirer has no "real and living" doubt concerning the truth of such claims.

To be sure, some other inquirer Y may be convinced that the inquirer X's corpus $\underline{K}$ contains error and might even identify some specific locus of error. But X is committed to rejecting Y's judgment (made relative to Y's corpus) pending X's having good reason for opening up his or her mind to Y's view. The mere fact that Y disagrees is not, in general, sufficient reason for X's coming to question his or her own view. Such disagreement would be a good reason for X's opening up his or her mind if X's goal were to avoid disagreement with others; but not if it were to gain new error-free information.

This means that the inquirer is committed to avoiding error in expansion *on the assumption that every sentence in $\underline{K}$ is true.*

No one should be completely opinionated – that is, have a maximally consistent corpus in $\underline{L}$. And even if someone were, one could enrich $\underline{L}$ to reveal and represent domains of ignorance. When the inquirer takes for granted that everything in his or her corpus $\underline{K}$ is true in $\underline{L}$, he or she is not committed to the converse claim that everything true in $\underline{L}$ is in $\underline{K}$.

As long as an expansion is nondegenerate (i.e., avoids remaining in suspense between the elements of U), then from the inquirer's point of view relative to $\underline{K}$, there is a serious possibility that such expansion will import a false claim into the corpus initially judged to be error free. However, some expansion strategies will be less likely to import error than others. In particular, if the rejection set associated with one expansion strategy is contained in the rejection set associated with another, the first strategy should be at least as likely to avoid error as the second. Risk of error may be measured by probability of error where probability is understood to be credal, belief, or judgmental probability relative to $\underline{K}$. This credal probability measure defined over the algebra generated by U may be represented as follows:

> Let P be a real valued probability measure defined over all potential corpora in K. That is to say, $P(\underline{K}_2/\underline{K}_1)$ is taken to be well defined for every consistent $\underline{K}_1$ in K and every $\underline{K}_2$ in K and the measure obeys the requirements of the calculus of (finitely additive) probability. P represents a (strictly Bayesian) *potential confirmational commitment* for K.
>
> If $\underline{K}$ is the current corpus and $\underline{K}_d^+$ is a potential expansion where d is the disjunction of some subset of U, the probability of avoiding error in expanding from $\underline{K}$ to $\underline{K}_d^+$ is $Q(d) = Q(\underline{K}_d^+) = P(\underline{K}_d^+/\underline{K}) = Q(d/\underline{K})$. Q is the *credal probability function* relative to P and $\underline{K}$.

In my opinion, anyone who rates the expansion strategy that is more probable preferable to other potential expansion strategies relative to his or her cognitive goals is committed to avoidance of error as a component of such cognitive goals.

Consequently, in assessing expansion strategies, inquirers who regard avoidance of error as a component of the cognitive goals they are promoting need to evaluate the probability that no sentence added by implementing an expansion strategy is false and the probability that some such sentence is false. Such evaluation is determined by a credal probability distribution Q defined over the elements of the Boolean (and possibly sigma) algebra generated by the elements of U. Hence, in evaluating potential expansion strategies, we need to consider, in addition to the corpus $\underline{K}$ and the ultimate partition U, the credal state as represented by the credal probability Q over the algebra generated by elements of U'.

6.5 Informational Value

Regarding avoidance of error as a component of one's cognitive goals does not mean counting it as the sole component. If the inquirer sought to minimize risk of error alone – that is, to maximize the probability of the expansion strategy – the best option would be complete suspense of judgment between the elements of U. That is the degenerate expansion strategy, which is the decision not to expand at all.

The point is a simple one. If avoidance of error is a concern of efforts at expansion, rational agents should never expand unless there is some incentive to risk error. That is to say, the goals of efforts at expansion must include some desideratum that can induce us to risk, to some degree, importing error. Such a desideratum may be found in the demand for more specific (i.e., logically stronger) information.

Specificity or logical strength will yield only a partial ordering of the expansion strategies. It will fail to yield comparisons, for example, of the elements of U. Other desiderata such as simplicity (in some sense or other), explanatory adequacy may on some occasions be invoked to fill in the slack.

Perhaps, sometimes these comparisons fail to be made. But when one conjecture is considered a simpler theory than another, more explanatory, or more effective in satisfying the constraints of some research program, it need not be well correlated with the risk of error to be incurred. The partial ordering with respect to strength extended by appeal to such considerations represents an evaluation of inducements to risk error. Specificity, explanatory power, simplicity, and other such factors are constituents of the aims of inquiry which may be aggregated into an assessment of potential expansions with respect to how well they gratify what I call *demands for information*. The assessments of the expansion strategies made are, in this respect, assessments of informational value, and it is the informational value carried by an expansion strategy that can compensate for the risk of error it incurs and may sometimes warrant its implementation.

In discussing contraction strategies in sections 2.1 and 2.3, we also needed to consider the assessment of informational value. In contraction, one loses information and, hence, informational value. The cognitive aim where contraction is required is to minimize the loss of *damped* informational value. Damped assessments of informational value call for the evaluation of saturatable contractions with respect to undamped informational value. The undamped informational value of a corpus or potential state of full belief reflects the same kinds of desiderata that are relevant in inductive expansion.

Assessment of informational value can be represented by an informational value determining M-function defined over potential corpora of the sort initially discussed in sections 2.1 and 2.3. Given the corpus $\underline{K}$ to be expanded, the informational value of a potential expansion is the informational value added to that already in $\underline{K}$.[2]

Illustration 1: The problem is to estimate the mean of a normal distribution (to simplify, let us suppose the variance is known). The issue could be to ascertain the difference between the mean weights of fifteen-year-old boys in Somalia and Croatia, which, given suitable modeling assumptions, would be the mean of a normal distribution for a random variable representing differences in weights for matched pairs of Somalian and Croatian boys. Here the ultimate partition would be all the point estimates of the unknown mean. This might be any real value on the real line. Then the ultimate partition would be nondenumerably infinite. Or it could be a set of interval estimates of fixed Lebesgue measure partitioning the real line. The ultimate partition would then be denumerably infinite. If each element of the ultimate partition is considered equally informative, the informational value determining probability M would assign each element 0 M-value and, hence, equal content $1 - M$ of 1. Or the ultimate partition could be finite, consisting, for example, of three hypotheses: The unknown mean is less than some value $-x$, between $-x$ and $+x$, and greater than $+x$. The first and third alternatives could carry M value of 0.5, the middle one M-value of 0. In terms of content or informational value, the middle alternative is clearly superior to the others.

Illustration 2: It is well known that in choosing between rival theories, those factors I have called determinants of informational value play a large role. Simplicity and explanatory power in a domain, whatever these may mean, illustrate the point well. It is also well known that when an inquirer faces a choice between rival theories, the rival theories do not exhaust the logical possibilities. Sometimes they are exhaustive relative to the initial corpus (the background knowledge); but sometimes they are not. When they are not, the ultimate partition will consist of the rival theories specified together with a residual or "catchall" hypothesis that none of the theories specifically singled out is true. Typically, the residual hypothesis will carry less informational value than the specific alternatives. This means that it may be rejected at a higher probability level than the alternatives. Indeed, as indicated in illustration 1, it may be rejected even though it is "almost certain" – that is, carries credal probability 1.

The informational values carried by the specific theories in the ultimate partition may depend on the research programs of the inquirers just as they do for authors like Kuhn or Lakatos. But such informational values are not the sole determinants of the merit of potential answers. Risk of error and, hence, credal probability needs to be taken into account as well; and such assessments do not depend on the "paradigms" or "research programs" but on the background information and confirmational commitments of the inquirers.

My thesis is that insofar as such informational value so construed can be represented by a numerical index (I do not claim that it can always be so represented), the index should be a positive affine transformation of a function $\text{Cont}(d) = \text{Cont}(\underline{K}_d^+) = 1 - N(\underline{K}_d^+) = 1 - M(\underline{K}_d^+/\underline{K}) = 1 - M(d/\underline{K})$.

Here M is a probability measure over the potential belief states or corpora of the (at least Boolean) algebra generated by the elements of U. The function M obtained in this way is formally analogous to the master probability measure (confirmational commitment) P. N is analogous to Q.

N should not, however, be confused with Q. The latter is a measure of degree of probabilistic belief relative to $\underline{K}$. Such credal or subjective probabilities are used in computing the expected values of options in decision problems. By way of contrast, N functions are used to assess informational value added to $\underline{K}$ by expansion. They are not *expectation-determining* probabilities but *informational-value-determining* probabilities.

A philosophical benefit of understanding such familiar desiderata as explanatory and predictive power or simplicity as components of informational value that can induce an inquirer to risk error is that it eliminates the need for a priori presuppositions claiming that nature is simple or is likely to be simple. Everything else being equal, inquirers prefer coming to believe answers that are simple rather than answers that are complex; but they do not assume a priori that nature will gratify their desires. They want to believe simple and explanatorily powerful answers and are prepared to risk error in order to obtain them. Inquirers should not come to believe fully explanatorily powerful or simple answers merely because they desire answers with those features. The urge to relieve doubt in a gratifying way should be constrained by the concern to avoid error. Consequently, it is possible that an inquirer is justified in coming to believe a simple hypothesis not because nature is simple or likely to be simple but because the risk of error in adopting that hypothesis is sufficiently low and the informational value of the hypothesis sufficiently attractive to warrant adding the hypothesis to the set of full beliefs.

The N-function is construed here as a means of evaluating potential expansion strategies relative to U and $\underline{K}$. We have, however, already invoked the use of an informational-value-determining probability function in Chapter 2 when defining optimal contractions. Whether the (unextended) Ramsey test criterion is RTR or RTGL, optimal contraction plays a critical role in formulating the criterion. That is why I called these Ramsey tests "informational value based." The question to be faced, however, concerns the relationship between the M-function used in defining optimal contraction where losses in informational value incurred by various contraction strategies are to be assessed and the one just introduced where gains in informational value incurred through expansion are compared.

It is clear that the evaluation of contraction strategies discussed in Chapter 2 cannot be the same as the evaluation of expansion strategies. Potential contractions removing h from $\underline{K}$ are evaluated with respect to *damped* informational value. That is to say, the set of saturatable contractions removing h from $\underline{K}$ are appraised with respect to M-value or the *cont*-values defined relative to it. The damped informational value of the intersection of a subset of saturatable contractions removing h from $\underline{K}$ is the smallest *cont*-value assigned to a saturatable contraction in $\underline{K}$ (or 1 – the highest M-value).

Potential expansions adding some h to $\underline{K}$ assess informational value using the probability N so that $N(\underline{K}_d^+) = M(\underline{K}_d^+/\underline{K})$. The *cont*-function thereby determined remains undamped in its application to expansion.

In spite of these differences between assessments of informational value in expansion and contraction, they can both be determined by the same M-function. Both the evaluation of contraction strategies and the evaluation of expansion strategies reflect demands for information relative to $\underline{K}$. The assumption of identity is the assumption that these demands reflect the same research program in the contexts of both expansion and contraction. In expansion, evaluations of potential expansion of $\underline{K}$ are of concern. In contraction, evaluations of potential contractions of $\underline{K}$ are required. Clearly the domains being evaluated are different. But they can both be seen as fragments of an overall evaluation of the space of potential corpora. To be sure, one might change one's overall demands for information in shifting from expansion to contraction. But we may reasonably assume that such a change in overall goals calls for some justification. Pending a good reason for the change, we may suppose them to be the same. And when we are evaluating the acceptability of conditionals whether inductively extended or not, we should assume that the demands for information of value remain fixed.

With this identification in place, the contextual feature responsible for assessing informational value can be seen as the same as the contextual feature responsible for identifying optimal contractions and, hence, informational-value-based Ramsey tests. As we have seen, an informational-value-based Ramsey test operates relative to a pair $\langle K, M \rangle$ consisting of a potential corpus and an informational-value-determining probability M. In addition, account needs to be taken of a basic partition that delimits the set of potential contraction strategies just as account needs to be taken in inductive expansion of an ultimate partition constraining the set of potential expansion strategies. The inductively extended version of an informational-value-based Ramsey test requires reference to at least two other contextual parameters additional to those just mentioned: the ultimate partition U and the probability P defining the expectation determining probability Q relative to $\underline{K}$ over the algebra generated by U. There is, however, yet a third such parameter that should be considered: the degree of boldness or caution exercised by the inquirer.

6.6   Boldness and Inductive Expansion

I am supposing that the cognitive aim of inductive expansion in terms of which the legitimacy of inductive expansions is to be assessed is to obtain new error-free information. Such an aim should be regarded as a trade-off between the two desiderata of avoiding error and obtaining new information or, when risk of error is taken into account, as a trade-off between risk of error and new information. I have assumed elsewhere (Levi, 1967; 1984a, ch. 5) that such a trade-off for a given expansion strategy $\underline{K}_x^+$ is to be represented as a weighted average $\alpha Q(\underline{K}_x^+) + (1 - \alpha)N(\underline{K}_x^+)$ where $0 \leq \alpha \leq 1$. Moreover, on the plausible assumption that no error is to be preferred to any case of avoiding error, $\alpha \geq 0.5$.

With this in place, the ordering of potential expansions just obtained is preserved by the following positive affine transformation of the weighted average:

$$EV(\underline{K}_x^+) = Q(\underline{K}_x^+) - qN(\underline{K}_x^+).$$

Here $q = (1 - \alpha)/\alpha$. q ranges from 0 to 1 because $\alpha \geq 0.5$. The higher q is, the higher the relative importance attached to the acquisition of new information as compared to avoiding error. In this sense, the inquiring agent's *boldness* increases with increasing q. *Caution* decreases with increasing q.

On the model of inductive expansion I am proposing, the inquiring agent is supposed to maximize the value of EV. Where U is finite, the EV-value is equal to the sum of the EV-values of the expansion strategies represented by the elements of U that are themselves expansions of the given expansion strategy. These are the elements of U that are not rejected in implementing the expansion strategy. In order for an expansion strategy to carry a maximum EV-value, it is necessary and sufficient that none of these unrejected elements of U carry negative EV-value and that the set of such elements of U includes all those carrying positive EV-value.

However, two or more expansion strategies can meet these requirements for optimality. If there are elements of U carrying 0 EV-values, their inclusion or exclusion from the set of unrejected elements makes no difference to whether an expansion strategy carries maximum EV-value. However, when two or more expansion strategies are optimal, there must be a weakest optimal expansion. To break ties, I have suggested (1967; 1984a, ch. 5) adopting the weakest optimal expansion strategy. That is tantamount to adopting that expansion strategy for which the set R of rejected elements of U consists of all and only those elements carrying negative EV-value.

This, however, is tantamount to endorsing the following rule for expansion:

Inductive Expansion Principle:

If U is finite, adopt that expansion $\underline{K}_d^+$ of $\underline{K}$ relative to U, M, P, and q where d is the disjunction of elements belonging to the complement in U of the rejection set R defined as follows:

$x \in R$ if and only if $Q(\underline{K}_x^+) < qN(\underline{K}_x^+)$

In addressing cases where U is infinite, there are technical modifications that need to be made in order to obtain a rule. Some of these modifications are discussed in the appendix and presuppose reliance on techniques discussed in section 6.10. For the most part, we shall focus on finite U.

This proposal assumes that credal probabilities and informational-value-determining probabilities are numerically determinate. In my judgment, this is a defect. I have made proposals for remedying the defects elsewhere (Levi, 1980; 1984a, ch. 7). Although technical details of these proposals require further adjustment, focusing on them does not modify the main claims to be made. Discussing them here would complicate an already complex argument.

We are now in a position to define the inductive expansion of $\underline{K}$ relative to the set of potential corpora or belief states K, an ultimate partition U, the expectation-determining probability P, the informational-value-determining probability M, and the index of boldness q as $\underline{K}_d^+$ where this latter expansion is the one recommended by the inductive expansion principle.

The inductively extended expansion $\underline{K}_h^{+i}$ of $\underline{K}$ by adding h relative to $\langle K,U,P,M,q\rangle$ is the inductive expansion of $\underline{K}_h^+$ relative to $\langle K,U,P,M,q\rangle$ according to the inductive expansion principle.

Inductively extended AGM revision *i and inductively extended Ramsey revision *ri can then be defined as before.

6.7   Inductively Extended Belief Revision Systems

Gärdenfors defined a belief revision system as a pair $\langle K,*\rangle$ where * is taken as a primitive characterized by his revision postulates. My procedure has been to take * to be defined in terms of contraction and expansion. Such revision systems are Gärdenfors belief revision systems (brsg's). These systems can also be characterized by the pair of contextual parameters $\langle K, M\rangle$.

Ramsey belief revision systems (brsr's) are representable by pairs $\langle K, *^r\rangle$. Once it is understood that we are dealing with Ramsey and not Gärdenfors revision systems, they can also be characterized by pairs of the type $\langle K, M\rangle$.

Inductively extended Gärdenfors belief revision systems are representable by pairs $\langle K, *i\rangle$. But now a fuller characterization requires appeal not to a pair of contextual parameters but to a 5-tuple $\langle K,U,P,M,q\rangle$.

Similarly, inductively extended Ramsey belief revision systems are representable by pairs $\langle K, *ri\rangle$ where again this pair may be replaced by the same 5-tuple.

We have already reviewed postulates for revision appropriate to all four kinds of belief revision system. In the case of inductively extended AGM and Ramsey belief revision systems, we have done so without appealing to any specific proposal for principles of legitimate inductive expansion. Just appealing to the presystematic plausibility of the ampliativity and nonmonotonicity of inductively extended expansion has been sufficient to support certain important conclusions about the extent to which inductively extended AGM revision and Ramsey revision satisfy postulates for * and for *r.

On this basis, certain conclusions were drawn about inductively extended conditional logic (tacitly) using IERTGL and IERTR and taking interpretations to be inductively extended brsg's of the type $\langle K, *i \rangle$ or brsr's of the
type $\langle K, *ri \rangle$.

Now, however, the use of inductively extended brs's (of both varieties)
can be given more structure by appeal to the criteria for inductive expansion
just sketched and the 5-tuples of contextual parameters relevant to such criteria. An inductive expansion function *i or *ri is defined relative to a space
K of potential corpora, an ultimate partition U, an informational-value-
determining M-function, a credal probability P, and index of caution or boldness q. For some 5-tuples of parameters, the function will be exportable, for
others importable, for others variable but constant, and for others fully
ampliative.

I shall not rehearse the conclusions already reached in Chapter 5 specifying that (K*1), (K*2), (K*i4) for inductively extended brsg's (K*ri4) for
inductively extended brsr's, (K*5), and (K*6) are satisfied. However,
Chapter 5 left the status of (K*i7), (K*i3m), (K*i8), (K*i4m), (K*ri8), and
(K*ri4m) unsettled. We know, for example, that an inductive expansion
function satisfies all of these principles if and only if inductive expansion is
ampliative but constant. In this chapter, necessary and sufficient conditions
for the principles favored here to secure ampliative but constant inductive
expansion will be identified. It will also be argued that when inductive
expansion fails to satisfy these conditions, inductive expansion fails to be
exportable in all cases or importable in all cases. Some inductive expansion
functions will fail to be exportable, some to be importable – indeed, both
can fail. Consequently, none of the principles just cited or the associated
principles of conditional logic and suppositional inference are valid. Indeed,
limited importability (see section 5.7) of the sort needed to validate the Cut
condition for inductively extended suppositional inference fails. However,
when the method of bookkeeping (see section 6.10) is instituted and inductive expansion is redefined utilizing this method, Cut will be restored.

Consider the following variant of the taxicab example. R stands for "Red
Cab," B for "Blue Cab," C stands for "carries the Mark of Cain." The percentages of cabs in the city of each category runs as follows:

RC	R~C	BC	B~C
20%	65%	10%	5%

Observe that 85 percent of the cabs are Red and 15 percent are Blue.
two-thirds of the cabs with the Mark of Cain (involved in accidents) are Red
and one-third blue.

Suppose the M-function assigns values of 1/4 to each of the four cases but
the ultimate partition U consists of the hypothesis that the cab Jones selects

is Red and the cab selected is Blue. These two alternatives carry equal N-values of 1/2. An element of U is rejected if and only if its expectation determining probability is less than q/2. h = Jones selects a cab at Central Square. f = the cab selected has the Mark of Cain. g = the cab is red. Relative to $\underline{K}^+_h$, the probability of Jones selecting a Red Cab is 0.85 and of selecting a Blue Cab is 0.15. As long as q is greater than 0.3, the inductively extended revision $\underline{K}^{*i}_h = \underline{K}^{+i}_h$ contains g. But it does *not* contain ~f. (It does not contain f either.) The reason is, given U, f cannot be rejected. The ultimate partition is too coarse grained to recognize the claim that the cab carries the Mark of Cain to be rejected. Yet $\underline{K}^{*i}_{h\wedge f}$ does not contain g provided that q is less than 2/3. Consequently, for the given M- and P-functions, for U as the two-element partition Red and Blue and q greater than 2/3, (K*ri8) and a fortiori (K*i8) are violated.

It may, perhaps, be objected these principles are violated because of the use of an ultimate partition that ought not to be allowed. Perhaps we should require that ultimate partitions be used such that all propositions involved in the specific application of (K*i8) or (K*ri8) should be eligible for inductive acceptance and rejection.

The objection is not compelling. Someone interested in whether Jones selects a Red or a Blue Cab would not regard an expansion by adding the claim that the cab carrying the Mark of Cain was selected to be an answer to the question. But if the agent took U' consisting of the four hypotheses RC, BC, R~C, and B~C as his ultimate partition, the expansion strategy involving the addition of the disjunction of RC and BC would be a potential answer. What dictate of reason forbids him or her from adopting U and refusing to countenance such potential answers as offering information because they fail to answer his or her question?

Such an inquirer could still expand for the sake of the argument by adopting R in $\underline{K}^{*i}_h$, refusing to add either C or ~C, and suspend judgment between R and ~R in inductively extended expansion $\underline{K}^{*i}_{h\wedge f}$.

Nonetheless, even though using U' is not obligatory, it is not forbidden either. Hence, it is worth examining the results of using U' rather than U. I shall suppose that M assigns equal value of 1/4 to each element of this partition relative to $\underline{K}$ and, hence, relative to $\underline{K}^*_h = \underline{K}^+_h$. The inductive acceptance rule applied given $\underline{K}^+_h$ and $0.4 < q \leq 0.8$ yields rejection of BC and B~C so that the disjunction of RC and R~C is accepted. This means that g is in $\underline{K}^{*i}_h$ but neither f nor ~f are.

Consider now the inductive expansion of $\underline{K}^*_{h\wedge f}$. U' reduces to RC and BC in this case with expectation determining probabilities of 2/3 and 1/3 respectively. The M-values are equal and, hence, equal to 0.5. As long as q is less than or equal to 2/3 neither alternative will be rejected. Putting these two results together, let us restrict ourselves to values of q in the interval from

0.4 to 2/3. Even though g is in $\underline{K}^{*i}_h$ while neither f nor ~f is, neither g nor ~g is in $\underline{K}^{*i}_{h\wedge f}$.

This result shows that there is a a 5-tuple $\langle K,U,P,M,q \rangle$ that violates (K*ri8) and, hence, (K*i8). This is so even though the finer-grained partition U' that met the objection was employed. Not only is inductive expansion variably ampliative but it is not exportable.

It is not importable either. $\underline{K}^{+i}_h$ contains g when U' is the ultimate partition and q ranges between 0.4 and 0.8. Expansion of this corpus by adding g brings nothing new in. Consider, however, $\underline{K}^{+i}_{h\wedge f}$. For q greater than 8/17, this leads to the rejection of RC. Thus, there is a range of values of q for which inductive expansion is not importable. (K*i7) and (K*i3m) fail to be satisfied.

Moreover, for $8/17 < q \leq 2/3$, inductive expansion lacks both exportability and importability. (K*i7), (K*i8) and (K*ri8) all fail.

If we could insist that q should be restricted somehow in a manner that rules out inductive expansion rules that fail to be importable and also fail to be exportable, q would have to greater than 0.8 or less than 0.4. One might seek to restrict the choice of values of q in a manner that secures that q falls in one of these ranges. If we set aside the difficult question of how such a restriction might be motivated, it will not help; for by changing the values of the M- or the P-functions, failures of either exportability or importability can be secured. There does not appear a useful non ad hoc way to restrict ourselves to 5-tuples $\langle K,U,P,M,q \rangle$ satisfying both importability and exportability.

Cut or cumulative transitivity may also be violated.[3] On the other hand, as we shall see in section 6.10, when bookkeeping is deployed, Cut will automatically be restored.

6.8 High Probability, Partition Sensitivity, Deductive Closure

One kind of objection that has been leveled at the proposals just summarized for inductive expansion is the sensitivity of inductive inference to the ultimate partition. Investigators might disagree in the conclusions they reach via induction because they adopt different ultimate partitions. A closely related objection complains about the dependency of the conclusions reached on the N-function.

It is widely thought that the objectivity of inference is compromised unless the contextual parameters relative to which the legitimacy of inductive expansions is assessed are restricted to the initial corpus $\underline{K}$ (i.e., the "evidence"), the credal probability judgments relative to the initial corpus, and some index of boldness or caution. Even these contextual parameters are thought to require further constraints in the name of objectivity. For example, the credal probabilities ought to conform to some objective standard and the index of caution or boldness ought to be rigorous enough to gratify everyone but the resolute skeptic. Everyone sharing the same "evidence" ought to reach the same conclusions.

I think these demands unreasonable. An inquirer is entitled to retain the state of full belief, state of credal probability judgment, and index of boldness he or she currently endorses unless there is good reason to change. Others may disagree with the inquirer but unless that disagreement is one that warrants the inquirer's reconsideration of some element of his or her current point of view, the inquirer should ignore the dissent. The mere existence of disagreement is not sufficient to justify reconsideration of one's viewpoint.

Epistemologists ought to take seriously an exploration of conditions under which disagreement with others should occasion modification of one's own view and when it should not. This, however, is not the topic of this book. The point to keep in mind is that one does not have to standardize levels of boldness or credal probability judgments or other contextual parameters so that the legitimacy of inductive expansions is uniquely determined by the evidence in order to guarantee the objectivity of inquiry in any humanly relevant sense. When inquirers differ with respect to any contextual parameter including the available evidence (i.e., their corpora of full belief), we should ask whether and how they should modify the disputed parameter values. To the extent to which we can offer answers to the very difficult questions that then arise, the demands of humanly relevant objectivity will have been satisfied.

These remarks apply not only to the three contextual parameters just cited but to the the choice of an ultimate partition and informational-value-determining probabilities. The choice of an ultimate partition reflects (1) the ability of the inquirer to identify potential answers to the question under study and (2) the demands for information that characterize the reasons for which the inquiry is undertaken. Informational-value-determining probabilities are a technical representation of the demands for information insofar as they impact on the evaluation of potential answers relative to an ultimate partition.

The abilities of inquirers do, indeed, vary; but when differences in the choice of an ultimate partition for a common problem emerge, inquirers can pool their abilities to come up with a sufficiently fine-grained partition to accommodate all potential answers one or the other recognizes. And when disagreements concerning the choice of an ultimate partition arise not due to differences in inquirer's abilities to construct potential solutions to problems but to differences in the demands for information adopted, inquirers can also identify ways and means of moving to a common ultimate partition and a way of assessing informational value represented by M-functions that avoids prejudging in favor of one or the other of different views (see Levi, 1984, ch. 7).

To the extent that inquirers can understand one another and recognize that they are in disagreement, they have the requisite abilities to perform the maneuvers just mentioned. Whether they should resolve their differences is another matter. That is an issue that cannot be considered here.

Thus, I deny that the context sensitivity of the criteria for inductive expansion endorsed here betrays demands for objectivity in favor of some politicized postmodern epistemology.

In any case, it is clear that alternative approaches to what looks like inductive expansion often do seem sensitive to the choice of partitions. Sometimes when statistical hypotheses are under consideration, one identifies a privileged partition into "simple" statistical hypotheses. Simple statistical hypotheses specify numerically definite probability distributions over "sample spaces" of possible outcomes of experiments so that given some outcome of experiment, the "likelihood" of a hypothesis (the conditional probability of the data given the hypothesis and $\underline{K}$) is well defined. One may then seek to add to $\underline{K}$ a claim that the true hypothesis is one of the simple hypotheses carrying maximum likelihood among the alternatives. Alternatively, one might add a hypothesis rejecting all but those hypotheses the ratios of whose likelihoods to the maximum likelihood are above a given threshold k'. Or one might take a set of exclusive and exhaustive hypotheses relative to $\underline{K}$ and add the claim rejecting all but those hypotheses, the ratio of whose expectation-determining probabilities to the maximum probability is over a certain threshold k" and form the closure.

These proposals and variants on them do yield an account of inductively extended revision satisfying (K*1) and avoiding inconsistency as (K*5) requires. Like the proposals I have made, they are sensitive to the partition used and they depend on an index of caution or boldness. To be sure, according to some proposals U is required to meet some standardized condition such as consisting of simple statistical hypotheses. Whatever the merits of this restriction, it cannot be satisfied in many situations where some sort of principled way to evaluate expansions is called for. According to other proposals, there are no serious restrictions on how U is selected. All of them either use a credal probability function (Q-function) relative to K or some surrogate for such a function such as likelihood.

There are entertainable proposals alternative to my own that suppress or ignore informational-value-determining probabilities as contextual parameters. In my judgment, the tendency to ignore the importance of information-determining probabilities as contextual parameters is a defect. But whether it is or is not a defect, the fact that such rules are relativized to surrogates for ultimate partitions and indexes of boldness or caution would seem to suffice to make a case that the failures of the standard AGM postulates for revision are not going to be removed if one adopts one of these alternative views.

Unreasonable though I think it to be, the distaste for the relativization of inductive acceptance to the available potential answers and demands for information is considerable and does to some very considerable degree drive many authors to take seriously the idea that high (expectation-determining) probability is necessary and sufficient for acceptance. Given a corpus $\underline{K}$ one should add to $\underline{K}$ all and only those sentences whose Q-values relative to $\underline{K}$ are at least as high as some threshold k. Typically k is required to be greater than 0.5 in order to avoid adding a hypothesis and its negation.

This view, like the one I favor, includes Q (or P) and an index of caution or boldness as parameters. The trouble with this simple idea is that the union of $\underline{K}$ and the set of sentences whose probability is greater than k is not going to be a theory – that is, a set closed under consequences – unless k is set equal to 1. To set k = 1 is tantamount to abandoning the use of inductive acceptance rules. So k must be less than 1.[4]

An immediate consequence of such high-probability criteria for inductive acceptance is that the set of sentences accepted via induction (a Kyburgian corpus) is not a theory and, hence, does not represent a state of full belief. Inductively extended expansion and revision must violate the AGM revision postulate (K*1). If a Kyburgian version of an inductively extended Ramsey test is deployed, axiom schema (A2) of conditional logic must fail.

The remedy is, of course, to require closure of the initial corpus together with what is inductively accepted under logical consequence. But the high-probability rule then recommends expansion into inconsistency.

Resisting the deductive closure requirement does not avoid the embarrassment of inconsistency. Inductivism claims that inquirers are sometimes justified in adding new information to a body of knowledge, full belief, or evidence that is not entailed by that evidence through ampliative inference. Once the new information is added, it becomes just as certain, from the inquirer's point of view, as items that were in the belief state prior to its admittance. And it may (though it need not) be less vulnerable to being given up than some such items.

If one supplements the high-probability criterion with the recommendation to form the deductive closure of propositions accepted via induction according to the high-probability rule, the result will be an inconsistent belief state. That is to say, high-probability rules recommend deliberately importing inconsistency into the belief state. Respect for avoidance of error should never warrant deliberate expansion into inconsistency.

Refusing to require deductive closure on the set of sentences accepted via induction does not help. Prior to expansion, the inquirer is sure that exactly one of the tickets in the million-ticket lottery will be drawn. To add all of the beliefs represented by sentences of the form "ticket i will not be drawn" for i between 1 and 1,000 but refraining from forming the closure will not prevent the inquiring agent from recognizing that he or she has added an error to his or her state of belief.

Thus, the advocate of abandoning the deductive closure requirement in order to save the high-probability rule must declare that avoidance of error does not matter as a goal of inductive expansion. Such a view might be rationalized by appeal to some form of extreme instrumentalism that insists that avoidance of error is not a desideratum in changing full beliefs or by abandoning inductivism (as probabilists and Popperians do) and declaring that the high-probability rule is not a criterion of legitimate inductive expansion of a state of full belief. Either way, we have every right to wonder what purpose there is in having such a high-probability rule.

That high-probability rules are not relativized to ultimate partitions and demands for information should not be an argument in their favor. They violate requirements of deductive closure or consistency. This consideration should trump sensitivity to the choice of a partition.[5] We shall return to some of the issues raised here in section 8.3.

6.9   Belief and Disbelief

Section 6.8 overlooks one kind of high-probability rule for inductive expansion that might be thought to meet the needs of a belief change theory accommodating inductive expansion while remaining faithful to the demands of deductive closure. Unlike, however, the first approach I mentioned, high probability does *not* refer to probability in a sense conforming to the requirements of the calculus of probabilities but to another notion of probability satisfying rather different requirements.

A formalism that seems well suited to the purpose was (to my knowledge) first suggested by G. L. S. Shackle (1949). He did not call his measure "probability" nor have most of its reinventors. However, L. J. Cohen, who is to be counted among those who have rediscovered Shackle's formalism, has called his version "Baconian probability" in contrast to "Pascalian probability" characterized by the standard calculus of probabilities (Cohen, 1977). I myself see no special advantage to calling Shackle measures of degrees of belief "probability."[6]

It would be utterly inexplicable why so many serious people have endorsed high-probability rules in spite of their blatant deficiencies unless there is some presystematic urge that they think can be gratified in only this way. To deny the urge totally is more than many philosophers can bear. But, perhaps, some of those wedded to high-probability rules can be persuaded to change their minds once it is pointed out that there is an alternative measure purported to calibrate "degrees of belief" or "evidential support" that has some presystematic precedent and which does not confront the difficulties that expectation-determining probability does when it is used in the high-probability rule.

If the results of using a high degree of support rule are to yield an inductive expansion relative to consistent $\underline{K}$, it must be the case that we are making an evaluation representable by some function b from the sentences in $\underline{L}$ to elements of some ordered domain S such that for every k in S those sentences in $\underline{L}$ that carry a b-value at least as great as k form a consistent theory containing $\underline{K}$.

Several consequences immediately follow for b-functions for sentences in $\underline{L}$ relative to high-support (i.e., b-value) rules for inductive expansions of $\underline{K}$:

(b1) If h and h' are logically equivalent, $b(h) = b(h')$.

(b1k) If $\underline{K}$ entails that h and h' have the same truth value, $b(h) = b(h')$.

(b1k), of course, entails (b1). For the purpose of comparison with other views, they are distinguished.

(b2) $b(h \wedge g) = \min(b(h), b(g))$.

$b(h)$ and $b(g)$ are both above the threshold of acceptance k if and only if both are accepted in the inductive expansion at that threshold. Hence, the minimum is above the threshold. If the set of accepted sentences via inductive expansion of $\underline{K}$ is closed under consequence, the conjunction h $\wedge$ g is accepted as well.

(b3) If h is in $\underline{K}$, it carries a maximum b-value in S.

Suppose h is in $\underline{K}$ but h does not carry maximum b-value. There is some k such that $b(h) < k$. If k is the threshold of acceptance, h would not be in the inductive expansion of $\underline{K}$. This contradicts the notion of an inductive expansion.

(b4) For every h in $\underline{L}$, either $b(\sim h)$ or $b(h)$ is the minimum b-value (in S).

Suppose $b(h) = k$ and $b(\sim h) = k'$ where $k' < k$. If k' were not the minimum value in the ordering of S, there would be a $k^* < k'$ such that both h and h' are in the inductive expansion of $\underline{K}$ when k^* is used as the threshold. But inductive expansions into inconsistency are disallowed.

Corollary: If h is in $\underline{K}$, $b(\sim h)$ = the minimum b-value in S.

The set S of values of the b-function could be any totally ordered set with a maximum and a minimum containing enough elements to cover the distinct b-values needed for sentences in $\underline{L}$. We can insure enough elements by taking S to be elements of some interval of real line. Let it be the unit interval from 0 to 1.

Given the b-function, no matter how it is obtained, it is possible to derive a d-function representing degree of disbelief or *potential surprise* as Shackle called it in the later 1940s. This measure is the dual of the b-function. d(h) = b(~h).

(b1)–(b4) spawn parallel assumptions for d-functions.

(d1) If h and h′ are logically equivalent, d(h) = d(h′).

(d1k) If $\underline{K}$ entails that h and h′ have the same truth value, d(h) = d(h′).

(d2) d(h ∨ g) = min(d(h), d(g)).

(d3) If $\underline{K}$ entails ~h, d(h) = 1 (or the maximum in S).

(d4) Either d(h) = 0 or d(~h) = 0.

Corollary: If $\underline{K}$ entails ~h, d(~h) = 0.

It should be noted that if neither h nor ~h is an element of $\underline{K}$, a b-function is constructible according to which both b(h) and b(~h) take the minimum value. This situation would arise if there is no threshold value warranting shifting from suspension of judgment between h and ~h.

I have assumed that degrees of belief and disbelief are representable by real valued functions with maximum and minimum values. For the purposes to which I shall be putting these measures, it is enough to regard them as ordinal in the sense that one positive monotonic transformation is as good a representation as another. I shall assume that the ordering of hypotheses with respect to degree of belief and of disbelief is representable in the reals and, hence, need not be a well ordering.

These requirements do not determine a unique b-function for $\underline{K}$. They do not even offer us an interpretation of the b-function. They merely reflect the formal requirements needed to insure that the high-b-value rule for inductive expansion yields a consistent expansion of $\underline{K}$.

The result obtained is the formalism for Shackle's theory of potential surprise (disbelief) and the dual theory of degree of belief (Shackle, 1949, 1961).[7]

As it stands, however, the theory is deficient. Given a corpus $\underline{K}$, we need to be in a position to specify the factors that generate a d-function or a b-function relative to $\underline{K}$. Otherwise we lack a criterion of legitimate inductive expansion. Once we have checked to see whether the formal requirements are satisfied, we recognize that these requirements are very thin. They guarantee that our inductive expansions will be consistent and deductively closed theories; but that is all they guarantee.

As we shall see, measures formally like Shackle measures have been interpreted in a variety of ways just as measures satisfying the formal requirements of the calculus of probability have. It is important, therefore, to devote some attention to the intended applications of Shackle measures in order to ascertain whether the applications are legitimate and to avoid confusion between two legitimate but distinct applications.

If Shackle measures are intended to be satisficing measures of degree of belief and disbelief, to interpret the b-functions and the d-functions is to provide inductive expansion criteria. Moreover, it is to provide criteria having certain properties over and above the features of ampliativity and nonmonotonicity mentioned previously. Three properties may be identified:

I) The criteria must be *deductively cogent*. That is to say, the set of propositions or sentences added to $\underline{K}$ and the set accepted by the criteria must be closed under logical consequence. The criteria are, in other words, criteria for legitimate inductive expansion.

II) The criteria must be *caution dependent* (boldness dependent). If one already has a b-function, the set of propositions or sentences added to $\underline{K}$ is one whose b-values are at least as great as some threshold value k. The higher k is, the higher the degree of caution. $q = b^* - k$ (b^* the maximum b-value) is the degree of boldness. Corresponding to each value of k, therefore, is an inductive expansion criterion parameterized to an index of caution.

III) The criteria are *partition sensitive*. Deductively cogent and caution dependent expansion rules expand $\underline{K}$ by rejecting some sentences consistent with $\underline{K}$ but not in it and failing to reject others. This means that given a partition U, for each level of boldness q, there will be a rejection subset R of U. For any h equivalent given $\underline{K}$ to a disjunction of elements of U, the b-value for ~h is equal to the degree of potential surprise, disbelief, or rejection of h. If we determine the b-value for ~h relative to a refinement U' of U using a set of exclusive and exhaustive alternatives (given $\underline{K}$) $e_1, \ldots e_i, \ldots$, we must insure that the b- and d-values are robust under such shifts. To determine the d-value for h, one must consider any refinement of h into a disjunction of alternatives of the type $h \wedge e_i$ where the e_i's are exclusive and exhaustive given $\underline{K}$. The d-value for h must be the minimum among d-values for such components. Hence, in order to guarantee the necessary robustness, we must either (1) relativize assignments of b- and d-values to some standard partition (perhaps, a space of possible worlds), or (2) merely assume that the rules we deploy and background information are such that robustness is insured. Neither of these alternatives is promising. The only other alternative is to maintain (as I do) that inductive expansion and, hence, judgments of potential surprise and degree of belief are relative to an ultimate partition reflecting the cognitive aims of the inquirer.[8]

The criteria I have proposed for inductive expansion are deductively cogent, caution dependent, and partition sensitive. Partition sensitivity is revealed in the role of the ultimate partition U in defining the potential expansion strategies. Caution dependence is revealed in the importance of the index q, which represents the trade-off between the concern to avoid error in expansion and the concern to increase informational value. Deductive cogency is secured by requiring that an expansion strategy $\underline{K}^{+}_{d}$ be the result of adding d to the corpus $\underline{K}$ and forming the deductive closure where d asserts that the true element of U belongs in the complement of the rejection set $\mathbf{R}$ relative to U.

The inductive expansion rules I have proposed are not unique in possessing the properties I have just described. However, they may serve to illustrate how d-values and b-values can be interpreted in the sense of being derived from such rules.

Notice that the index q takes all the real values between 0 and 1. It is not essential that q be bounded in this way. We could use any positive monotone function f(q) to index boldness or any positive monotone transformation of 1/q to index caution and still be in a position to derive a d-function and a b-function. But the index q is itself derived from a weighting of the utility of avoiding error and the utility of obtaining new information and the ratio of these weights for all modes of weighting does yield a real valued index in the unit interval. Hence the range of values q may take is a range of real values in some interval or other.[9]

Clearly there is a tendency for more elements of U to be rejected the higher the value of q. Of course, this is only a tendency. Consider, for example the infamous case of the million-ticket fair lottery. Let U be the set of hypotheses of the form "ticket i will win." Each such hypothesis receives the Q value of 10^{-6}. It receives the same N-value. No matter what value of q in the permitted range is used, no element of U is rejected even though the probability that ticket i will not win is extremely close to 1. A high credal probability relative to $\underline{K}$ is not sufficient for addition into $\underline{K}$.

By way of contrast, consider a fair coin tossed 1,000 times. U is the set of all hypotheses concerning the relative frequencies of heads. There are 1,001 of these. Assume that each of these carries equal N-value of 1/1001. The Q function is the binomial distribution $(n!/[r!(n-r)]!)0.5^{n}$ where n is 1,000 and r is the frequency of heads. If q = 0, no element of U is rejected. If q is greater that $(1,001)(0.5^{n})$, the prediction that all the tosses yield heads (tails) is rejected. As q increases, more elements of U representing frequencies near 0 and 1 are rejected. When q = 1, the unrejected elements of U predict relative frequencies in a relatively small interval around 0.5. Although it is true in this case that the total probability of no error in the recommended inductive expansion is very high, that is not the reason the expansion is recommended. It is recommended because it consists of all elements of U whose probabilities are not sufficiently low (although all elements of U carry low credal probabilities) to be rejected. And how low is low enough will depend on the value of the index q adopted.

Observe further that for each element x of U, there is a value q(x) of U such that q(x) is the maximum value of q at which x fails to be rejected. If x is not rejected for any value of q, that value is 1. If it is rejected for all values of q, the value is 0. Let y be any potential expansion in the algebra generated by the elements of U. In particular, let y be the disjunction of any subset W of U. y is rejected if and only if all elements of W are rejected. Hence, the maximum level of q at which y fails to be rejected is the maximum level of q(x) for x in W.

Consider then any z in $\underline{L}$ that is not the disjunction of a subset W of U. If $\underline{K}$ entails that z has the same truth value as a disjunction y of some such subset W, we can equate q(z) with q(y). But suppose this is not the case. In that event z is equivalent (given $\underline{K}$) to a disjunction of a subset W' of a refinement U' of U. When U is the ultimate partition, any element of U' is rejected if and only if the element of U it entails (given $\underline{K}$) is rejected. Hence, there will be a maximum level q(z) of q at which z fails to be rejected.

Thus, the function q(x) is defined for all sentences in $\underline{L}$. As I have constructed it, it is a function from the sentences in $\underline{L}$ to the unit interval; but any finite interval on the real line could be the range of the function. And any positive monotone transformation of q(x) (i.e., order-preserving transformation) will be equivalent to q(x).

The q-function for $\underline{K}$ relative to $\langle K,M,P,U \rangle$ has the following formal properties:

(q1)   If x and y are logically equivalent, for every consistent $\underline{K}$ in K the q-function relative to K satisfies q(x) = q(y).

(q1k) If $\underline{K}$ implies that x and y have the same truth value, q(x) = q(y).

(q2) $q(x \vee y) = \max(q(x), q(y))$.

(q3)   If x is in $\underline{K}$, q(x) = 0 (= the minimum q-value).

(q4)   If x is in $\underline{K}$, q(~x) = 1 (= the maximum q-value).

(q1k) implies (q1) as a special case. It will prove useful, however, to take note of both (q1) and (q1k).

Let y be the intersection of any pair w and w′ of expansion strategies in the algebra generated by U. Define d(y) to be $1 - q(y)$. Demonstrating that (d1) – (d4) hold is an easy exercise. The d-function is an index of the degree of confidence of rejection. If the expansion strategy y is representable as the closure under consequence of $\underline{K}$ and some sentence y (the expansion of $\underline{K}$ by y), d(y) may be regarded as the degree of disbelief that y relative to $\langle \underline{K}, M, P, U \rangle$. Then we may say d(w ∨ w′) = min[d(w), d(w′)]

Relative to the same quadruple, we may define b(y) to be d(~y). This represents the degree of belief that y. Degree of belief is the dual of degree of disbelief. It is easy to see that (b1)–(b4) are satisfied.

Thus we have derived a measure of degree of disbelief and degree of belief exhibiting the formal properties of Shackle measures of degrees of potential surprise or disbelief and of degrees of belief. The derivation involves the use of a family of deductively cogent, caution-dependent, and partition-sensitive criteria for inductive expansion.[10]

As in the case of the calculus of probabilities, however, it is important to distinguish between the formal properties of Shackle – measures and applications of the formalism. There have been many rediscoveries of Shackle's formalism in recent years. Many of them seem to begin with some primitive notion of "support," "degree of belief," "degree of possibility," "expectation," or the like and construct a formal structure. And even when there is some effort to indicate an intended application, it is clear that not all of them are intended to interpret the measures in terms of inductive expansion rules as I have done and among those that may be understood in this way, not all share an approach to inductive acceptance along the lines I have described.

However, any application that seems designed for inductive expansion would appear to share with my proposal reference to inductive rules that have the three properties I have previously identified.

Given such a rule (or parametrized family of rules), it is clear that when an inquirer endorses a degree of boldness to exercise, that is tantamount to saying that the expansion strategy to be endorsed is a consequence of the set of negations of elements of U whose degrees of belief in Shackle's sense are "high enough" where "high enough" is defined by reference to the function q(x) in the manner sketched previously.

L. J. Cohen (1977), who has noticed this property, has suggested calling degrees of belief in this sense "Baconian probability" and has observed that one can say that one's inductive conclusion is justified if one's "Baconian probability" is high enough. This claim is, indeed, a sound one, provided that an acceptable interpretation of Shackle's formalism is used.[11]

So the idea that one should "accept" a conclusion if its probability is sufficiently high makes good sense provided that "probability" is interpreted in a manner quite different from what is required by the calculus of probabilities. The calculus of probabilities cannot itself be given such an interpretation unless "high enough" is construed strictly as "with probability 1."

Shackle (1961) contemplated ways of modifying assessments of potential surprise in the light of new data so that the new assessments would somehow be a function of the old assessments and the data. Shafer (1976) and

Spohn (1988) among many others have sought to do the same. Unlike Shackle and Spohn, Shafer thought that in cases where evidence e_1 warranted positive b-value to a proposition h while e_2 consistent with background $\underline{K}$ and e_1 warranted positive b-value to ~h, $e_1 \wedge e_2$ warranted assigning positive b-values to both h and ~h. He elegantly showed how to do this using a method of combining data initially proposed by A. P. Dempster (1967) and elaborated by Shafer. The result is an index of support that violates either (b2) or (b4). Shafer acknowledges and, indeed, emphasizes the point. He claims that his view is a generalization of Shackle's needed to account for the fact that two logically consistent data reports may be dissonant in the sense that they by themselves warrant by conflicting conclusions.

If the measures of inductive support had been interpreted in terms of families of inductive acceptance rules having the properties described here, it would become apparent that data that are dissonant in Shafer's sense would not accord positive b-values to inconsistent hypotheses. Given an ultimate partition U, informational-value-determining measure N, and credal probability measure Q relative to initial corpus $\underline{K}$, Q and M would be updated via conditionalization and Bayes theorem in the usual manner relative to expansions of $\underline{K}$ by adding e_1, by adding e_2, and and by adding $e_1 \wedge e_2$. Application of the inductive expansion rules in these three cases could lead to assignment of positive b-value to h relative to the expansion by adding e_1 and 0 to ~h, positive b-value to ~h and 0 to h relative to the expansion by adding e_2. But under no circumstances would both h and ~h receive positive b-values relative to $e_1 \wedge e_2$.

Shafer could reasonably respond that he does not have in mind an interpretation of his support functions in terms of families of inductive expansion rules along the lines I have sketched here. However, it does seem fair to ask him to offer an alternative interpretation or family of applications. He rests content with appeals to intuitions, which I, for one, do not understand.

Spohn has offered an account of updating with b-values that never yields from an initial b-function and data a measure that fails to satisfy Shackle's requirements. However, like Shafer, Spohn has not offered an account of applications of his proposals yielding a clear interpretation of his measures in terms of other notions.

The scheme I have proposed does provide for an account of updating that does not take b-functions to be determined by prior b-functions and new data. And the account is the straightforward result of the proposed interpretation of Shackle measures.

It has been noted that the index of boldness can be eliminated by evaluating grades of confidence of rejection or degrees of potential surprise in Shackle's sense. But Shackle measures lack significance unless they are given an interpretation. The interpretation in terms of caution-dependent and deductively cogent criteria for inductive expansion calls for using the index of boldness or caution. Consequently, what may seem to be an elimination of a contextual parameter amounts to little more than the suppression of explicit mention of it. Of course, if one is not going to expand at all, one may rest content with reporting what the results of expansion would be at various levels of boldness. And this can be conveniently summarized by Shackle measures. However, if one is going to expand, one will have to be committed to expanding so that the elements of U rejected are those belonging to the set of those rejected at some level of confidence d* or greater. And this is tantamount to picking a degree of boldness q*.

Let us return to our main topic – inductively extended Ramsey test conditionals and suppositional reasoning.

In section 4.2, I pointed out that statements like "If this coin were tossed 1,000 times, it would probably land heads approximately 50 percent of the time" are not to be construed as revealing the degree of expectation-determining probability assigned the claim in the "then" clause relative to a revision (whether according to Gärdenfors or to Ramsey) of $\underline{K}$ by adding the information conveyed in the "if" clause. I suggested that "probably" is used here to indicate only that the conditional is an inductively extended conditional. Even so, there is a yearning among students of conditionals for an interpretation that does allow for grading inductively extended conditionals. We can introduce the expression "h $I>_r$ g" to express the claim that if h were the case, g would probably to degree r be the case. However, again the "probably to degree r" does not represent expectation-determining probability. What it does mean is that h I > g is acceptable according to either IERTR or IERTGL relative to a degree of boldness q such that $1 - q(\sim g) = b(g)$ relative to $\underline{K}_h^*$ or $\underline{K}_h^{*r}$ is equal to r.

Unlike expectation-determining probability, the use of b-functions here preserves the essential features of (K*1) and the axiom (A2) of conditional logic on Gärdenfors presentation.

Finally, this approach does explicate a sense of "conditional" surprise (d-value) and "conditional" degree of belief (b-value) that g given h. It is equal to that value r for which the inductively extended Ramsey test delivers h $I>_r$ g as acceptable.

6.10   Bookkeeping

Given a $\langle K, U, Q, N, q \rangle$, an investigator might reject elements of some subset of U and obtain an expansion $\underline{K}^i$ of $\underline{K}$ via induction. Relative to $\underline{K}^i$, there will be a truncated ultimate partition U^i, a new credal probability Q^i obtained by conditionalization via Bayes theorem, and a new informational-value-determining probability N^i derived in the same way. Assuming that the index of boldness remains the same, the inquirer can then reapply the inductive expansion rule to determine what new hypotheses are to be rejected.

This process of *bookkeeping* can be reiterated until the set of unrejected elements of U can be reduced no further. If U is finite, in a finite number of iterations, a point will be reached where no new elements of the original U are removed for the given value of q.[12]

From an inductivist point of view, there can be no objection to bookkeeping. If induction is expansion of the state of full belief, once one is in the new expanded state of full belief one might contemplate expanding again. To forbid such bookkeeping would be tantamount to denying that inductive expansion is a transformation from one belief state to another belief state relative to which further reasoning can take place. This attitude conflicts with the inductivist perspective explored in this book.

In the special case where q = 1, it can be shown that, for finite U, the conclusion recommended at the end of such bookkeeping iteration is to reject all but those elements of U for which the ratio $Q(h)/N(h)$ is a maximum.[13] Thus, if N assigns equal value to all elements of U, this result recommends accepting the disjunction of all elements of U bearing maximum credal probability.

The result for q = 1 just mentioned, however, argues against the general practice of being maximally bold. When q = 1, bookkeeping will tend to allow one to learn too fast from experience. If a fair coin is tossed 1,000 times, it is reasonable to conclude that it will land heads approximately 500 times. But it is rash to come to full belief through bookkeeping that it will land heads exactly 500 times. There are, perhaps, some contexts where bookkeeping with q = 1 makes sense; but in my judgment, most of the time one should be substantially more cautious than this.

Instead of considering as one inductive expansion rule the inductive expansion principle of section 6.6, one can adopt the rule that recommends adopting the results of bookkeeping iteration of the inductive expansion principle. The criterion that emerges is a caution-dependent, deductively cogent, and partition-sensitive rule.

It has one further property. When q = 1, it is ampliative but constant in the sense of Chapter 5 for all sentences equivalent given $\underline{K}$ to claims that some element of a subset of U is true. The reason is this. Suppose one begins with background corpus $\underline{B}$ and considers inductively extended expansions of it relative to U, Q, N, and q via bookkeeping when q = 1. Let $\underline{K}$ be the corpus that is the result of inductive expansion via bookkeeping with q = 1. It is the result of expanding $\underline{B}$ by the disjunction of those elements of U for which the ratio Q/N is a maximum. Let this disjunction be c.

Let us now expand $\underline{B}$ by adding d that is consistent with c – that is, that amounts to rejecting some set of the surviving unrejected elements of U. Since the surviving unrejected elements have equal Q/N ratios, the result will be tantamount to $\underline{K}^+_d$. But that is tantamount to claiming that the result is ampliative but constant.

Of course, if a given sentence in the language is not equivalent to the claim that a member of some subset of U is true, the inductive expansion function can fail to be ampliative but constant.

In Chapter 5, it was argued that if inductive expansion is ampliative but constant, (K*i7) and (K*i8) are satisfied by *i and (K*i7) and (K*ri8) are satisfied by *ri. We know that all ampliative but constant expansion functions are permutable so that $\underline{K}^{+i}_x = (\underline{K}^i)^+_x$. * satisfies (K*7) and (K*8). Substitute $\underline{K}^i$ for $\underline{K}$ in these postulates and use permutability to obtain (K*i7) and (K*i8) for *i. Similarly, *r satisfies (K*7) and (K*r8). Substitute $\underline{K}^i$ for $\underline{K}$ and use permutability to obtain (K*i7) and (K*ri8).

Main Thesis 1:

> If the family of caution-dependent, deductively cogent, and partition-sensitive rules deployed are bookkeeping iterations of the inductive expansion principle of section 6.6 and if q = 1, the inductive expansion rule that results is ampliative but constant (i.e., permutable).
> If q = 1, *i obeys (K*i7) and (K*i8) while *ri obeys (K*i7) and (K*ri8).

This conclusion may be further generalized. Take *any* family of deductively cogent, caution-dependent, and partition-dependent inductive expansion rules and construct a new family of such rules by bookkeeping. If the degree of boldness is at the maximum allowed, the resulting rule is ampliative but constant and all the ramifications for *i and *ri cited previously obtain.

Maximum boldness and bookkeeping also insure the validity of versions 2 of weak and of strong conditionalization for *i, version 2 of (NU), inductively extended Cut, and cautious monotony. The corresponding principles of conditional logic will be summarized in section 6.11.

If bookkeeping with q < 1 is used, inductive expansion functions will, in general, fail to be ampliative but constant. Some will fail to be exportable and some will fail to be importable. (K*i7), (K*i8), and (K*ri8) will be violated. Hence, so will versions 2 of weak and strong conditionalization, version 2 of (NU), and inductively extended cautious monotony. On the other hand, the following result can be proved:

Main Thesis 2:

> Using bookkeeping rules iterating use of the inductive expansion principle, for all values of q greater than 0, inductively extended Cut is satisfied.[14]

I have not been able to extend this claim to bookkeeping rules for arbitrary caution-dependent, deductively cogent, and partition-sensitive families of rules.

The coin-to?ssing example provides some basis for concluding that, in general, the use of the bookkeeping procedure with q = 1 is excessively rash. On the other hand, it is arguable that inductive expansion ought to be iterated until one has exhausted its resources via bookkeeping and reached a "fixed point." Doing so with q < 1 secures the validity of inductively extended Cut. But not much else.

6.11 What Is Left of Conditional Logic?

The focus of the previous discussion has been primarily on inductive expansion and inductively extended expansion even though the results were extended to inductively extended AGM and Ramsey revisions. Readdressing our attention to suppositional reasoning, when such reasoning is inductively extended, the main target is inductively extended revision. In effect, we have already summarized the results of the discussion in section 6.10 in terms of revision transformations and nonmonotonic principles of reasoning that remain and do not remain legitimate when revision is inductively extended. In this section, the results for conditional logic will be stated.

As in the previous discussion of conditional logic in Chapter 4, Gärdenfors's axiomatization of Lewis's VC shall be used as a baseline. The discussion will be restricted to languages where the conditional operator is not iterated even once.

Instead of beginning with brsg's of the type $\langle K,*\rangle$ or brsr's $\langle K,*r\rangle$, we must start with inductively extended brsg's of the type $\langle K,*i\rangle$ or inductively extended brsr's of the type $\langle K,*ri\rangle$. Inductive expansion is assumed to conform to the inductive expansion principle of section 6.6 both with and without bookkeeping and with maximum boldness and without. Whereas both of the unextended brs's could be determined from the pair $\langle K, M\rangle$, the inductively extended brs's involve five parameters $\langle K,U,P\ M,q\rangle$. Notice that, as the potential belief corpora in K, the Q and N functions used in inductive expansion vary. But how they vary is uniquely determined by the potential corpus adopted, P and M, which remain fixed. Given that U and q may be held fixed through variations of $\underline{K}$ that imply that exactly one element of U is true, satisfiability in a belief revision and positive validity in a system and in all systems of a given kind can all be well defined. Hence, the claims about validity of axioms can be investigated. I do no more than report results here.

The axioms of VC according to Gärdenfors used here are, of course, all inductively extended by replacing ">" by "I>."

Inductively extended versions of axioms (A1)–(A4) are positively valid in all inductively extended brsg's and brsr's.

The first important casualty is (A5), which is not valid even for the logic of noniterated conditionals (the only logic we are considering here). The inductively extended version of (A5) is the first version of (A5) stated in section 5.7. This fails to be positively valid for all inductively extended brs's.

The second version of (A5) in section 5.7, which is *not* a substitution instance of (A5) but of inductively extended (A7), is positively valid for all inductively extended brsg's and brsr's when bookkeeping is used and $q = 1$. When q is less than 1 or bookkeeping is not used, there are brsg's and brsr's for which it fails to be positively valid.

We have already discarded (A6). For brsg's, we have instead (A6t), which remains valid when brsg's are inductively extended. As before, (A6t) fails for brsr's both unextended and extended; but (A6r) is valid for brsr's even when extended. (A7) survives inductive extension for all brs's.

(A8) depends on the adoption of both (K*7) and (K*8). The inductively extended version of (A8) is valid if and only if both of the inductively extended versions (K*i7) and (K*i8) for inductively extended AGM revisions hold. (A8r) is positively valid if and only if (K*i7) and (K*ri8) for inductively extended Ramsey revisions hold. Where bookkeeping with $q = 1$ is used, (K*i7) and (K*i8) both hold and inductively extended (A8) is positively valid for the inductively extended brsg's thereby generated. Inductively extended (A8r) holds by similar reasoning for inductively extended brsr's.

When q is less than 1 or bookkeeping is not used, neither inductively extended (A8) nor inductively extended (A8r) is positively valid for all brs's in their respective domains.

Once more, when $q = 1$ and bookkeeping is deployed, inductively extended (K*7), (K*8), and (K*r8) all apply. Hence, inductively extended (A9) and (A10) are positively valid in the inductively extended brsg's generated thereby, and inductively extended (A9) and (A10r) are positively valid in the inductively extended brsr's generated thereby. As soon as q is less than 1 or bookkeeping fails to be employed, these claims must be withdrawn.

These results parallel corresponding results obtainable for the postulates of inductively extended revision and for the suppositional inference relation. They show that in the general case, inductively extended conditional logic becomes very weak. At the same time, they indicate how bookkeeping with $q = 1$ can recapture everything but inductively extended (A5).

None of this implies that we need discard the more substantial results applicable in the case of unextended conditionals. I have already contended that conditionals as couched in natural language often appear to be ambiguous. They are sometimes interpretable as unextended conditionals and sometimes as inductively extended conditionals. This corresponds to the need for ways to register conditional or suppositional judgments of serious possibility where the only inference in the strict sense is monotonic deductive inference and also to express conditional judgments of serious

possibility where inductive inference is involved as well. My aim throughout this discussion has not been to dismiss interest in unextended Ramsey
test conditionals but only to direct attention to the importance of considering inductively extended Ramsey test conditionals as well.

6.12   Induction as a Generalization of Deduction

In 6.10, I pointed out that if bookkeeping is used to construct a caution-
dependent, deductively cogent, and partition-sensitive set of inductive
expansion rules from an initial family of this kind, whether it is of the sort
I have proposed or is some alternative proposal, requiring maximum
boldness (q = 1) does restrict the legitimate inductive expansion rules to
ampliative but constant rules that can be used to characterize inductively
extended Gärdenfors or Ramsey revision as conforming to postulates to
unextended revision of one of these two varieties. I objected to imposing
such restrictions on the grounds that to require maximum boldness as a
constraint on legitimate inductive expansion yields awkward results.

This point could be conceded. Yet, there are some 5-tuples of contextual parameters with q less than 1 where bookkeeping can lead to ampliative
but constant rules. It may be argued that legitimate inductive expansion
should be restricted to such contexts on the grounds that principles like
(K*i7) and (K*i8) or (K*ri8) or, at least, principles like inductively extended Cut and inductively extended cautious monotony are reasonable
constraints to impose on revisions in general or on inferential relations
that deviate from the straight and narrow of monotonicity.

As it happens, bookkeeping using the criteria for inductive expansion I
favor will secure Cut at all positive levels of q. But I would deny that the
adequacy of my proposals hinges on results like that. Cautious monotony,
for example, fails. That is not a sign of weakness in the proposals I favor but
rather in inductively extended cautious monotony. Even in the case of
inductively extended Cut, I do not see how to insure that all families of
caution-dependent and caution-dependent expansion rules can yield with
the aid of bookkeeping the validity of inductively extended Cut except by
stipulating that Cut should be satisfied. But, when it comes to inductive
expansion, satisfaction of Cut by inductively extended revision and the
corresponding pseudoimplication relations lacks a self-standing motivation except as a reflection of a desire to make induction look as much like
deduction as possible.

My contention has been throughout that inductive expansion is dependent on five factors or on variants on them. It is at least arguable that one
might seek to impose constraints on the values these five contextual parameters can have, such as insisting that q = 1. Any such proposal needs to be

examined on its merits. I have raised an objection to adopting q = 1 with bookkeeping. The proposal now being considered would simply stipulate that no package of 5-tuples where q < 1 is legitimate unless it yields an ampliative but constant inductive expansion function.

One can formulate a restriction of this sort. But to defend it on the grounds that it yields results that are attractive for nonmonotonic inferential relations that are "natural" generalizations of monotonic inferential relations of the sort associated with deductive logic is to rate the demand to obtain such natural generalizations of deductive logic as having a priority over the requirements of inductive inference.

The issue here is in part a matter of research program. If one thinks of nonmonotonic reasoning as a variant of monotonic inference, specifies in advance how far nonmonotonic reasoning is allowed to vary from monotonic reasoning, and then seeks to accommodate inductive as well as belief-contravening nonmonotonicity within that framework, the stipulation that inductive expansion functions should be permutable (= ampliative but constant) might seem natural.

To someone who thinks of deduction and induction as the two main types of inference and who focuses on the differences between the two that arise in applications, it will be desirable to appeal to independently formulated principles of inductive expansion so that comparisons of the formal properties of deductive and inductive inference can be made in a systematic manner.

Of course, the principles regulating inductive inference are far more controversial than those regulating deductive inference. I have relied on proposed principles of inductive expansion that are controversial in some respects. However, I have also tried to indicate that most of the main results obtained are robust with respect to weaker and less controversial assumptions about inductive expansion. Thus, ampliativity was shown to insure nonmonotonicity with no belief contravention. And whether or not one wishes to use caution-dependent, deductively cogent, and partition-sensitive principles, the legitimacy of some constraints on inductively extended revision operators depends on the role of bookkeeping and the value of the boldness parameter. To this extent, the proposed criteria for inductive expansion offer a reasonably adequate tool for making an independent assessment of the respects in which induction is and is not a generalization of deduction.

Of course, I am a partisan of efforts to derive nonmonotonic inferences from principles of inductive inference rather than characterizing them by postulates specifying how they deviate from deductive reasoning. I do not know how to refute decisively those who adopt this latter approach. Refutation, in any case, is not the main issue. One can learn a great deal from the work of those who study nonmonotonic consequence relations. I merely wish to emphasize the importance of pursuing the study of principles of inductive inference as well.

The matter is of some importance. The current interest in nonmonotonicity on the part of students of artificial intelligence is not, in point of fact, focused on belief-contravening supposition. Suppositional reasoning is of interest. But as I shall argue in Chapters 7, 8, and 9, the focus is often on jumping to conclusions – that is to say, on ampliative inference whether it is suppositional or involves genuine change in view. Yet, there is a widespread tendency to try to avoid consideration of statistical and inductive reasoning and to address the topic of nonmonotonic ampliative reasoning in a framework that sees it as a variant on deductive, explicative, monotonic reasoning. Given the diversity of views held on the topics to be considered and the hermeneutical difficulties involved in understanding what is intended by the authors to be discussed, I cannot claim to be in a position to document this tendency with perfect accuracy. However, I hope to be able to support the view that insofar as my interpretations do not do full justice to the views to be considered, they are sufficiently on target to establish the relevance of the proposals I have been sketching to current discussions of nonmonotonic reasoning.

Appendix

In section 6.6, the inductive expansion principle was formulated for finite U. In section 6.10, the bookkeeping procedure was combined with the inductive expansion principle for finite U. As promised in section 6.10, the inductive expansion principle with bookkeeping will be extended to infinite U of primary interest. The proposals made are elaborations of ideas proposed previously (Levi, 1980). Three types of cases will be considered:

> *Case A:* U is the set of point estimates of parameter whose possible values are the real numbers in a given finite interval (except at most for a subset of Lebesgue measure 0).

The domain of case A predicaments could be extended to estimating n-dimensional parameters in a hypervolume of n-dimensional Euclidian space; but the effort will not be undertaken here. It is assumed that the Q and N distributions over U are countably additive. This restrictive assumption could be removed at some additional complication in my exposition. I shall not do so.

Case B: U is countably infinite.

In case B, I shall not assume that the Q and N distributions are countably additive. They are, of course, finitely additive.

Case C: U is the set of point estimates of parameter whose possible values are the real numbers on the real line from $-\infty$ to $+\infty$. In each finite subinterval at most a set of points of Lebesgue measure 0 is removed.

In case C, it is not assumed that the Q and N distributions over U are countably additive. However, the corresponding distributions conditional on the true estimate falling in a finite subinterval of positive Lebesgue measure are countably additive (for any such subinterval).

Consider first *Case A*. A cumulative credal probability distribution $Q(x \leq k)$ is given over the interval and a corresponding informational-value-determining cumulative probability distribution $M(x \leq k)$. For any seriously possible real value point estimate h_x asserting that x is the true value, consider the ultimate partition consisting of $h_{x,\epsilon} = x_{\pm\epsilon}$ and its complement in the interval. Reject the exact point estimate h_x for index of boldness q if and only if there is a value $\epsilon *$ such that for every $\epsilon \leq \epsilon *$, $Q(h_{x,\epsilon}) \leq qM(h_{x,\epsilon})$. Bookkeeping can then be instituted to reiterate this process.

In cases where the cdf's for the Q- and M-functions are continuous and differentiable so that densities are defined, the point estimate x will be rejected if and only if the credal density f(x) is less than qm(x) where m(x) is the informational-value-determining density. Bookkeeping with q = 1 implies that the point values carrying maximum value f(x)/m(x) alone survive rejection if such a maximum exists.

A maximum will always exist unless the maximum value is ruled out by the initial corpus $\underline{K}$ as not a serious possibility. In that case, all elements of U that are serious possibilities will be ruled out via induction. Expansion into inconsistency will be recommended. In such cases, bookkeeping with maximum boldness is forbidden. When q is less than 1, at least one point estimate must survive rejection both in the case where distributions are continuous and when they are not. This is another consideration arguing against bookkeeping with maximum boldness additional to the one offered in section 6.10. The objection to adopting q = 1 with bookkeeping in the case of finite U does not claim that exercising maximum boldness leads to inconsistency in that case but merely that it yields inductive conclusions that are presystematically too bold. In case A, maximum boldness can lead to inconsistency when there is no element of U that is maximally "preferred" according to the assessments of Q and N values.

Turn now to *Case B*. Let h be any element of a countably infinite U. Any finitely additive probability distribution over U can be represented as a weighted average or mixture of a countably additive distribution P_c and a "purely finitely additive" distribution P_d (i.e., a finitely additive probability P_d such that if a nonnegative countably additive set function S satisfies the condition that for every x in U, $P_d(x) \geq S(x) \geq 0$, then for every x in U, $S(x) = 0$) (Yosida and Hewitt, 1952 cited in Schervish, Kadane, and Seidenfeld, 1984, and used there to state important results concerning breakdowns of "conglomerability" when countable additivity is not satisfied). In less technical terms, a purely finitely additive probability distribution over a countably infinite U is a uniform distribution assigning 0 unconditional probability to each element of U. A uniform distribution over U is also representable by a sigma finite, countably additive set measure (i.e., the measure assigns positive finite real values to all elements of U while the total measure assigned all elements of U is infinite). Sigma finite measures can be used to determine the conditional probability of elements of a finite subset F of U given that some element of F is realized. A conditional distribution may also be derived for F from P_c. The actual conditional distribution for F will be derived from the weighted average of the additive component and the sigma finite or uniform distribution.

The Q and N distributions over U are of these kinds. There is one requirement that will be imposed on the N distribution but not the Q distribution. It is mathematically possible to construct a probability – indeed, a countably additive one – that orders the elements of U in an infinitely descending chain. The sequence of hypotheses h_i carrying probability $(1/2)^i$ illustrates the point. Every such scheme secures an ordering of the elements of U whose probability values are bounded by 0 – the probability of a contradiction. If the probability is an informational-value-determining probability reflecting the inquirer's demands for information, the inquirer countenances elements of U whose informational values are arbitrarily close to the informational value of a contradiction. I suggest that this attitude ought not to be countenanced. Assessments of informational value ought to be bounded away from inconsistency. This means that countably additive measures over countably infinite alternatives cannot assign to the elements of U more than finitely many distinct probability values and, hence, that finitely but not countably additive N-functions cannot do so either.

This proposal is to be understood as a constraint on demands for information. It insists that all elements of U should carry less informational value than a contradiction by some fixed positive amount. In that setting, it seems attractive, although demonstration seems out of the question. On the other hand, this argument does not carry over to credal probability distributions where no such restriction is imposed.

Consider any finite subset of n members of U containing h and determine the conditional Q and N distributions over that subset.

The inductive expansion principle can then be applied at the given level of boldness and iterated via the bookkeeping procedure relative to the given n-fold finite subset of U. h will be *n-rejected* if and only if it is rejected for at least one such subset of n-elements of U. h is *rejected* if and only if it is n-rejected for every n greater than some n^*.

Note that this procedure requires that bookkeeping must be applied to each n-fold truncated ultimate partition. This is required in order to take care of cases such as the following. Let the Q distribution assign probability $(1/2)^i$ to h_i, $i = 1, 2, \ldots\ldots$. Let M assign equal 0 probability to all elements of U (and equal positive sigma finite measure). h_n will be rejected relative to the n-fold partition consisting of $h_1, \ldots, h_n$ but there will an m^* such that for all $n + m$ fold finite subsets of U containing h_n, h_n will not be rejected if bookkeeping is not used. The upshot is that no element of U will be rejected at any value q. However, if bookkeeping is used at each stage, it will be $n + m$ rejected if q is sufficiently large and will stay rejected for still larger subsets of U.

The same example explains also why it is not feasible to apply the inductive expansion principle directly to U. In this case, the N-function assigns 0 probability to all elements of U and positive Q-values so that the rule would fail to reject any element of U. If we used a sigma finite measure for N, all elements of U would receive an equal positive value that can be arbitrarily selected. We could arrange matters so that all elements of U are rejected by the rule or only some of them depending on the positive value selected. On the other hand, the ratio of the Q-value of elements of U to their sigma finite M-values (whatever that value might be) completely orders the elements of U so that one might formulate a rule recommending rejecting all elements of U except those for which the ratio is greatest (if such a maximum exists). This corresponds to a case of maximum boldness. In the kinds of examples under consideration, the boldness parameter is not well defined relative to U for values less than a maximum. Consequently, we cannot specify levels of boldness less than the maximum directly. The technique proposed here uses well-defined indices of boldness relative to finite subsets of U to supply the required index in conformity with the decision theoretic rationale offered for the inductive expansion principle for finite ultimate partitions.

The question arieses: Are there cases where there is no maximum for the ratio of credal probability to informational-value-determining probability?

Consider a situation where the Q-distribution is uniform but the N-function assigns h_i the value $(1/2)^i$. In that case, every element of U will be rejected no matter how cautious the inquirer is according to the procedure proposed here so that induction yields expansion into inconsistency.

In case A, we have already seen that this eventuality can occur when $q = 1$. But in this case, requiring the agent to exercise a less than maximum but positive degree of boldness is of no avail. Only q = 0 which mandates no inductive expansion at all will do.

Observe that in the example just described the least upper bound of informational values assigned elements of U is precisely the informational value assigned to a contradiction. I have previously suggested that this kind of assessment of informational value is unacceptable. If the restriction is acceptable, the ordering with respect to informational value can assign no more than finitely many distinct values to elements of U. The ordering with respect to Q/N must have a maximum as must the ordering conditional on the true element of U falling in any subset.

Finally turn to *Case C*. This case may be treated by combination of the previous two methods. Divide the real line into positive intervals of length L. These intervals constitute a countably infinite partition of the real line. Apply the methods used in case B to this situation. An interval of length L is *L-rejected* if and only if it is rejected by the case B methods relative to such a partition at the given level of caution q.

When an interval of length L is L-rejected, all the point estimates in the interval are also L-rejected. But some of these point estimates may fail to be L'-rejected for L' < L. A point estimate is rejected if and only if there is an L* such that for all L < L*, that point estimate is L-rejected.

Notice that the requirement that the N-function or its sigma finite representative should be such that for any L, the N-values of the elements of L-partition should have no more than finitely many distinct values. With this observation in place, when q < 1, some elements of U must go unrejected so that consistency is preserved. When q = 1, no such guarantee can be provided for all cases.

These techniques remain in the spirit of the decision theoretic approach that rationalizes the inductive expansion principle with bookkeeping. They understand the injunction to maximize expected epistemic utility for infinite U as limiting cases of maximizing expected epistemic utility for finite U under specific understandings of what such limiting cases should be.

It is worth mentioning that in all three cases, the ratio $Q(h)/N(h)$ where Q and N represent either probability distributions over U, their sigma finite representations or the appropriate densities when such exist give an ordering of the elements of U. If the agent is maximally bold, he or she will suspend judgment between the most "preferred" elements according to that ordering. If there are no such optimal elements, maximal boldness leads to contradiction. To avoid that untoward result, the inquirer should be less than maximally bold and suspend judgment between elements of U that are sufficiently well preferred according to the ordering. What this means has been sketched for the three cases under consideration.

7 Defaults

7.1 Defaults and Nonmonotonic Implication

I have taken for granted that nonmonotonic pseudoinferential relations of the sort represented by expressions like "h |~ f @ $\underline{K}$ and E" or nonmonotonic strictly inferential relations represented by "h |≈ f @ $\underline{K}$ and E" are related to transformations of the corpus $\underline{K}$ in the manner indicated in section 5.4. This approach extends the interpretation of nonmonotonic implication in terms of belief revision proposed by Makinson and Gärdenfors (1991). On the basis of this construal, some alleged principles of nonmonotonic reasoning have been examined in both inductively unextended and extended settings.

Although the expression, *nonmonotonic logic* is of very recent origin, the phenomena addressed by students of nonmonotonic logic have been considered in one way or the other for centuries. This is true both of the nonmonotonicity due to the ampliativity of inductive inference and the nonmonotonicity due to the belief-contravening character of counterfactual suppositional reasoning. One advantage of extending the Makinson–Gärdenfors approach in the manner I have suggested is that it facilitates understanding of the contrast between nonmonotonicity due to belief contravention and due to ampliative inference.

Nonmonotonic reasoning is not customarily represented, however, in the fashion elaborated in the previous chapters.

Interest in nonmonotonicity under that name arose initially in computer science. Given some specified body of information (knowledge cum data base) and the kinds of questions that need to be answered, students of computer science sought to model, for the given case, the drawing of conclusions from the given information.

The default systems of Reiter (1980), Poole (1988), or Brewka (1991), circumscription according to McCarthy (1980), autoepistemic logic of Moore (1983), and the like eschew efforts to identify a general criterion for inductive expansion like the inductive expansion principle of section 6.6 in terms of which an inductive inferential relation can be defined for the specific case under consideration when supplemented by values for pertinent contextual parameters. Instead, a standardized method of representing legitimate inferences (or an inferential relation) is devised to be applied in all applications in a certain domain. No guidance is offered concerning the inferences from the given knowledge cum data base other than what is built into the constraints on the standardized method of representation.

To explain the difference, I shall first illustrate how the inductive expansion principle applies to an example and then consider the way in which a variant on the default representation of Reiter (1980) may be used on the same example.

Example 1: Let r(x) = x is a radical, f(x) = x is a feminist, and b(x) = x is a bank teller. Suppose that the percentages of individuals in a given population that belong in the eight categories derivable from these predicates are given in the following table:

Table 7.1

rfb	rf~b	r~fb	r~f~b	~rfb	~rf~b	~r~fb	~r~f~b
2	15	1	2	5	10	10	55

Suppose further that Linda is selected at random from this population. By this, I mean that some procedure is used to select her from the population that results in my assigning credal probabilities to hypotheses as to which category she belongs equal to these percentages.

If I were then to ask in which category Linda belonged so that U might be regarded as the partition relative to which I generated potential answers and I regarded each element of U as carrying equal M-value of 0.125, the inductive expansion principle gives definite recommendations for each value of q as to which elements of U to reject and what conclusion to reach by inductive expansion. The following table gives this information by specifying the degrees of disbelief (d-values) in the sense of section 6.9 for each element of U.

Table 7.2 d-values for elements of U

rfb	rf~b	r~fb	r~f~b	~rfb	~rf~b	~r~fb	~r~f~b
.84	0	.92	.84	.6	.2	.2	0

From Table 7.2 one can compute the d-values for all disjunctions of subsets of elements of U. Take the minimum d-value in the given subset. One can also obtain b-values for such disjunctions. Take the minimum d-value assigned an element of the complement in U of the given subset.

Table 7.3 gives the d-values for elements of U when bookkeeping is used.

Table 7.3

rfb	rf~b	r~fb	r~f~b	~rfb	~rf~b	~r~fb	~r~f~b
.84	.56	.92	.84	.74	.56	.56	0

All of these judgments are relative to a background corpus $\underline{B}$, which shall be held fixed.

Consider now $\underline{K} = \underline{B}^i$ for the case where q = 1 and the inquirer is maximally bold. Without bookkeeping, the agent will suspend judgment between rf~b and ~r~f~b. The expansion of $\underline{B}$ by adding the disjunction of these alternatives is identical with $\underline{K}$.

When bookkeeping is used with q = 1, $\underline{K}$ consists of the consequences of $\underline{B}$ and ~r~f~b.

We can characterize the results of these two forms of inductive expansion by specifying the maximally consistent expansions or extensions of $\underline{K}$ (not $\underline{B}$) expressible as elements of U.

Notice that there is a distinction to be drawn between specifying the set of maximally consistent expansions or extensions $\underline{K} = \underline{B}^i$ (which are, of course, a subset of the consistent extensions of $\underline{B}$) and the use of the inductive expansion principle to derive the set so specified from $(\underline{B}, U, Q, M, q)$. Reiter (1980) suggested a procedure for obtaining consistent and closed extensions of $\underline{B}$ from (a subset of) $\underline{B}$ and a set of default rules by requiring the extensions of $\underline{B}^i$ to be maximal consistent sets of sentences default derivable from $\underline{B}$ and the set of defaults.

A default derivation starts with $\underline{B}$. (Reiter allows the initial set to be a subset of $\underline{B}$ that is not closed under logical consequence – a point to which I shall return in section 7.2. I shall presuppose that the initial set is closed under logical consequence.)

Let $\underline{B}'$ be a set that is default derived from $\underline{B}$ and x be consistent with $\underline{B}'$. Then y may be added to $\underline{B}'$ as the next step in a default derivation from $\underline{B}$ provided that there is a default rule of the form $\mathbf{M}x/y$. Suppose that $\underline{B}^*$ is default derived from $\underline{B}$, z is in $\underline{B}$, and x is consistent with $\underline{B}^*$. Then y may be added to $\underline{B}^*$ as the next step in a default derivation from $\underline{B}$ provided that there is a default rule z; $\mathbf{M}x/y$ conditional on the "prerequisite" z. A maximal default derivable extension of $\underline{B}$ is the closure under logical consequence of a set of sentences derivable from $\underline{B}$ relative to the given set of defaults to which no more sentences can be added by default derivation.

Thus, if we wanted to obtain the set of extensions of $\underline{B}$ whose elements are rf~b and ~r~f~b, the following defaults would be employed: $\mathbf{M}r/r$ asserts that since it is possible (i.e., consistent with $\underline{B}$) that Linda is a radical, Linda is a radical. $\mathbf{M}$~r/~r. In addition, there is a default conditional on a "prerequisite." One might say that radicals are normally feminists. This default is represented as r; **M**f/f. Given that Linda is a radical and that it is possible that she is a feminist, to derive the claim that she is a feminist: ~r;**M**~f/~f. **M**~b/~b (i.e., since it is possible that Linda is not a bank teller, she is not a bank teller).

Given these five defaults, two maximal default-derivable extensions of **B** may be shown to exist: rf~b and ~r~f~b, if **B** is as specified in example 1. There r is consistent with **B** so by the first default, it may be added. Once it is added, it can be used as the prerequisite of the third default along with the consistency of f with what is already available to obtain f. The fifth default can then be used to add ~b. Nothing more can be added by default. So take the deductive closure of rf~b, and a maximal default-derivable extension of **B** has been obtained. Using the second, fourth, and fifth defaults, ~r~f~b may also be added to **B** along with logical consequences to form a maximal default-derivable extension.

Thus, rather than using the inductive expansion principle with q = 1 along with a specification of the other contextual parameters to derive a set of "extensions," Reiter uses a set of defaults and default derivability. If bookkeeping had been deployed along with the inductive expansion principle and q = 1, only the extension associated with adding ~r~f~b to **B** is obtained. In that case, the defaults in the set identified before would have to be reduced to the set consisting of the second, fourth, and fifth.

The method just sketched offers a system of rules (default rules) which, together with the initial corpus **B**, determines the inductive expansion **K**. Understood as principles of inductive expansion, default rules make explicit reference to one contextual feature relative to which inductive expansion takes place – to wit, the initial background information **B**. Yet no one seems to think that the default rules are principles of reasoning that are context-independent principles of inductive expansion. In example 1, the default "radicals are normally not bankers" is not taken to be a principle that is of universal applicability. One can think of contexts where this default would not be endorsed and of other situations where there may be differences of view as to whether it is an acceptable default or not. One can suppose for the sake of the argument that the default is not acceptable and that "radicals are normally bankers" is acceptable.

The default rules are themselves information of a certain kind distinct from the information contained in **B** – although, according to other approaches such as that of Poole (1988), this information is represented as part of the belief state **B**. By way of contrast, the principles of inductive expansion proposed in Chapter 6 seek to formulate criteria that specify explicitly a larger set of contextual parameters relevant to inductive expansion so that the principles themselves are relatively speaking context independent and, hence, applicable to a broader range of situations. It would be hubris to suggest that the proposals made have specified all the relevant contextual parameters or that there are not alternative ways to parse relevant contextual features and formulate general principles that will yield equivalent characterizations of inductive expansion. It would be just as pretentious to insist that the proposals could not be modified so as to yield a better systematic account of inductive expansion. Nonetheless, it is arguable that characterizing inductive expansions by means of systems of defaults fails to provide a systematic account of all the relevant contextual parameters, whereas proposals like the one based on the inductive expansion principle can do better.

The account of default reasoning I have just sketched does not precisely capture Reiter's procedure in ways some of which will be explained in section 7.2. My purpose here is to explain what appears in a rough and ready way to be the relation between the methods computer scientists have used in nonmonotonic reasoning and principles of inductive expansion as I have conceived them.

Reiter's approach or the alternative methods that have been favored by others have rested content with supplying defaults designed for a given application or family of applications. There have been several other methods proposed for deriving $\underline{K} = \underline{B}^i$ from $\underline{B}$ cum defaults and it has not always been clear how these methods relate to one another. For example, as Kraus, Lehmann, and Magidor (1990) observe, the deliverances of circumscription, autoepistemic logic, and default logic with regard to the assertions of nonmononotic implication $h \wedge f \vdash g$ and $T \vdash \sim f \vee g$ are different from one another and differ from verdicts Kraus, Lehmann, and Magidor regard as sensible.

Many logicians and computer scientists have been discontent with the seeming chaos that has attended these approaches and have sought to address the topic of nonmonotonic reasoning from a different angle. In 1985, D. Gabbay recommended looking at the nonmonotonic inference relations determined by a given system of nonmonotonic principles and exploring various properties that it might or might not be desirable for such relations to have. That is to say, the project was to consider the relation between input premises and output conclusions as some sort of inferential relation represented in the fashion of Tarski or Gentzen but lacking the property of monotonicity exhibited by deducibility. This approach was taken up by Makinson (1989) and Shoham (1987) and subsequently by Kraus, Lehmann, and Magidor (1990) and Lehmann and Magidor (1992).

These approaches have been usefully illuminating; but they are, as they stand, coherence conditions on what Kraus, Lehmann, and Magidor (1990) call "assertions" – that is, specific claims as to which inferences from premises to conclusion are licensed. Moreover, although these authors have explored quite clearly the properties of various systems of coherence conditions, many of their judgments concerning which systems of coherence conditions ought to be instituted remain little more than obiter dicta. We may despair of ever identifying a perfectly noncontroversial set of coherence conditions.

The source of difficulty here seems to me to be found in the circumstance that such investigations, like those of the default reasoners, ignore the contextual parameters other than the initial corpus $\underline{B}$ and the set of defaults or the nonmonotonic inferential relation adopted in a given situation that are relevant in the given context. Identifying connections between these coherence conditions on "assertions" and general but context-sensitive principles of inductive inference ought to contribute to clarifying our judgments as to what coherence conditions ought to be imposed on systems of assertions characterizing a nonmonotonic inferential relation.

Makinson and Gärdenfors (1991) sought a remedy for this deficiency by connecting the alleged inferential relations with belief revision. The account of belief revision they deployed is a theory of AGM revision based on the classic work of Alchourrón, Gärdenfors, and Makinson (1985). In addition to the initial corpus $\underline{B}$, it appeals to the revision operator or to the underlying assessment of entrenchment or, as I understand it, assessment of damped informational value.

As it stands, however, the Makinson–Gärdenfors proposal of 1991 does not accommodate inductively extended revision in either the AGM or Ramsey sense.

Gärdenfors and Makinson (1993) can be understood as an attempt to do just that by supplementing the earlier proposal with the use of a notion of expectation in a manner that relates entrenchment to expectation.

Authors like Kyburg (1961, 1983) and Pearl (1992) resort to different contextual parameters. They invoke considerations of probability in addition to the initial corpus to address the question of inductive expansion.

My dissent from Gärdenfors and Makinson, on the one hand, and from Kyburg and Pearl, et al., on the other, concerns the identification of the relevant contextual parameters appropriate to an account of inductive expansion. In the last three chapters of this book, my aim is to elaborate on the relations between the proposals I have made and alternative views and to explore the extent to which such views can be accommodated within the framework I favor.

The rest of Chapter 7 will focus on the ideas found in Reiter's pioneering article of 1980 and the work of Poole (1988). I shall argue that a modified version of Reiter's approach is rationalizable as the results of inductive expansion with bookkeeping. Poole's approach seems to advocate the use of inductively extended nonmonotonic inferential relations based on ampliative but constant inductive expansion transformations that I have rationalized in terms of families of caution-dependent, deductively cogent, and partition-sensitive inductive expansion rules when bookkeeping is exploited with maximum boldness.

I shall also suggest that many of Reiter's informal remarks register sympathy with a view of default reasoning that requires ampliative but constant inductive expansion – even though his own formal suggestions deviate from this requirement.

I cannot and do not claim that the views of these authors as stated precisely conform to the results of using bookkeeping with maximum boldness. Where there are deviations, I shall argue from my own philosophical perspective and sometimes from theirs that their views would be improved by instituting modifications that secure precise conformity. Hence, my arguments will be enmeshed, to some extent, in philosophical controversy. But even those who disagree with my views will, I hope, agree that the connection I claim to hold between assumptions and results is in order.

I will not consider alternative approaches such as circumscription and autoepistemic reasoning, although I am inclined to think that arguments to a similar conclusion may be reached.

Chapter 8 will review and survey various kinds of assessments of the conclusions of nonmonotonic reasoning including probability, entrenchment, degree of belief in the Shackle sense, degrees of possibility, and the like and take a critical look at their use in characterizing nonmonotonic reasoning. Within this setting, the critique of high-probability views of inductive expansion of the sort favored by Kyburg and Pearl among others will be pursued further.

Chapter 9 will consider some of the ideas found in Kraus, Lehmann and Magidor (1990), Lehmann, and Magidor (1992), and Gärdenfors and Makinson (1993). In the same spirit of critical hermeneutics adopted in this chapter, I shall argue that these authors are also advocating the use of ampliative but constant inductive expansion functions of the sort obtainable via bookkeeping with maximum boldness.

Because, as I have already indicated, bookkeeping with maximum boldness is a practice that, at best, has limited legitimacy and that bookkeeping with less boldness will satisfy Cut but will violate many salient requirements imposed by these authors on nonmonotonic implication relations, my arguments, to the extent that they are cogent, ought to raise serious questions concerning the domain of legitimate applicability of the theories of nonmonotonic reasoning that tend to be favored in the literature.

7.2   What Is a Default?

It is well known that most birds fly but that some birds do not. Penguins are a notorious exception. We come to believe via the testimony of our senses or of competent witnesses or by some previous inductive inference that the animal Tweety is a bird. Or we are invited to suppose, for the sake of the argument, that this is so. From this information, we seek to answer the question whether Tweety flies.

It is contended that the force of the so-called default principle that most birds fly or that birds normally fly is to prescribe a rule for drawing conclusions about flying from information that the individual involved is a bird. In the absence of counterindications, one should conclude by default that Tweety flies from the claim that Tweety is a bird.

It is far from obvious that "Most birds fly" and "Birds normally fly" as understood presystematically express default principles of inference in the sense just indicated. "Most birds fly" is usually employed to make a statistical claim to the effect that more birds belong to flying species than to nonflying species or that the chance of an individual selected at random from the population of birds belonging to a flying species is greater than the chance of such a selection failing to belong to a flying species. Yet, a leading student of default reasoning denies that "Most birds fly" is such a statistical claim when interpreted as a default principle (Reiter, 1980).

Reiter subsequently elaborated by explaining that under the interpretation of "Most birds fly" as an expression of a default, it is a claim about "prototypical" birds (Reiter and Criscuolo, 1981). This remark suggests that "Most birds fly" is better rendered by "Birds normally or typically fly" or by "On average, birds fly." The prototypical bird is then the normal or average bird.

Reiter's elaboration on what he meant by "Most A's are B's" is a marked improvement on his initial formulation. However, according to this reading, defaults remain statistical claims. They are no longer claims about what happens for the most part but about what happens on average or, if not that, the most likely among a range of alternative outcomes. What is normal is what is either the mean or the mode of a probability distribution. (Originally, it was taken for granted that the distribution is a normal distribution but the idea may be extended to any unimodal distribution.) Prototypes might on this understanding be quite rare. But prototypicality remains a statistical or probabilistic concept.

Because statistical notions are not plausibly represented within a first order framework, Reiter wishes to avoid them. Instead of avoiding them, however, he appears to be using them without incorporating their statistical features in the formal framework.

Perhaps, however, Reiter means to suggest that his project is to abstract from the details of a probability distribution over possible outcomes or states and consider only some features of the distribution. This interpretation concedes that prototypicality or normality is a statistical or probabilistic idea while dispensing with the need in the set of applications of interest to Reiter to consider the details of probability distributions that may require going beyond the use of a first-orderizable language.

According to this reinterpretation of Reiter's intent, Reiter (1980) gives explicit instructions concerning how systems of default rules are to operate on sets of sentences in a first-order language representing a set of initially given assumptions from which inferences are to be drawn. For present purposes, we may focus on closed normal defaults. These are rules that take as inputs a prerequisite α that is a closed first-order formula and a consequent w. The rule has the form: α; $\mathbf{M}$w/w. Thus, "Birds fly" as instantiated for Tweety is the closed normal default with prerequisite "Tweety is a bird" = b and consequent "Tweety flies" = f. "$\mathbf{M}$Tweety flies" means "it is consistent to assume that Tweety flies."

Suppose "Birds fly" instantiated by Tweety is the only default in D. Given a consistent set $\underline{A}$ of sentences representing full beliefs of the agent, one can "extend" $\underline{A}$ by considering all logical consequences of $\underline{A}$. Defaults are used to characterize a different notion of a consistent extension. Roughly speaking, an extension is a consistent, deductively closed set of sentences in the given language containing the initial set that is default derivable from the sentences in $\underline{A}$.

For example, if $\underline{A}$ contains b = "Penguins are birds" and "No penguins fly" but does not contain either p = "Tweety is a penguin" or its negation, the default rule b;$\mathbf{M}$f/f entitles us to extend $\underline{A}$ to the deductive closure of the sentences in $\underline{A}$ and f. This is an extension of $\underline{A}$ by the set of default rules and is, indeed, the only extension of $\underline{A}$.

Reiter's scheme seems to suggest that the set of defaults defines a generalized or inductive expansion transformation. This is the construal I offered in discussing example 1 of section 7.1. Strictly speaking this construal does not agree with Reiter's own account of the use of defaults.

In section 7.1, I required the set of sentences to be a corpus $\underline{B}$ – that is, a deductively closed theory. Reiter does not do so. The initial set $\underline{A}$ need not be logically closed even though it is consistent. The extensions generated by the defaults are, however, logically closed. This aspect of Reiter's proposal yields two interesting results.

(a) An extension of $\underline{A}$ according to Reiter need not contain the logical closure $Cn(\underline{A})$ of $\underline{A}$.

(b) Given that $Cn(\underline{A}) = Cn(\underline{A}')$ so that $\underline{A}$ and $\underline{A}'$ represent the same state of full belief $\underline{B}$, the set of extensions determined from $\underline{A}$ by a given set of defaults need not be the same as the set of extensions determined by $\underline{A}'$.

Makinson (1994, sec. 3.2, observations 2 and 3) recognizes these two ramifications of Reiter's scheme. He points out that when the nonmonotonic implications of $\underline{A}$ are taken to be members of the set of logical consequences of the intersection of the consistent extensions of $\underline{A}$ (as I, in effect, did in discussing example 1 in section 7.1), (a) is tantamount to the violation of the logicality requirement that the nonmonotonic implications of $\underline{A}$ include the logical consequences of $\underline{A}$.

Makinson suggested that this peculiarity of Reiter's scheme could be avoided by a "touch of the screwdriver." One could amend Reiter's definition of an extension of $\underline{A}$ by adding the stipulation that a Reiter extension be enlarged by adding all of the logical consequences of $\underline{A}$ and then be closed under logical consequence.

Makinson also pointed out the following counterexample to cautious monotony supported by Reiter's system. Let $\underline{A} = \{a\}$ and $\underline{A}' = \{a, a \vee b\}$. The defaults are (a; **M**c/c) and (a ∨ b; **M**~c /~c).

$\underline{A}$ has one extension: $\underline{E} = Cn(\{a, c\})$. Notice that $Cn(\underline{A}) \subset \underline{E}$.

$\underline{A}'$ has two extensions: $\underline{E}$ and $\underline{E}' = Cn(\{a, \sim c\})$. Both extensions contain $Cn(\underline{A}') = Cn(\underline{A})$.

According to our understanding that the nonmonotonic implications of $\underline{A}$ and $\underline{A}'$ are $\underline{E}$ and the intersection of $\underline{E}$ and $\underline{E}'$ respectively, the first touch of the screwdriver requires that a |~ c and a |~ a ∨ b. Nonetheless, it is not the case that a ∧ (a ∨ b) |~c. Cautious monotony (section 5.8) is violated.

The example also satisfies ramification (b) of Reiter's proposal. $\underline{A}$ and $\underline{A}'$ represent logically equivalent bodies of information and yet yield different sets of extensions.

Ramification (b) of Reiter's scheme seems to me to be as objectionable as ramification (a) when an account of default reasoning is taken to be an account of ampliative inference – as the examples illustrating its use suggest it is intended to be. If the total information available to agent 1 is logically equivalent to the total information available to agent 2 so that their closures under logical consequence are identical, any differences in the conclusions they reach via ampliative reasoning should be due to failures in one way or another to recognize the logical equivalence between their evidence sets or due to differences in the other contextual factors such as the ultimate partition U, credal probability Q, informational-value-determining probability M, or index of boldness q. Reiter does not appear to think that the legitimacy of the default reasoning depends on contextual features such as those specified here. Consequently, we have been offered no reason why, when the logical closures of two sets are the same, the inductive inferences licensed should not be the same.

This consideration suggests that Reiter's default theory should be modified so that defaults are required to operate on deductively closed sets of sentences as I did in section 7.1. This modification precludes $\underline{A}$ and $\underline{A}'$ yielding different sets of extensions from the same defaults. The emendation of Reiter's approach just suggested also has the virtue of allowing us to regard Reiter's approach as providing transformations of a deductively closed theory to another such theory where the transformation contains the initial theory if the data are consistent with the initial theory.

These considerations prevent the appearance of Makinson's counterexample to cautious monotony from emerging. However, for quite different reasons failures of cautious monotony can emerge, as shall be explained in section 7.3. My contention here is that Reiter's proposal is, nonetheless, improved by requiring that his defaults be applied to logically closed theories and not to arbitrary consistent sets of formulas.

As defaults are customarily presented, claims like "Birds normally fly" appear to license nonmonotonic inferences from "Tweety is a bird" to "Tweety flies." But then the presence of multiple extensions that are mutually inconsistent appears to present a difficulty. This difficulty was finessed in section 7.1 by taking the extensions to represent serious possibilities relative to the inductive expansion $\underline{K} = \underline{B}^i$ of $\underline{B}$. This finesse follows the suggestion of Makinson (1994) and Gärdenfors and Makinson (1993). This reading does not, however, seem to be the one adopted by Reiter.

Reiter has explicitly acknowledged the phenomenon of multiple extensions relative of a given set $\underline{A}$ relative to a fixed set of defaults. For example, one default states that if it is consistent to assume that a person's hometown is that of one's spouse, do so. Another states that if it is consistent to assume that a person's hometown is at the location of place of employment, do so. Suppose we know that Mary's spouse lives in Toronto while her employer is located in Vancouver. Here we have two extensions incompatible with each other. One states that Mary lives in Toronto, the other in Vancouver.

Reiter acknowledges that "from the point of view of conventional logical systems this is a perplexing, seemingly incoherent situation" (1980, p. 86). And, indeed, it *is* incoherent and downright inconsistent to infer incompatible conclusions from the same evidence in the same context. Inductive expansion into inconsistency ought to be forbidden. But to Reiter, it only *seems* incoherent.

From our intuitive perspective that default assumptions lead to beliefs, the example makes perfect sense; one can choose to believe that Mary's hometown is Toronto or that it is Vancouver, but not both. It would seem that defaults can sanction different sets of beliefs about the world. (1980, p. 86)

Thus, Reiter is not urging us to form inconsistent beliefs via induction or in any other deliberate manner. But the method for avoiding inconsistency is not the one favored by Gärdenfors and Makinson. According to Reiter, we need to pick one extension as the expansion rather than another and inductively expand B to that extension. Gärdenfors and Makinson suggest instead that we remain in suspense between the several extensions.

How does Reiter think we should decide which extension to choose? Reiter elaborates on his point of view as follows:

Imagine a first order formalization of what it is we know about any reasonably complex world. Since we cannot know everything about that world – there will be gaps in our knowledge – this first order theory will be incomplete. Nevertheless, there will arise situations in which it is necessary to act, to draw some inferences despite the incompleteness of the knowledge base. That role of a default is to fill in some gaps in the knowledge base, i.e., to further complete the underlying incomplete first order theory so as to permit the inferences necessary to act. (1980, pp. 86–7)

Reiter does not explain why one should favor one of the consistent extensions of the "default theory" (e.g., that Mary lives in Toronto) rather than the other (that she lives in Vancouver). Rather he justifies his refusal to countenance suspension of judgment between these alternatives on the evidence given. (This interpretation is supported by the footnote on p. 86 where Reiter reveals the possibility of suspense but refers back to the text that supports ruling suspense out of court.) We face situations where time is short and we must make decisions as to which option to exercise or policy to implement. Peremptoriness of choice as to what to do is taken to require a corresponding demand that the inquirer decide what to believe – that is, "fill a gap in the knowledge base."

I take it to be noncontroversial that decisions an inquirer makes ought to be constrained by his or her beliefs (both full beliefs and probabilistic beliefs). Nonetheless, a gambler can bet a substantial sum of money on the outcome of a toss of a coin being heads while remaining in suspense as to whether the coin lands heads or lands tails. Someone who must decide today whether to list Mary's residence in a professional handbook as Toronto or Vancouver without an opportunity to check with Mary might have to make an arbitrary choice. But the decision to *report* her residence as Toronto is not to be confused with a *belief* that Mary lives in Toronto. That is to say, the editor of the handbook makes a report that does not sincerely reflect his or her convictions.

It is not utterly unthinkable that the editor of the handbook is honest both with himself or herself and with the public. The honest editor would declare that putting down Vancouver as Mary's residence is just a possibility and that there is another possibility, Toronto. In that case, the editor is not registering a belief that Mary lives in Vancouver.

Telling the truth as one believes it or acknowledging doubt when one has it may not always be the best policy even from the vantage point of intellectual morals. Good teachers, for example, often insincerely feign an open mind in order to engage in nonthreatening dialogue with dissenting students in the service of a good pedagogical cause. But such dissimulation should be understood for what it is. Similarly one might pretend to a definite opinion where one is in doubt; but such pretense should not be confused with genuine conviction.

Poole (1988) has offered a different view from Reiter's of the conclusions of default reasoning. Following Reiter, he avoids considerations of statistical or probabilistic reasoning, resting content with first-order representations of the information deployed. Unlike Reiter, however, Poole takes the position that the default reasoning with which he is concerned is merely "hypothetical." The inquirer (Poole's "user") provides "the forms of possible hypotheses" the user is "prepared to accept in an explanation" (p. 28). According to Poole, such forms of possible hypotheses are representable by open formulas or schemata such as "x is a bird ⊃ x flies," substitution instances of which are the possible hypotheses the inquirer is prepared to accept in an explanation (Poole, 1988, pp. 28–9). What Poole means by "is prepared to accept in an explanation" remains to be elucidated. I shall examine it more closely shortly.

Given such a roster Δ of forms of possible hypotheses and, in addition, given background information $\underline{F}$ or knowledge of "facts," Poole calls the union of $\underline{F}$ with a set $\underline{D}$ of substitution instances of Δ, where the union is consistent, a *scenario*. (In his more general formulation, the scenario includes in the union a set of "constraints" on substition instances of Δ as well.) Any sentence g that is a logical consequence of such a scenario is potentially explained by the scenario.

Poole contends that his approach can be deployed to characterize default reasoning (1988, p. 31). In the context of explanation, Poole contends that g describes something "we are actually observing." In default reasoning, g represents a prediction. One problem with this approach is that if g represents something we are observing and we are seeking to explain it, g is taken to be known. g is trivially a consequence of $\underline{F}$ alone so that $\underline{D}$ is not needed in the derivation. To show that g is explained by the scenario, it is necessary first to contract $\underline{F}$ by removing g before constructing the scenario. But relative to the contracted $\underline{F}$, the explanatory task reduces to the predictive one. Explanation becomes showing that g was to have been expected according to the system $\underline{D}$ of possible hypotheses together with the contracted network of facts. The use of possible hypotheses as defaults becomes the primary application and the explanatory use is elucidated in terms of the model of default reasoning. I shall focus on the use of Δ and its instances in the setting of default reasoning.

In this setting, Poole says that a set $\underline{D}$ of instances of Δ consistent with $\underline{F}$ is a set of "possible hypotheses." Assuming that the deductive closure of $\underline{F}$ is the current corpus, he is not claiming merely that $\underline{D}$ is a set of serious possibilities relative to that corpus. $\underline{D}$ is, indeed, such a set. But there are many material conditionals consistent with $\underline{F}$ that are not instances of Δ and, hence, are serious possibilities relative to $\underline{F}$ but are not possible hypotheses. To be a possibility in Poole's sense, the conditional must be a substitution instance of an element of the default set Δ of open formulas.

I suggest that Poole's notion of a possible hypothesis can be understood to be the notion of a material conditional that is a serious possibility relative to an ampliative inductive expansion of the closure of $\underline{F}$ generated by the set of defaults Δ. In this sense, all logical consequences of the scenario are serious possibilities relative to that inductive expansion.

Poole characterizes an extension of $\underline{F}$ and Δ as the logical consequences of a maximal scenario of $\underline{F}$ and Δ. Any such extension is a seriously possible extrapolation from $\underline{F}$. The set of sentences contained in all such extensions will then constitute the inductive expansion of $\underline{F}$ according to Δ. This interpretation of Poole is the one adopted by Makinson (1994) and Gärdenfors and Makinson (1993). It is called the "skeptical" interpretation of default reasoning. As in the case of Reiter, a system of defaults can allow for multiple extensions of what is known; but the new beliefs warranted by the defaults on the basis of the initial information do not require the choice of one of the extensions so that all the gaps are filled in the information available but allow for suspension of judgment.

I shall adopt this skeptical interpretation in discussing not only Poole's version of default reasoning but Reiter's as well, even though, in the case of Reiter, this clearly goes against Reiter's explicit pronouncements.

With these understandings in place, we are in a position to interpret a system of defaults Δ as defining a generalized expansion function E designed to operate on consistent corpora. As we shall see, there are some important differences in the way in which Reiter (as amended) and Poole understand extensions. But given a characterization of the set of extensions of an initial corpus, we can define inductively extended expansions, AGM revisions and Ramsey revisions, corresponding conditionals, and pseudoimplication relations along lines already elaborated in Chapter 5. Because the discussion in the literature has focused on nonmonotonicity, I shall restrict attention to the pseudoimplication relations generated by systems of Reiter and of Poole defaults.

7.3   Simple Normal Defaults and Nonmonotonicity

In comparing his defaults with Reiter's, Poole begins with a restricted class of Reiter's closed normal defaults of the sort mentioned in section 6.2. These are closed normal defaults without extralogical prerequisites. They have the form **M**w/w. Any such default is an instance of a Poole default schema where w is a substitution instance of an open formula. He shows (theorem 4.1) that Reiter extensions of the logical closure of $\underline{F}$ using such defaults coincide with extensions in his sense of the corresponding Poole defaults.

Poole then proceeds to explain how, according to his scheme, one can embed a set of defaults Δ into the first-order object language by a process he calls "renaming" defaults. He associates with each default (e.g., "Birds fly") a primitive predicate (e.g., "Birdsfly") together with a postulate stating that everything that is a birdsfly flies if it is a bird. This first-order sentence exhibits the form of one leg of a Carnapian reduction sentence (Carnap, 1936) that serves as an axiom for a primitive disposition predicate. Poole replaces the set of defaults Δ by a set Δ' of instantiations of associated default predicates such as "Tweety is a birdsfly" and augments $\underline{F}$ by adding for each default the associated reduction sentence serving as a default postulate. He shows that the notion of a scenario and maximal extension relative to $\underline{F}$ and Δ is equivalent to the notion of a scenario and maximal extension relative to $\underline{F}'$ and Δ' (Poole, 1988, theorem 5.1).

Poole worries about the following problem: The default birdsfly, together with the information that Tweety does not fly, yields a scenario where Tweety is not a bird. After rehearsing several examples of such situations, Poole considers evenhandedly the views of those who see nothing wrong and the views of those who do. He modifies his characterization of scenario by introducing the notion of "constraints" in order to accommodate such worries (definition 7.4, p. 39). However, he does not seem to take a position on the merits of using constraints or dispensing with them. Following Makinson (1994), I shall focus on the approach that does not appeal to constraints.

With this understood, Poole's formal apparatus is sufficiently clear for us to deploy it to make the comparisons with Reiter he exploits. He undertakes to extend his comparisons to generalized defaults, including closed normal defaults that contain prerequisites of the form α;**M**w/w. According to Poole, such a default cashes out in the first-order formula $\alpha \supset$ w.

This, however, does not quite conform to Reiter's conception of a default. To be sure, if α is in $\underline{F}$ or its closure, w will be in an extension. That is common ground between Reiter and Poole. However, if α is not in $\underline{F}$, $\alpha \supset$ w will be in the Poole extension of $\underline{F}$ but not in the Reiter extension.

Thus, if "Birds fly" is a default but "Tweety is a bird" is not in $\underline{F}$, "Tweety is a bird $\supset$ Tweety flies" is in the Poole extension but not in the Reiter extension.

Suppose then we begin with a corpus $\underline{B}$ containing neither b (= Tweety is a bird) nor its negation and neither f (= Tweety flies) nor its negation and have "Birdsfly" as a default. According to Reiter, $\underline{B}^i$ might not contain b $\supset$ f even though $\underline{B}_b^{+i}$ contains f. In such a case, $(\underline{B}^i)_b^+$ would not contain f and would not be a superset for $\underline{B}_b^{+i}$. Inductive expansion via Reiter defaults is not importable in the sense of section 5.6. (K*i7) and inductively extended conditionalization are violated.[1] On the other hand, as Makinson (1994) has shown, Poole's version of inductive expansion is importable provided we dispense with constraints. (K*i7) is satisfied. Inductive expansion rules derived from Poole defaults fail to be importable when constraints are allowed. Without the constraints, however, Poole's proposal derives inductive expansion functions from defaults that satisfy permutability and, hence, are ampliative but constant. (K*i7) is satisfied by both *i and *ri. (K*i8) is satisfied by *i and (K*ri8) by *ri.[2]

Thus, Poole's proposal seems to generate an ampliative but constant inductive expansion function. This means that inferences licensed by Poole defaults can be rationalized by bookkeeping with q = 1 but not with q < 1.

Reiter's theory clearly fails to be ampliative but constant. It fails to be importable and satisfy (K*i7). Indeed, Reiter-like inductive expansion functions fail to satisfy cautious monotony and, hence, (K*i8). Inductive expansion is not exportable any more than it is importable.[3]

Nonetheless, Reiter's defaults must insure the satisfaction of Cut by the nonmonotonic implication relation. Suppose that h $\mid\!\sim$ f and h $\wedge$ f $\mid\!\sim$ g. If these results are obtained by Reiter defaults from $\underline{B}$, there must be a default h; Mf/f so that f is in at least one extension of $\underline{B}_h^+$. Since h $\mid\!\sim$ f, f must be in all the extensions. By similar reasoning, there must be a default h $\wedge$ f;Mg/g to guarantee that g is in an extension of $\underline{B}_{h\wedge f}^+$. Again, since by hypothesis h $\wedge$ f $\mid\!\sim$ g, g is in all extensions. But then g must be in all extensions of $\underline{B}_h^+$.

In Chapter 6, we established that Cut is not always satisfied by inductive expansion rules but is obeyed when bookkeeping is deployed for all values of q. Reiter's defaults do generate nonmonotonic inferential relations that are rationalizable by bookkeeping rules. On the other hand, because Reiter's defaults do not guarantee importability or exportability of the corresponding inductive expansion function, the bookkeeping rules are for values of q that may be less than 1.

As I have already indicated, using bookkeeping with q = 1 as is required in order to rationalize the use of Poole's defaults yields inferences that seem too strong. Provided Reiter's method of generating extensions from defaults is amended in the ways indicated in section 6.2, his method does seem superior to Poole's in this respect.

To be sure, Reiter's method leads to violation of cautious monotony. The conjunction of Cut and cautious monotony are characteristic of "cumulative inference relations" (Makinson, 1994). The importance of Cut is that it allows us to add conclusions reached to our premises without generating new conclusions that we could not reach before. Cautious monotony allows us to add conclusions to premises without losing conclusions that we could reach before. These requirements turn out to be less compelling than they seem to be in the case of ampliative inference. Both can be violated when the inductive expansion principle I have proposed is used without bookkeeping. Only Cut is preserved when bookkeeping is used but q is less than 1. To obtain cumulativity, bookkeeping with q = 1 is required. But as I have argued, this leads to dubious recommendations in inference. In the light of these considerations, the failure of Reiter's defaults to secure cautious monotony may not appear to be a defect – especially since default reasoning does appear to be designed to characterize inductive expansions.

7.4   The Default Assumption

Reiter's defaults (construed as operating on initial information closed under logical consequence and with the "skeptical" interpretation of the role of extensions of the initial information) yield inductive expansion transformations defining inductively extended revision operators obeying inductively extended Cut but failing (K*i7), (K*i8), and inductively extended cautious monotony. Not only do Reiter's defaults fail to yield an ampliative but constant inductive expansion function; but they do not yield a "cumulative" pseudo implication relation.

In contrast to Makinson (1994), I do not think this to be unfortunate. My concern here, however, is to argue that given Reiter's own informal remarks concerning default reasoning, he should agree with Makinson and not with me. Indeed, the spirit of his ideas as expressed in his informal remarks seems to require that inductive expansion should be ampliative but constant.

To conclude that Tweety does fly is an ampliative and, hence, nonmonotonic inference from the information that Tweety is a bird and the default rule that birds fly. If, prior to making the inductive expansion, additional information had been given such as the information that Tweety is a penguin, the conclusion that Tweety flies would have been undercut and, hence, undermined.

The presence of nonmonotonicity is not due to belief contravention. The supposition that Tweety is a bird does not require a contraction of the initial background information. And the supposition that Tweety is a penguin does not contradict the claim that Tweety is a bird. What is true is that if, in addition to supposing that Tweety is a bird, we had supposed that Tweety is a penguin, then the inference to the conclusion that Tweety is a flying bird would have been undermined and, indeed, undercut by the information that Tweety is a penguin. It is only if it is *not known* that Tweety belongs to a non-flying species of bird (such as penguins) that it is legitimate to infer that Tweety is a flying bird from the supposition that Tweety is a bird and the background information. It is no part of the supposition or the background to which it is added that Tweety is not a penguin any more than it is that Tweety is a penguin.[4] That Tweety is a flying bird is *inferred inductively*. That Tweety is not a penguin is a logical consequence of this inductive conclusion and the background assumption that no penguin flies. It is not part of the background information to begin with. Hence, supposing that Tweety is a penguin is not belief contravening. The nonmonotonicity in the suppositional inference from "Tweety is a bird" to "Tweety flies" is revealed by comparing this inference with the suppositional inference from "Tweety is a penguin" to "Tweety does not fly."

To be sure, if the agent actually came to full belief that Tweety is a bird rather then merely supposing it true for the sake of the argument and then inductively inferred that Tweety is a flying bird, the agent might inadvertently expand into inconsistency subsequently via routine expansion by adding the claim that Tweety is a penguin. But here the conflict is not between the claim that Tweety is a penguin and the initial background information or between the background information cum claim that Tweety is a bird and the information that Tweety is a penguin. The tension is between the claim that Tweety is a penguin and the conclusion reached by induction that Tweety flies and, hence, is not a penguin.

In that case, the conflict should be removed. One might do so by questioning the reliability of the program for routine expansion deployed in acquiring the information that Tweety is a penguin. One might also question part of the initial background (e.g., no penguins fly) or the conviction acquired by induction that Tweety belongs to a flying species of bird. Or one might question all of these things. The inquiring agent faced with inadvertent inconsistency should contract from inconsistency in one of these directions. A sketch of a decision theoretic approach to this matter is given elsewhere (Levi, 1991, ch. 4).

No such issue is relevant to default reasoning as a species of inductive inference. If the initial supposition is "Tweety is a bird," the inference is to the conclusion "Tweety flies." If the initial supposition is "Tweety is a bird that is penguin" the inference is to the conclusion "Tweety does not fly." There is no question of retreating from an inconsistency in belief. The second inference does not inject conflict into a state of full belief containing the information that "Tweety flies."

By way of contrast, when the inductive conclusion that Tweety flies and is not a penguin has already been drawn from the information that Tweety is a bird and then new information reveals Tweety to be a penguin, the decision as to what to contract is not an inference from premises judged true to a conclusion judged true. Because it is not an inference, it is not a nonmonotonic inference. And it is surely not an extrapolation from the information given in the background together with the specifically stated premises.

I have already conceded that certain forms of suppositional reasoning are nonmonotonic because they are belief contravening. But the "Tweety" example and others like it that are used to exemplify default reasoning do not illustrate this belief-contravening type of nonmonotonicity. They appear to involve a nonmonotonicity due to some form or other of ampliative reasoning that is the hallmark of inductive inference. The consistent inductively extended expansion of $\underline{B}$ by adding "Tweety is a bird" yields "Tweety flies." The consistent inductively extended expansion of $\underline{B}$ by adding "Tweety is a bird that is also a penguin" yields "Tweety does not fly." These are the marks of nonmonotonicity in ampliative reasoning without belief contravention.

Reiter (1980, sec. 1.2) is far from clear on this point. He contends that reasoning in cases such as the Tweety example is nonmonotonic precisely because subsequent discovery that Tweety is a penguin calls for withdrawing the claim that Tweety flies. This fallacious idea seems to be so widespread among students of nonmonotonicity in the artificial intelligence community that it is sometimes taken to be definitive of the concept of nonmonotonicity.

If Reiter's view were sound, default reasoning could not be a form of inference at all. Its nonmonotonicity would be the result of belief contravention. In that event, neither the case of Tweety nor other examples of default reasoning offered by Reiter illustrate default reasoning.

Perhaps Reiter's view is not the uncharitable one I have attributed to him but a somewhat different one. Perhaps Reiter means to be suggesting that the nonmonotonicity of default reasoning as a species of ampliative inductively extended inference can be used to prescribe how inquirers who face a problem of belief contravention due to the fact that an inductive inference has already been made should retreat from inconsistency.

To fix ideas, consider the following two variants of the Tweety example:

Scenario 1: The inquirer begins with background information $\underline{B}$ about which types of birds do and do not fly.

First Stage: The inquirer then engages in an extended expansion by first expanding <u>B</u> by adding b = "Tweety is a bird" and then inductively expanding. The net result is a corpus <u>K</u> that is the expansion of <u>B</u> by adding b and the claim f = "Tweety is a bird of a flying species."

Second Stage: The inquirer comes to believe true "Tweety is a penguin" = p and is thrown into contradiction because <u>B</u> and, hence, <u>K</u> contains "No penguins fly." Hence, the expansion of <u>K</u> by adding p is inconsistent. The inquirer retreats from inconsistency by either (a) retreating from the expansion adding p to the first-stage belief state <u>K</u>, (b1) retreating to the expansion of <u>B</u> by adding p (so that the new corpus contains ~f), (b2) removing the assumption that "No penguin flies" so that f is retained, or (c) retreating to the intersection of two or three of the corpora just listed.

Scenario 2: At the first and only stage, the inquirer expands <u>B</u> by adding both b and p (or, in effect, p alone). ~f is in the new corpus.

The default reasoning in scenario 1 is the *first stage* inductively extended expansion of <u>B</u> by adding b, which contains the claim f, and, because <u>B</u> contains the claim that no penguins fly, the claim ~p as well. The subsequent addition of the information p at stage 2 is not default reasoning (i.e., inductive expansion) at all but is a revision of the corpus obtained via default reasoning at stage 1.

Scenario 2 is an inductive inference at stage 1 from different premises from those deployed at stage 1 of scenario 1. Instead of adding b alone to <u>B</u>, both b and p are added together. As a consequence, the conclusion f of the default reasoning at stage 1 of scenario 1 is undercut in the sense that with the new information in scenario 2 the conclusion that ~f is justified.

Reiter seems to think that, because the scenario 2 default reasoning from the conjunction of b and p yields a corpus containing ~f, the revision obtained at stage 2 of scenario 1 ought to be the option (b1). In this sense, the nonmonotonicity of default *inference* can yield prescriptions concerning how to modify a belief state in the face of conflicting data.

Reiter assumes that the position that should be reached in the revision at stage 2 of scenario 1 of the result of stage 1 should be the same corpus as would have been obtained had scenario 2 taken place rather than scenario 1. In scenario 2, there is no belief contravention at all. However, because the new information added to <u>B</u> includes p as well as b rather than b alone, the conclusion reached is different than it is at the first stage of scenario 1. And this indicates how the agent ought to respond at the second stage of scenario 1.

On this interpretation of Reiter's view, the nonmonotonicity of default inference is not belief contravening as he sometimes appears to be saying but is rather a guide prescribing how to address situations where belief contravention is required. Belief-contravening nonmonotonicity is alleged to recapitulate default nonmonotonicity in the following sense:

The Default Assumption:

> When the extended inductive expansion adding proposition h to initial belief state $\underline{B}$ that is licensed by default rules of inference conflicts with new information g obtained subsequently, the modification made in the belief state should agree with the inductively extended expansion adding h ∧ g to $\underline{B}$.

The transformation I have called an inductively extended expansion $\underline{B}_b^{+i}$ by adding b was formally characterized as the result of two transformations. It is an inductive expansion of an (unextended) expansion of $\underline{B}$ by adding b. Formally $\underline{B}_b^{+i} = (\underline{B}_b^+)^i$. This second representation does not mean that the inquirer who engages in inductively extended expansion literally changes his or her belief state or corpus in two steps. The transformation could represent a doxastic trajectory through two corpora or it could represent a shift in belief state from the initial state to its inductively extended expansion.

The same is, of course, true of $(\underline{B}_{b\wedge p}^+)^i$.

Consider then the transformation that results from starting with the belief state represented by corpus $(\underline{B}_b^+)^i$ and receiving the information that p. This is the predicament at stage two of scenario 1. Since ~p is in the corpus $(\underline{B}_b^+)^i$, the addition of the information is in conflict with that corpus. To avoid inconsistency, I suggested that (roughly speaking) four alternatives are available to the inquirer:

- **(a)** Contract from inconsistency to $(\underline{B}_b^+)^i$. That is to say, remove the new information p.
- **(b1)** Contract from inconsistency to $(\underline{B}_{b\wedge p}^+)^i$. That is to say, give up the information obtained inductively from $\underline{B}_b^+$, add p, and then perform an inductive expansion. This prescription recommends calling into question the information obtained by inductive expansion initially before adding p.
- **(b2)** Do not question the results of inductive expansion (in particular, do not question f). Rather question some other background assumption such as "No penguins fly."
- **(c)** Shift to the intersection of two or three of these alternatives.

According to the default assumption, a recommendation as to how to proceed can be obtained by considering the content of the corpus $\underline{B}_{b\wedge p}^{+i}$.

This corpus cannot, however, be one representing the option of throwing out the information that Tweety is a penguin as in (a), throwing out other background information as in (b2), or suspending judgment according to some version of (c). (b1) is recommended.

Thus, the default assumption does impose a very definite restriction on the response that should be made when new information comes into conflict with old. It implies that the net effect of adding conflicting new information to a corpus obtained by inductive expansion from an initial corpus B should always be an AGM revision or an inductively extended AGM revision where the new information is retained and information obtained via inductive expansion removed.

Gärdenfors among many authors has, at least by implication, favored a requirement closely resembling the default assumption:

The Revision Assumption:

> If new information conflicts with a belief state, the new information should be retained and elements of the old information removed in eliminating inconsistency.

Advocates of the default assumption (who seem to include Reiter among their number) should presumably regard the demands of the revision assumption as too weak. It prohibits endorsing contraction strategies (a) and (c). Choice between (b1) and (b2) is left open. The clear intent, however, is to leave the corpus $\underline{B}$ and, indeed, $\underline{B}_b^+$ unsullied and to give up information from $\underline{B}_b^{+i}$ that is added by inductive expansion.

The Default Restriction:

> If new information x conflicts with a belief state $\underline{B}^i$ obtained previously via inductive expansion from $\underline{B}$ and does not conflict with $\underline{B}$ and if new information x is to be retained in contracting from inconsistency, then the contraction from inconsistency should be to the contents of the inductively extended expansion $\underline{B}_x^{+i}$

The revision assumption rules out strategies (a) and (c). The default restriction does not do so. However, it does assert that if (a) and (c) are ruled out, (b2) should be ruled out as well. It says that if x is to be retained, follow (b1). If x is not retained, follow strategy (a) or (c). The two of them jointly mandate (b1). The default assumption is, therefore, the conjunction of the revision assumption and default restriction.

The default restriction has another consequence. The result of adding p to B_b^{+i} and contracting from inconsistency in a manner retaining p is a revision of B_b^{+i} by adding p both according to AGM and according to Ramsey. The default restriction imposes a constraint on such revisions additional to those imposed by $(K*1)$–$(K*8)$ or the corresponding postulates for Ramsey revision whether or not the revision assumption holds.

Let x be consistent with $\underline{B}$ and, indeed, suppose that neither x nor $\sim$x is in $\underline{B}$ but x is inconsistent with $\underline{B}^i$. Adding x to this corpus yields the revision $(\underline{B}^i)_x^*$. The default restriction requires this revision to be identical with $\underline{B}_x^{*i} =$

$\underline{B}_x^{+i}$ since x is supposed to be consistent with B. And this should mean that when x is consistent with $\underline{B}^i$, $\underline{B}_x^{+i} = (\underline{B}^i)_x^+$. This requirement is nothing else but our old friend the permutability condition on inductive expansion. As we have seen, requiring permutability is equivalent to requiring inductive expansion to be ampliative but constant.

Thus, the default assumption, which Reiter appears to endorse, implies the default restriction, which in turn entails the permutability of inductive expansion and, hence, that inductive expansion should be ampliative but constant. But Reiter's account of default reasoning allows for inductive expansion functions that are not permutable. The two claims cannot be consistently maintained.

I have, to be sure, helped myself to some assumptions that Reiter might resist. In order to claim that defaults generate an inductive expansion function, I have assumed that Reiter defaults are required to operate on deductively closed theories or corpora to derive extensions. And I have followed Makinson and Gärdenfors in adopting the "skeptical" interpretation according to which the default reasoning should yield the intersection of the Reiter extensions obtained from the initial corpus via the defaults. For this reason, it cannot be claimed that my interpretation of Reiter's views is entirely accurate. However, I think it is a charitable interpretation. Two sets of sentences $\underline{A}$ and $\underline{A}'$ whose deductive closures are identical ought to yield the same inductive or ampliative extrapolations, or so I have argued. And the adoption of the skeptical interpretation seems to me another exercise in charity as well. Once these points are conceded, it seems to me that Reiter's commitment to something like the default assumption does lead to the conclusion that his proposals are unsatisfactory according to his own position. Either his scheme should be modified in the direction of Poole's version of default reasoning or he should abandon the default assumption.

Observe that the components of the default assumption are formulated so that neither presupposes the other. The revision assumption does not presuppose the default restriction and vice versa.

Thus, someone could endorse the default restriction but reject the revision assumption and, hence, the default assumption. And the reverse is also entertainable. Reiter seems to endorse both.

And he does not seem to be alone. It seems to be a commonplace in the literature on nonmonotonic reasoning that if one has reached a conclusion by default reasoning that one would not have endorsed had one had additional information at one's disposal initially, the subsequent acquisition of that additional information should lead to withdrawal of the conclusion.

This idea has received able defense from philosophers. Indeed, ever since 1961, H. E. Kyburg has endorsed a notion of defeasibility ingredient in the default assumption. Kyburg, unlike Reiter and many others in computer science and artificial intelligence, is explicit in insisting that the nonmonotonic inference involved in default reasoning is ampliative and inductive (see, e.g., Kyburg, 1990b, p. 319). And he explicitly takes statistical considerations into account. However, Kyburg thinks that such ampliative, inductive and probabilistic inference conforms to high-probability rules and bites the bullet on the question of abandoning logical closure of the results of inductive inference. Consequently, one cannot conclude from his endorsement of the default assumption that he is committed to ampliative but constant inductive expansion rules.

We cannot saddle Reiter with such rules either unless we modify his account of defaults along the lines already indicated. But those who think that ampliative and, hence, nonmonotonic inference transforms deductively closed sets of assumptions to deductively closed sets of conclusions containing the assumptions and who also endorse the default assumption are committed to the use of ampliative but constant inductive expansion rules.

We have seen that Reiter's approach to defaults can be easily adjusted so as to understand nonmonotonic inference as transforming deductively closed sets to deductively closed sets containing them. Yet, when so modified they fail to use inductive expansion rules that are ampliative but constant.

In my judgment, this is not a defect. The default assumption is extremely questionable. It reflects a kind of foundationalism in epistemology that ought to be rejected. I shall now turn to this issue.

7.5 Foundationalism

On the construal of Reiter's view just suggested, default rules generate inductively extended expansion rules that are ampliative and nonmonotonic. Reiter also endorses the default assumption. However, an inquirer need not be committed to the default assumption in order to endorse criteria for inductive expansion. As I have argued in section 7.4, Reiter's own account of default rules yields nonmonotonic inference relations that conflict with the strictures of the default assumption.

To be sure, the scenario 1 first-stage conclusion via inductive expansion that Tweety flies could be open to modification in subsequent inquiry according to inquirers who reject the default assumption. There are entertainable circumstances in which the inquirer would be justified in giving up this conclusion. If we set aside the status of logical and mathematical truths or whatever else one thinks to be a conceptual truth, every belief is liable to being abandoned under some circumstances or another. This includes not only the products of theoretical inquiry and inductive inference but also the testimony of the senses. In this sense, every substantive belief is believed by default.

The default assumption is more substantial than this general assumption of the corrigibility of full belief. The default assumption prescribes how one should proceed when one's belief state goes into conflict depending upon the default procedures originally deployed in acquiring the beliefs producing the conflict.

Endorsing the default assumption as a constraint on all inductive expansion is tantamount to a commitment to some sort of epistemological foundationalism as Reiter acknowledges in one brief passage.

This last example also demonstrates the need for some kind of mechanism for *revising beliefs* in the presence of new information. What this amounts to is the need to record, with each derived belief, the default assumptions made in deriving that belief. Should subsequent observations invalidate the default assumption supporting some belief, then it must be removed from the data base. The Truth Maintenance System of Doyle (1979) is a heuristic implementation of just such a system for belief revision. (1980, p. 86)

In this passage, Reiter explicitly allies himself with the so-called foundationalist view of belief change as characterized by Harman (1986). According to this view, when an inquirer obtains beliefs in conflict, it becomes necessary to look to the derivations of these beliefs.[5] Beliefs derived from other beliefs via default reasoning are especially vulnerable to being given up due to their nonmonotonicity in the undermining sense when the default assumption is made. The default assumption is then seen to be a keystone of at least one version of a foundationalist view.

Harman (1986) and Gärdenfors (1990b) have both objected to this sort of foundationalism on the grounds that it requires the inquirer to keep a record in memory of the defaults used to obtain his or her beliefs. In the context of devising inquiring automata, this seems an excessive demand on limited storage capacity.

Although I share Harman's and Gärdenfors's opposition to foundationalism of this sort as well as other varieties, my own focus has not been on artificial intelligence. But, in any case, the objection does not seem to me compelling.

Foundationalism is objectionable because there is no way to satisfy demands that all full beliefs that an agent has at a given time have a justification. Given this impossibility, foundationalism leads to skepticism that requires rational agents to withhold all belief.

On my view (Levi, 1980, ch. 1), we should turn our back on any "pedigree" epistemology requiring that all beliefs be justified or acquired by some process that yields reliable beliefs. There is no need to justify beliefs already held. In the absence of any reason for contemplating whether to give up settled assumptions, the law of doxastic inertia sets in. This law exhorts us not to scratch where it does not itch.

The position that emerges is not "coherentist" in any useful sense. Of course, belief states ought to be coherent. But the coherence of a belief state is not a reason for endorsing it. It is true that the incoherence of a state is a good reason for retreating from it. But there are many coherent states. In a context where the inquirer is deciding which of several such states to move to, given his or her current state, a question of justification arises. But it cannot be settled by considerations of coherence alone since all the moves under consideration are coherent.

Doxastic inertia is not to be confused with Harman's "principle of conservatism" that states that one is justified in continuing fully to accept something in the absence of any special reason not to.

Previously (Levi, 1991), I argued that the demand for justification arises in a setting where the inquirer is concerned to decide between either rival expansion strategies or rival contraction strategies.

In the context of expansion, one of the expansion strategies available is to refuse to expand at all and to remain with the status quo. It may be that in such a context (e.g., of deciding between rival expansion strategies via induction), the best expansion strategy will be to remain with the current corpus or belief state. But given that one is deliberating as to whether to expand and if so how to expand, the decision not to expand *does* require a reason. The reason for continuing in the belief state one is in is that the current belief state represents the best trade-off between risk of error and information gained among the available expansion strategies. Harman's principle of conservatism plays no role.

Inquirers may be compelled to contract to escape from incoherence or inconsistency. In such cases, they need to choose among alternative contraction strategies. On the view I favor, they should minimize loss of informational value. Once more, Harman's principle plays no role.

Sometimes, however, inquirers contract without being compelled to do so in order to escape from inconsistency. Perhaps some interesting new hypothesis incompatible with the initial corpus has been presented for consideration (Levi, 1991, ch. 4). One of the options available is not to contract. The decision not to contract and to remain with the status quo in that context calls for a reason. Roughly speaking, the inquirer needs to decide whether the expected benefits in the way of informational value gained from contraction are better or worse than the expected benefits of remaining with the status quo. Once more, Harman's principle of conservatism plays no role.

Of course, agents are not always deliberating as to how to modify their beliefs states. And when they are not, there is no need to worry about justifying changes of belief states rather than remaining with the status quo. Harman's principle of conservatism is supposed to provide a reason in such cases. But that is supplying a nonexistent demand.

Doyle (1992) rightly sees that Harman's principle of conservatism can be incorporated into his foundationalist approach precisely because Harman's principle provides a reason for continued belief. But Doyle's procedure does not accommodate the Peircean version of antifoundationalism and anticoherentism I favor. The Peircean view requires no reasons for current belief but only for changes of belief.

When an inquirer does modify his or her belief state, the inquirer does not, in general, have good reason for giving up the beliefs he or she had about what his or her erstwhile belief state was. There is no good reason for any kind of change in this respect. The demands on storage capacity may be as demanding for my kind of antifoundationalism as for foundationalists like Reiter or Doyle.

Of course, storage capacity limits the capacity of human inquirers and social institutions as well as automata to fulfill their doxastic commitments. And there is no doubt that agents not only forget some of their former views. They also scrap written and other records of their views. Indeed, sometimes such jettisoning of records is done coolly and with good reason. This does not mean, however, that the inquiring agent has abandoned his doxastic *commitment* to fully believing that he or she had the views to which the records testify. Rather it means that he or she has reduced his or her ability to fulfill these commitments upon request. Everything else being equal, one should seek to extend one's capacity to fulfill doxastic commitments (such as recognizing the logical consequences of what one fully believes or remembering what one once believed). But such ability enhancement and maintenance is costly and decisions under scarcity of resources need to be made. The point is that the management of storage ought to be a problem for everyone – foundationalist, coherentist and advocate of the Peircean belief–doubt model of inquiry alike.

Thus, the merits of foundationalism cannot be settled by considering the advantages of this or rival views of belief change in designing expert systems or intelligent automata. The controversy is over prescriptive issues in epistemology.

By saying this, I do not mean to suggest that there are authorities in philosophy departments to whom students of artificial intelligence ought to defer. Epistemology is a game anyone can play. But since we know at the outset that no ideals of cognitive rationality are implementable or even approximately so, criticisms of epistemological ideals in terms of implementability in machines does not appear to be a promising line to take.

Rather than becoming drawn into a discussion of the endless and tedious attempts to defend foundationalism with straightedge and compass, I prefer a more positive approach – to wit, developing an alternative view of belief change based on the attitude registered by Peirce in his "Fixation of Belief." Foundationalism is an epistemological program that has failed in spite of several centuries of effort. I see no reason to continue the debate over its merits more than is necessary.

With this understood, my first objection to Reiter's endorsement of the default assumption is that this assumption rests on a bankrupt epistemological tradition.[6]

7.6   Foundationalism and Defaults

Reiter and others do, indeed, seem committed to some variant of foundationalism or other. A version of foundationalism seems sufficient to secure the permutability of inductive expansion functions. Is it necessary?

Recall that the default assumption that entails the commitment to foundationalism consists of two parts: the revision assumption and the default restriction.

The revision assumption does carry a whiff of foundationalism with it to the following extent. It presupposes that when observations or the reports of expert witnesses are added to an inquirer's corpus in conflict with it, the new information is to be retained and the background information modified. This requirement suggests that the testimony of the senses has some priority epistemologically that invariably gains it admission into the modified belief state. It is tempting to conclude that this priority is due to the foundational character of observation reports.

This whiff of foundationalism does not, however, stipulate which elements of the background beliefs are to be given up to save consistency. It is conceivable that the background contains beliefs that are the outcome of observation, which are given up to make way for the new information. The inquirer might have initially come to believe that Tweety is not a penguin because of observations he had made of Tweety. Subsequently, he looks again and observes that Tweety is a penguin after all. One observation competes with another. The revision assumption stipulates in that case that the new observations are favored over old. That this should be so seems questionable. Still there is no contradiction in taking this view. The foundationalism presupposed by the revision assumption does not require the foundation to be incorrigible.

Foundationalism is implied more powerfully when the revision assumption is combined with the default restriction. In that case, it becomes clear that the fruits of previous inductions must be surrendered to make way for observations. Such a view suggests that the results of induction have an epistemologically less secure grounding than the fruits of observation.

The default restriction without the revision assumption does not, however, imply this result automatically. Without the revision assumption, one is not always prohibited from retaining the background beliefs intact when recalcitrant observations appear.

One might attenuate the default restriction so as to squeeze out any suggestion of foundationalism altogether. In formulating the default restriction, it has been taken for granted that the revision transformation involved is intended for use in situations where the agent supposes for the sake of the argument or comes to believe some proposition conflicting with his or her background information, as in stage 2 of scenario 1, and the new information is to be retained. The default restriction then requires the result of such revision to be identical with the result of an inductively extended expansion of an initial background corpus $\underline{B}$.

But suppose that x is acquired and added to $\underline{B}$ prior to inductively expanding $\underline{B}$ to $\underline{B}^i$. The default restriction still requires that inductively extended expansion by adding x to $\underline{B}$ should proceed *as if* $\underline{B}^i$ were being revised by x. I shall call this a *default pseudorevision*. According to the *weak default restriction*, no assumption is made as to whether default pseudorevisions are or are not identical with genuine revisions made in cases where x is acquired as new information after the inductive expansion B^i is implemented.

Thus, the default restriction has been parsed into the weak default restriction and an assumption of the identity of pseudorevisions with revisions. The vestigial foundationalism can be eliminated by abandoning the identity of pseudorevisions with revisions and resting content with the weak default restriction.

The result turns out to be nothing more or less than imposing a condition of permutability on inductive expansion functions. Hard-core foundationalists may be happy to defend permutability by appeal to the default assumption. But skeptics who want to endorse permutability for whatever reason may do so while remaining free of the charge of being closet foundationalists.

For my part, I want no part of a foundationalist perspective. At the same time, I doubt whether permutability is, in general, a reasonable constraint on inductive expansions. Although it does appear to be the case that many computer scientists who have studied default reasoning are committed to some form of foundationalism and also to permutability, it is important to realize that refuting foundationalism will not be sufficient for refuting permutability.

7.7   Defaults and Bookkeeping

In Reiter and Criscuolo (1981), Reiter seems to acknowledge that there is a statistical connotation to "Most birds fly." He insists, however, that there is a "prototypical" interpretation as well, which, as I suggested previously, seems to correspond to something rather like Weber's notion of an ideal type. But the notion of an ideal type is itself redolent with probabilistic connotation. To say that an ideal type bird flies is not necessarily to imply that more birds fly than don't. Ideal type birds might be very rare. A partial approximation to the idea of an ideal type is obtained by focusing on a probability distribution over a range of alternative species of birds and identifying the prototypical species as one whose probability of being present in a random selection is as great as that of any other species. That is to say, the ideal type is the one that is at the mode of the probability distribution. This does not quite capture Weber's notion. There may be no instances of a Weberian ideal type. Hence, a Weberian ideal type is not always a most commonly occurring type. Still one might claim that most commonly occurring types approximate in some sense ideal types. This understanding of prototypicality agrees well with the expression of defaults in the form "F's are normally G's" (e.g., "Birds normally fly"). In the special case where the probability distribution ranges over birds that fly and birds that don't fly, such prototypicality collapses into the claim that more birds fly than do not – that, is into the claim that most birds fly.

I suspect that the remark in Reiter and Criscuolo (1981) really is an unintentional concession that probabilistic considerations really do matter in default reasoning. More to the point, however, the construal of prototypicality as being at or near the mode of a probability distribution captures excellently an important feature of ampliative but constant inductive expansion that we have already suggested is implicated in the reconstructed version of default reasoning as using permutable inductive expansion transformation. This is especially true when prototypicality is understood as normality. A maximally normal individual (with respect to weight) would be one whose weight is the mode (= mean) of a normal distribution of weights. At least, this idea is, as Ian Hacking (1990) has observed, ingredient in the genesis of the use of the term *normal* in modern times.

In section 6.10, it was shown that when bookkeeping based on the inductive expansion rules I propose is exploited, the result of bookkeeping when $q = 1$ is a rule recommending adding the disjunction of all elements of the ultimate partition U for which the ratio Q/N is a maximum. When all elements of U carry equal N-value, this means that the rule recommends suspending judgment between all elements of U carrying maximum probability.

In the sense of normality as prototypicality, the recommendation is to expand inductively by adding the information that the most normal type (i.e., the most probable) is present.

The result of bookkeeping when $q = 1$ was shown to be an ampliative but constant inductive expansion rule. Since default reasoning under the default pseudorevision assumption is reasoning using ampliative but constant inductive expansion criteria, the suggestion that default reasoning is indeed inductive inference with bookkeeping and maximum boldness seems worth considering.

The plausibility of the suggestion is reinforced when it is observed that when q is less than 1 or bookkeeping is not used, inductive expansion will fail in general to be ampliative but constant, and, as a consequence, will fail to satisfy the demand that inductive expansions defined by defaults be simulatable by default pseudorevisions (section 7.6). As I argued there, this appears to be the surviving content of the default assumption when foundationalism is squeezed out. The point I am now making is that this observation coheres well with the Reiter–Criscuolo remarks about prototypicality.

In any case, it has been shown that a plausible reconstruction of the sort of nonmonotonic reasoning based on defaults can be developed, which sees such default reasoning as a species of inductive inference of the ampliative but constant variety and identifies such ampliative but constant inference with inductive expansion using the rules I have proposed when $q = 1$ together with bookkeeping.

7.8   Defaults and Partition Sensitivity

Both sets of Reiter defaults and sets of Poole defaults define extensions of an initial corpus representing a state of full belief. A Poole extension may be understood to be the set of deductive consequences of the initial corpus $\underline{B}$ and some further sentence or set of sentences generated by the defaults. Strictly speaking this is not quite true of a Reiter extension. However, in the special case where the Reiter defaults operate on the deductive closure of the "data base," Reiter extensions are also ordinary expansions of $\underline{B}$ by adding some sentence.

The skeptical interpretation of default reasoning defines the inductive expansion of $\underline{B}$ to be the intersection of all default extensions of $\underline{B}$. From this perspective, a default extension is a consistent expansion of $\underline{B}$ that remains a serious possibility not only relative to $\underline{B}$ but relative to the inductive expansion $\underline{B}^i$.

Thus, even though the union of the set of default extensions will often be logically inconsistent (so that the closure of this union is the set of all sentences in the language $\underline{L}$), the inductive expansion will not be.

For example, one might adopt the default rule: Mticket i will not win/Ticket i will not win for all substitution instances where i is replaced by the numerals 1 to 1,000,000. One could argue that this default allows one to say that, given that ticket i might not win relative to the background $\underline{B}$, it should remain a serious possibility relative to $\underline{B}^i$. Given that $\underline{B}$ implies that some ticket in the million is drawn, we can identify a million extensions, each of which is self-consistent but none of which is consistent with any other. The set of logical consequences of this set of extensions is inconsistent.

Kyburg (1990b, p. 321) has argued that it is implausible for an extension to specify one winner and 999,999 losers. It is implausible if, as Reiter suggests, one is called on to select one of the million extensions as the inductive expansion of $\underline{B}$. But Poole's suggestion is that one should take as the inductive expansion the intersection of all the default extensions, which, in this case, is no expansion at all. This recommendation urges suspension of judgment between alternative predictions as to the outcome of the lottery. In the case of a fair lottery where each ticket has the same chance of being picked, this recommendation is the presystematically recommended one.

Thus, a default extension is a consistent expansion of the initial point of view that remains a serious possibility after expansion. On this reading, there is nothing remarkable in there being several default extensions or in the union of the extensions being inconsistent. But what is accepted via induction is a new deductively closed theory that is an inductive expansion of the initial corpus.

Given an initial corpus $\underline{B}$ and an ultimate partition U, we may associate each element h of U with the deductive consequences of $\underline{B}$ and h – with $\underline{B}_h^+$. We might call such an expansion, relative to U, a potential or entertainable extension.

Given an inductive expansion rule, it will recommend removing some elements of U from their status as serious possibilities. A default extension may be thought of as the intersection of one or more entertainable extensions. The members of the default extension are seriously possible relative to the inductive expansion. Of course, the conjunction of members of two distinct default extensions may turn out to be inconsistent. That is not too surprising. It does represent a warning, however, against believing the union of the extensions to represent the belief state obtained via induction from $\underline{B}$. The inductive expansion $\underline{B}^i$ is the intersection of the set of extensions.

In effect, a set of defaults, when applied to a belief state relative to an ultimate partition U, determines which elements of U are to be eliminated via induction from the defaults and which are not.

Neither Poole nor Reiter do, of course, relativize their use of defaults to ultimate partitions. But it is easy to see that such relativization can be offered relative to several different such partitions.

Consider the lottery default and the ultimate partition consisting of the pair {ticket 1 wins or ticket 2 wins, neither ticket 1 nor ticket 2 wins}. Given the same set of defaults concerning the lottery as before, the new default extensions are the intersection of all extensions relative to the original partition in which the first element of the ultimate partition is true and the intersection of all extensions in which the second element is true.

The ultimate partition can be taken to be the set of maximally consistent extensions of $\underline{B}$ in the language $\underline{L}$. But I see no reason why inquirers are committed to determining inductive expansions and the corresponding expansions in this fashion.

It may, perhaps, be claimed that the use of defaults allows one to formulate inductive expansion rules without relativizing them to ultimate partitions. To my way of thinking this is no more and no less a virtue than the fact that default reasoning dispenses with explicit appeal to credal and informational-value-determining probabilities and degrees of boldness. Abstracting away from these contextual factors may sometimes be considered desirable; but when one seeks a systematic perspective on inductive expansion (and default reasoning is inductive, ampliative reasoning), taking these factors (or factors that an improved version of the proposals I am making may require) into account is mandatory.

Neither Reiter nor Poole attends to probabilistic considerations in his discussion of defaults and Reiter positively resists the idea. As I have just suggested, there may be contexts where it is desirable to abstract away from relevant contextual features. But in a systematic discussion of induction, an examination of the relevance of probability to induction is unavoidable. The neglect of probability by students of default reasoning hobbles their approaches when addressing this question.

One can readily appreciate the hostility to probabilistic imperialism registered by those who insist that there is more to legitimate inference than can be explicated in terms of probability. I share the general skepticism of the adequacy of probabilist approaches to inference. But resistance to such probabilist imperialism ought not to be reduced to denial of the relevance of probability to induction without closer argument. The view I have taken insists on the relevance of probability to induction without maintaining that probability is the whole story.

Not all participants in the discussion of nonmonotonic reasoning have neglected the importance of statistical considerations. And many of those who do not focus on statistical or probabilistic considerations do seek to take considerations of uncertainty and evidential warrant or support into account in one way or another. Chapter 8 will attend to some of these efforts.

8 Matters of Degree

8.1 Default Conclusions as Degrees of Belief

The conclusion of an inductive inference is a coming to full belief.[1] It entails a change in the agent's state of full belief and, in this respect, in one's doxastic commitment. Or, if the reasoning is suppositional, it is a full belief conditional on an inductively extended supposition. In both cases, the induction is a transformation of a corpus of full belief into another one. If default reasoning is a species of nonmonotonic inductive inference, the conclusion of a default inference must be representable by such a transformation.

It is far from clear, however, that students of default reasoning always understand the conclusions of default inference to be held so confidently. The inquirer might judge the conclusion as worthy of belief to some degree without coming to endorse it fully. In this sense, nonmonotonic reasoning need not involve inductive expansion to a new corpus of full beliefs.

What, however, is to be meant by a degree of belief or beliefworthiness? The currently fashionable view is that degrees of belief are to be interpreted as credal probabilities used in evaluating the expected values of options in decision problems.

There are alternative proposals. In particular, there is the interpretation as degrees of belief of measures having the formal properties first identified by G. L. S. Shackle (1949) and embedded in the b-functions described in section 6.7. Moreover, there are diverse proposals for ordinal or quantitative indices of beliefworthiness or inductive support such as the difference between a "posterior" probability and a "prior" probability, the ratio of a posterior and a prior, or the logarithm of such a ratio.

Confronted with this embarrassment of riches, those who insist that default reasoning assesses the degree to which the current corpus (expanded, perhaps, by adding a supposition for the sake of the argument) supports a proposition need to specify which of these evaluations of warranted degree of belief is appropriate to the assessments involved in default reasoning.

Prima facie the conclusion of a default inference cannot be an evaluation of the beliefworthiness of a proposition relative to an initial state of full belief. It is widely held that default reasoning in particular and nonmonotonic reasoning in general involves "leaping to conclusions" by adding information not entailed by the initial state of full belief to that state. Whether the reasoning involves a genuine change in state of full belief or a suppositional addition of information, there is a transformation of the current state of full belief adding new information to that state. Assigning h a degree of credal probability or any alternative candidate assessment of degree of beliefworthiness involves no transformation of the current state of full belief adding new information to the current state of full belief. The assessment of beliefworthiness of a hypothesis h that is a serious possibility relative to the current corpus $\underline{K}$ (supplemented when appropriate by suppositions) is a finer-grained characterization of the doxastic state than the distinction between seriously possible and impossible propositions provided by $\underline{K}$ itself. It provides more detailed information about the doxastic state but does not add new information to the doxastic state. It does not involve the kind of transformation in doxastic commitment that is characteristic of an ampliative inference.

This observation will not disturb antiinductivists who deny that ampliative inference is ever legitimate.

Antiinductivists of probabilist persuasion, such as R. Carnap and R. C. Jeffrey, advise against inductive expansion and urge us to rest content with judging hypotheses to be probable to varying degrees. They sometimes suggest, however, that judging hypotheses to be probable is a surrogate for the conclusion of an inductive inference.[2]

Antiinductivists of the Popperian persuasion also counsel against inductive expansion. They deny, however, that judgments of credal probability (or, for that matter, any other assessments of degrees of belief or support) are surrogates for conclusions of inductive inference.

The Popperians are right to object to probability judgments as ersatz inductive conclusions; but Popperian arguments are not needed to support the claim that probability judgments do not mimic and are not substitutes for the conclusions of inductive inferences. Bayesian doctrine itself argues in the same direction. Judgments of credal probability relative to the agent's current state of full belief are not the conclusions of ampliative inferences from the information contained in the state of full belief. To be ampliative, an "inference" must entail a change in doxastic commitment. But the current state of full belief and the state of credal probability judgment are two components of the agent's doxastic commitment at a given time. We may ask whether the two components cohere with one another; but the "transition" from one to the other requires no change in doxastic commitment.[3]

Because antiinductivist probabilists have mistakenly thought of probability judgment as the conclusion of an inference alternative to inductive inference, the antiinductivism implicit in their view has not always been appreciated.[4]

Popperians are right to oppose probabilist imperialism. They have tended, however, to overdo their opposition to the emphasis on credal probability judgment in inquiry. The concept of credal probability is a useful tool for representing important features of the attitudes of rational agents. Degrees of belief in the sense of credal or expectation-determining probability play a critical role in evaluating options in decision problems by providing a critical component of evaluations of expected value. But judgments of credal probability serve this function precisely because they are not the conclusions of inferences that are surrogates for inductive expansions. To the contrary, they supplement the judgments of serious possibility determined by the state of full belief by making distinctions between serious possibilities with respect to expectation-determining probability.

Thus, those who advise us never to come to full belief that h via inductive inference but merely to judge it probable that h to some degree or other are not urging us to replace inductive inferences with alternative ampliative inferences. They are urging us to eschew ampliative inference altogether. Unfortunately probabilists often obfuscate the clarity of this view by claiming that they can somehow mimic the results of ampliative inductive inference with the aid of probability judgments. The charge of obscurantism is a fair one given the absence of any clear standard for successful mimicry. This allows probabilists to pick and choose the features of inductive inference they seek to imitate in a manner immune to critical scrutiny.

I do not think any version of antiinductivism, be it Popperian or probabilistic, is tenable. In inquiry, we seek new error-free information. There are, broadly speaking, two ways to obtain it: by routine expansion (via the testimony of the senses or reliable witnesses) and by inductive expansion. Neither method can alone succeed fully in gratifying our curiosity. If inductive expansion is never considered legitimate, the acquisition of new error-free information will be seriously frustrated.

When Shackle introduced his notion of potential surprise or degree of disbelief and the dual notion of degree of belief, he thought of these modes of appraisal as alternative measures of uncertainty to be used in a decision theory that is not based on the injunction to maximize expected utility but is rather a generalization of the optimism–pessimism approach (see Levi, 1984a, ch. 14, sec. 10). Thus, he intended it as a rival to probability as a representation of judgments of uncertainty. Since maximizing the optimism–pessimism index violates even the attenuated version of the expected utility principle I have defended elsewhere and, for that reason, yields unattractive results in some contexts, Shackle's application of his theory of potential surprise does not appear to me to be tenable.

Earlier (Levi, 1966, 1967), I proposed an alternative interpretation of Shackle's formalism in terms of deductively cogent, caution-dependent, and partition-sensitive inductive expansion rules along the lines sketched in section 6.9. Unlike Shackle's interpretation, the notions of degrees of belief and disbelief that result are not surrogates for credal probability in a theory of decision making. Credal probability serves one function in an account of rational deliberation – to wit, as the uncertainty component in a determination of expected value. Shackle-like appraisals of degrees of belief and disbelief serve quite a different function. An inquirer using caution-dependent criteria for inductive expansion who has not fixed on the degree of caution or boldness to exercise can use Shackle measures to determine the set of sentences to be added to the initial corpus of full belief at various levels of boldness. Such an inquirer will not have actually implemented an inductive expansion strategy; but he will have made an appraisal of hypotheses currently considered seriously possible relevant to assessing the potential expansion strategies appropriate to a given context of inquiry. Example 1 of section 7.1 is a rudimentary example of how such information can be reported using d-values or degrees of potential surprise (disbelief).

Thus, the inquirer might identify the set of hypotheses carrying positive (Shackle-type) degrees of belief (i.e., whose negations carry positive degrees of potential surprise). This set of hypotheses would include all items in the current corpus as well as the items that should be added were the inquirer to be maximally bold (given the inquirer's current state of full belief, ultimate partition, credal state, and assessment of informational value). It is arguable that the "conclusion" of default reasoning has been understood to be this particular expansion strategy.

Strictly speaking, of course, making such an identification is not implementing an inductive expansion strategy. The agent who makes the identification may not be maximally bold. The default conclusion may not be the conclusion the agent is actually prepared to draw. Even so, default reasoning so construed presupposes the inductivist assumption that there is a distinction to be made between legitimate and illegitimate inductive expansions. This distinction is represented by the family of deductively cogent, caution-dependent, and partition-sensitive expansion rules deployed. There would be no point in making appraisals of hypotheses with respect to degrees of belief and disbelief according to the interpretation of Shackle measures as b-functions and d-functions proposed in section 6.7 unless induction were sometimes legitimate.

Shackle measures as I have interpreted them are *satisficing* measures of beliefworthiness or inductive support. One can say that h should be added or accepted into the corpus of full belief if and only if b(h) is sufficiently high (see section 6.8). The set of sentences added to the initial corpus via induction consists of all the sentences whose b-values are above a given threshold.[5]

Suppose that an inquirer has fixed on a level of boldness or caution given his state of full belief, ultimate partition, credal state, and appraisal of informational value. The inquirer's task is then to assess potential expansion strategies with respect to expected epistemic utility as represented by $Q(h) - qN(h)$ (see section 6.5). This assessment may also be considered an assessment with respect to inductive support. Now, however, the support is in the *maximizing sense*. It applies to expansion strategies. An expansion strategy relative to $\underline{K}$ may be represented as the deductive closure of $\underline{K}$ and some sentence d to be accepted into the corpus via induction as strongest.[6]

If we use measures of inductive support in the maximizing sense, a default conclusion may be said to be the sentence accepted as strongest via induction from the corpus of full belief carrying maximum support – relative to the given corpus, ultimate partition, credal state, demands for informational value, and the maximum degree of boldness.[7]

Assessments of degrees of beliefworthiness in the maximizing sense and satisficing sense can both be made prior to implementing an inductive expansion strategy. Moreover, if the agent adopts a level of boldness short of the maximum and implements it, he or she can still identify the expansion strategy that would have been adopted were the level of boldness the maximum. This judgment can be made appealing to either the maximizing or satisficing measure of inductive support. Both methods should yield the same expansion strategy. There is no conflict between maximizing and satisficing measures of inductive support. There is only a difference in the factors emphasized.

Thus, it is possible to recognize default reasoning as a form of appraisal of hypothesis distinct from the change in commitment involved in ampliative inductive expansion. The conclusions supported by default reasoning might be those that an agent should reach were he or she maximally bold. Or they might be the conclusions the agent should reach were he or she more cautious to some degree. And these conclusions may be identified using either measures of inductive support in the maximizing or in the satisficing sense.

But even if one construes the conclusions of default reasoning in this fashion rather than as inferences to full belief, default reasoning remains parasitic on ampliative inductive reasoning and the nonmonotonicities it manifests are due to this circumstance. Antiinductivists cannot gain any comfort from the concession.

8.2   Probability as Maximizing Support

Probabilists have tended to be blind to the importance of interpretations of beliefworthiness other than explications purely in terms of probability. This is partially due to inhospitability of induction to the reductionist tendencies of probabilistic imperialism. As a consequence, probabilists treat nonprobabilistic accounts of beliefworthiness or evidential support dismissively or with complacent neglect. Worse yet, they promote confusion by claiming that credal probability can perform functions in inquiry that one might have thought would have been done better by nonprobabilistic indices of beliefworthiness.

I have illustrated the difference between inductive support in the maximizing sense and in the satisficing sense (and the corresponding senses of beliefworthiness) using my own proposals. I think my proposals preferable to the few rival suggestions – such as Niiniluoto's indices of expected epistemic utility deploying measures of verisimilitude (See Niiniluoto, 1986; Levi, 1986c) – that merit serious consideration. However, neither Niiniluoto's proposals nor mine recognize probability to be a measure of inductive support in either the maximizing or satisficing sense.

Nonetheless, because of the sociological fact that so many writers have been bemused by probabilist ideas, it may be worthwhile digressing from the main line of my argument to summarize some of the more obvious defects with probabilist proposals.

Probabilists have sometimes understood probability to be inductive support in the maximizing sense. Given a list of potential answers generated by an ultimate partition, however, the potential answer that carries maximum probability is refusal to expand at all – that is, refusal to reject any element of the ultimate partition. That option incurs no risk of error from the inquirer's perspective (see Popper, 1962; Levi, 1967).

Of course, a probabilist who is also an antiinductivist need not be dismayed by this point. If probability is inductive support in the maximizing sense, then ampliative inductive inference is never legitimate.

If that is so, however, there is no point in having an index of inductive support in the maximizing sense. Probabilists should not be interested in identifying it with probability.

A good example of the confusion of probability with maximizing support is found in the reports of experimental results published by A. Tversky and D. Kahneman concerning the so called "conjunction fallacy" (Tversky and Kahneman, 1983). According to the calculus of probability, probability of a conjunction of two "events" x and y is no greater than the probability of x and the probability of y. Tversky and Kahnemann report that experimental subjects violate this requirement in estimating relative frequencies and determining betting rates where it is clear that the conjunction fallacy is an error.[8] They also report that experimental subjects violate the requirement when asked to rank hypotheses with respect to probability after having been given some background information. A well-known example of the kind of question Tversky and Kahneman consider is the following:

> Linda is 31 years old, single, outspoken and very bright. She majored in philosophy. As a student, she was deeply concerned with issues of discrimination and social justice, and also participated in anti-nuclear demonstrations.

The experimental psychologists then invite the experimental subjects to rank a series of propositions about Linda including the following:

> Linda is active in the feminist movement. (f)
> Linda is a bank teller. (b)
> Linda is a bank teller and active in the feminist movement. (b&f).

Tversky and Kahneman report that most experimental subjects regard b&f as more probable than b.

In this exercise and in problems concerning medical diagnosis where a similar result occurs, even though the experimental subject is invited to rank the hypotheses with respect to probability, we should not assume without experimental evidence that the experimental subjects understand "probability" in the sense of the calculus of probability.

Indeed, there is substantial reason to doubt that they do so. The setting of the problem seems to invite them to ascertain which of the rival hypotheses is more beliefworthy or better supported on the basis of the "evidence" supplied in the brief description of Linda. It is a reasonable conjecture that they interpret the question to be: Which of these sentences would it be better to add to the background information via induction? That is to say, the experimental subjects might well interpret the request to rank the propositions with respect to probability as asking for a ranking with respect to evidential support in the maximizing sense. As we have seen, it is arguable that probability cannot be inductive support in the maximizing sense *precisely because probability invariably ranks weaker answers as at least as good as stronger ones.* Experimental subjects in the Linda and cognate cases may have some presystematic understanding of this and, hence, need not be suffering from cognitive illusions, making a mistake, or committing a fallacy.

My aim is not to defend experimental subjects as free of all fallacies. Tversky and Kahneman appear to have a convincing case to make that experimental subjects commit the conjunction fallacy in some tasks. This is so in estimating frequencies and assessing fair betting rates. However, they commit no such fallacy in the Linda problem, the medical diagnosis problem, and cognate tasks.

It may be that experimental subjects use the same "heuristics" for making probability judgments in these cases that they do for making judgments of evidential support in the maximizing sense in cases of medical diagnosis and in problems like the Linda problem. As Kahneman and Tversky have observed, heuristics often give correct answers. They are, after all, rules of thumb for rough-and-ready answers to questions that one uses in lieu of exact rules. That they will go wrong in some cases is to be expected. It is important, however, to recognize when they go wrong and when they do not.

Tversky and Kahneman have disavowed any prescriptive or evaluative intent when they charge the presence of a fallacy. But it seems quite clear that they are claiming that experimental subjects are making mistakes or are subject to cognitive illusions in the Linda case and their readers have understood this to mean that steps should be taken to develop methods for preventing such illusions from impacting on more serious contexts of judgment. If the experimental subjects are not making mistakes in the medical judgment cases and are being urged to think that they are, this is a serious matter.

Tversky and Kahneman have not, to my knowledge, designed experiments that establish that in problems such as the question about Linda, experimental subjects understand "probability" in the sense of the calculus of probability. They have shown that conjunction fallacies occur in determining fair betting rates and in estimating frequencies; but these observations are not the kind of control relevant to the problem under consideration. Consequently, the results of their experiments leave it unsettled as to which of the following three possibilities hold:

(a) Experimental subjects are committing fallacies in the calculus of probabilities.

(b) Experimental subjects are failing to clearly distinguish maximizing support from probability. In that case, experimental subjects might be guilty of the fallacy of equivocation but not the conjunction fallacy.

(c) Experimental subjects may mean to evaluate inductive support in the maximizing sense and are not confused, committing a fallacy, or suffering from cognitive illusion.

Deploying a principle of charity as a guide to interpretation cannot help us here. If (c) is endorsed and a charitable interpretation of the experimental subjects' responses is adopted, Tversky and Kahneman and not their experimental subjects are confusing evidential support in the maximizing sense with probability. We are being uncharitable to the experimental psychologists. If alternative (a) is adopted, the experimental subjects are making mistakes in calculus of probabilities. No charity is granted to them. In case (b), experimental subjects are confusing evidential support in the maximizing sense with probability while the experimental psychologists are misdiagnosing what has gone wrong. We cannot be charitable to everyone.

The moral of the story seems to be that everyone, experimental psychologist and experimental subject alike, should be reminded of the fallacy of conflating inductive support in the maximizing sense with probability.[9]

Of course, whether the conflation is a fallacy is predicated on the claim that probability cannot be evidential support in the maximizing sense. My argument for this claim is not based on experimental data. It is, as already explained, based on a commitment to inductivism that rejects the antiinductivist presupposition shared by Popperians and probabilists that ampliative inductive inference is never legitimate.

In interpreting experimental subjects along the lines I am suggesting, they are made out to be tacit inductivists as well. Antiinductivists can and do try to offer rival interpretations. Tversky and Kahneman, however, want to remain above both the hermeneutical and philosophical fray. They should, therefore, remain neutral with respect to the prescriptive and philosophical issue as to whether one should be an inductivist or antiinductivist and avoid interpreting experimental subjects as evaluating probabilities where there is a well-known alternative interpretation available. This means that in running their trials they should control for whether experimental subjects are evaluating inductive support in the maximizing sense or whether they are evaluating expectation determining probabilities.

In point of fact, many Bayesians anxious to provide a measure mimicking measures of inductive support have not appealed to probability. The favorite (but not the sole) measure of inductive support proposed is $P(h/x) - P(h)$ where P is a measure of credal probability relative to some initial state of ignorance B and the current state of full belief is $\underline{B}^+_x$. Such measures are quite capable of ranking b&f over b in Linda-type problems.

Consider example 1 in section 7.1. Linda is selected at random from the population described and found to be a radical. The prior probability of her being a bank teller is 0.18. The probability of her being a bank teller given that she is a radical is 0.15. The difference between the conditional probability and the prior probability is -0.03. The prior probability of her being a feminist bank teller is 0.07. The conditional probability of her being a feminist bank teller given that she is a radical is 0.1. The difference between the conditional probability and the prior probability is 0.03. On the proposed measure of inductive support, Linda's being a feminist bank teller is better supported by the evidence of her being a radical than is the hypothesis that she is a bank teller whether or not she is a feminist.

Bayesians who favor such measures of maximizing inductive support seem to be committing what Tversky and Kahneman are calling a fallacy.

There is some difference between authors advocating such measures of inductive support in the maximizing sense as to what the state relative to which the "prior" probability is to be assessed should be. Fortunately, it is unnecessary to explore this matter.

Let h and h' be two hypotheses that are *not* equivalent in truth value relative to $\underline{B}$ but *are* equivalent in truth value relative to $\underline{B}_x^+$. In our example, the claim that Linda is a feminist is not logically equivalent to the claim that she is a radical feminist. Yet, conditional on her being a radical, the two claims are equivalent. Expanding $\underline{B}_x^+$ by adding h and by adding h' yields the identical inductive expansion strategy. Relative to $\underline{B}_x^+$ they should receive the same inductive support in the maximizing sense. The difference between a posterior and a prior does not guarantee this result. It cannot, therefore, be a satisfactory measure of inductive support in the maximizing sense. A similar remark applies to variants on this proposal such as $P(\text{h}/x)/P(\text{h})$ or $log(P(\text{h}/x) - log P(\text{h})$.[10]

The index of expected epistemic utility $Q(\text{h}) - qN(\text{h})$ that I have proposed is an index to be maximized in choosing between inductive expansion strategies so that it is a candidate measure of inductive support in the maximizing sense. Taking Q to be the conditional or posterior probability, the informational value determining probability N replaces the prior probability. When q = 1, the formula is otherwise identical with the difference between the posterior and the prior. Notice, however, that even though h and h' are not logically equivalent, as long as they are equivalent relative to $\underline{B}_x^+$, they carry the same inductive support in the maximizing sense.

The measures of expected epistemic utility proposed by Niiniluoto (1986) are also intended to be measures of inductive support in the maximizing sense. They satisfy the equivalence condition just mentioned and can yield deductively cogent, caution-dependent, and partition-sensitive inductive expansion rules. The difference between Niiniluoto's proposals and mine concerns differences in our conception of the common features of the aims of scientific inquiry. Discussion of this disagreement would entail an even more substantial digression from the main line of argument (see, however, Levi, 1986c). What is clear, though, is that neither probability nor the difference between a posterior and a prior can serve as a measure of inductive support in the maximizing sense.

8.3 Probability as Satisficing Support

Probabilists are far more prone to consider probability as an index of satisficing support than as an index of maximizing support. A broad cross-section of the community of epistemologists and students of scientific methodology adhere to the view that a conclusion is to be adopted on the basis of evidence if its probability is high enough or, what amounts to the same thing, that a proposition is to be rejected if its probability is low enough. The view is endorsed by experts in probability theory and statistical inference of the stature of Borel, Kolmogoroff, and Cramer, to mention but a few.

One reason for adopting this view seems to be the thought that if the probability of h on the evidence is close enough to 1, it is certain enough for practical purposes. Consequently, the inquirer is justified in adding h to the corpus of full belief.

There is less to this argument than meets the eye. Even if it is conceded that propositions virtually indistinguishable from full beliefs are legitimately added to the state of full belief via induction, propositions that carry probability near 1 relative to a given state of full belief need not be virtually indistinguishable from propositions that are fully believed according to that state.

As is well known, assigning positive credal probability to a proposition h is sufficient but not necessary for judging h to be a serious possibility. h can be a serious possibility even if it is assigned 0 probability and its negation probability 1. In such a case $P(x/h)$ is well defined. It is not well defined if h is judged impossible.[11] For certain practical purposes, therefore, impossible propositions carrying probability 0 are distinguishable from serious possibilities carrying probability 0.

Thus, the probability that an unbiased coin will land heads on every one of infinitely many tosses or will land tails on every one of those tosses is 0. Yet, if it is settled that the coin is to be tossed infinitely many times, the truth of this disjunction is a serious possibility. Moreover, the probability of the coin landing heads every time conditional on its landing heads every time or landing tails every time could very well be 0.5.[12]

Not only do arguments for adopting probability as a satisficing measure of beliefworthiness or inductive support based on the virtual indistinguishability of "almost certain" and "certain" propositions fail to carry conviction; but the idea of using credal probability in this way is fraught with well-known difficulties.

Satisficing measures of inductive support or beliefworthiness typically presuppose that when a proposition carries positive support, it is at least coherent to add it to one's corpus of full belief if one is bold enough. Probability measures fail to satisfy the demands of this presupposition.

The index of probability ranges from 0 to 1. If propositions carrying positive credal probability very close to 0 are supported or judged worthy of belief to a positive degree, a sufficiently bold agent should be in a position to add it and all propositions carrying that degree of probability to the inductive expansion of his corpus and do so without contradiction. In the celebrated fair lottery of a million tickets, every hypothesis of the form "Ticket i will win" should be considered positively supported. A sufficiently bold agent should be justified in adding all such hypotheses to his corpus as well as their negations. But no advocate of high-probability rules would regard this inconsistent result as tolerable.

There is a good case for denying that the claim that ticket i will be drawn (carrying probability 0.000001) is worthy of belief or inductively supported to a positive degree at all. Those who have not been inculcated with the ideology of high-probability acceptance rules will often find it attractive to recommend remaining in total suspense as to which ticket in the lottery will be drawn and, hence, concerning whether ticket i will or will not be drawn (Levi, 1965, 1967).

The recommendation to suspend judgment is controversial (although that this is so continues to astound me); but those who dissent from the recommendation should and, I think, would concede that the claim that ticket i will be drawn is not supported at all and should be positively disbelieved even though it carries positive probability. Whether one thinks that suspension of judgment is in order or that one should disbelieve that ticket i will be drawn (and believe that it will not), one cannot equate degree of probability with degrees of inductive support in the satisficing sense.

But even if probability is not an index of inductive support in the satisficing sense, there may be a suitable function of probability that is. This is a point of view that H. E. Kyburg has advocated for many years (at least since Kyburg, 1961). Relative to a given theory K, Kyburg suggests that one can construct a new corpus or set of sentences including K whose probabilities are at least as great as some threshold value k above 0.5. This set of sentences is not closed under logical consequence.[13]

Since credal or epistemic probability for Kyburg is, in general, interval valued (and, hence, is a species of indeterminate probability of the sort the importance of which I emphasize in Levi, 1974 and 1980), the set of sentences "accepted" in the k-level corpus are those whose lower probabilities are greater than k. One could define an index of degree of belief ranging from 0 to 1 by defining the Kyburgian degree of belief of a sentence as $[k-0.5]/0.5$ where k is the lower probability of the sentence and k is at least 0.5. Otherwise, the Kyburgian degree of belief is 0. The degree of disbelief in a sentence could then be defined as the degree of belief in the negation.[14]

Kyburg (1961, 1974, 1983, 1990a, 1990b, 1992) has long insisted that inductive inference yields additions to the information "accepted" by the agent from initial information in opposition to the antiinductivist attitudes of the Popperians and probabilists. In his more recent publications, he has turned much of his attention to the literature on nonmonotonic reasoning. And among the themes he has aimed to discuss within the artificial intelligence community is the relevance of statistical considerations to nonmonotonicity.

Inductive inference yields additions to the information "accepted" by the agent from the initial information. As already noted, Kyburg contends that the criterion of acceptability is high probability. Relative to $\underline{B}$, h is acceptable inductively if and only if the credal or epistemological probability that h relative to $\underline{K}$ is above some threshold k at least as great as 0.5. That is why Kyburgian degrees of belief could be defined as I have suggested.

The calculus of probabilities implies that, given a probability function over sentences in a given language $\underline{L}$ relative to corpus $\underline{K}$ in $\underline{L}$, the set of sentences whose probabilities are greater than k will, in general, fail to be a deductively closed set. This point was forcefully emphasized by Kyburg in terms of his discussion of the so called "paradox" of the fair lottery (Kyburg, 1961). Kyburg argued that one can avoid deriving a contradiction from the background claim that some ticket in the million-ticket lottery will win and the conclusion that ticket i will win for each of the million substitutions of an integer for i by refusing to close this set of sentences under logical consequence. That is to say, one can have each of these million and 1 beliefs, but one is not entitled to accept any conjunction of a subset of these claims unless that conjunction also carries a probability greater than k.

There is, nonetheless, trouble in paradise. The conclusions licensed by Kyburg in the case of the fair-lottery paradox constitute an inconsistent set. Kyburg himself acknowledges that they do but is not disturbed. The set of sentences of the form "Ticket i will not win" together with the sentence "At least one of the tickets will win" are logically inconsistent. Kyburg qualifies this claim by suggesting that they are "strongly inconsistent" because every sentence in the language can be derived from them. But they are not "weakly inconsistent" because no single one of the n + 1 sentences just mentioned is inconsistent with the initial corpus (Kyburg, 1983, p. 233). According to Kyburg, all we require is a set of sentences representing beliefs that are weakly consistent.

Calling a set of accepted sentences that is inconsistent "weakly consistent" is not much consolation. As long as the inquiring agent recognizes that at least one of the sentences his criteria for acceptance recommend accepting is false, the inquiring agent has not avoided error in using the criteria for fixing beliefs.

It does not help Kyburg's case to take notice of analogues of lottery problems that arise in measurement, as he is wont to do. Kyburg invites us to consider cases of measuring magnitudes with apparatus possessing a known error distribution where one may state the value of the magnitude measured to be within three standard deviations of some single real number. Even though the probability of being in error is negligible in any single case, if such information is reproduced a large number of times, the probability of at least one such interval estimate of the real value being in error is considerable. Kyburg argues that we have reason to accept each and every single report of measurement. We have reason to accept the claim that at least one such report is in error. We have a lottery "paradox" of a sort Kyburg thinks arises frequently in scientific work. The set of inductively accepted sentences is inconsistent.

That the set of accepted sentences is strongly inconsistent obtains whether or not it is closed under logical consequence. However, the n + 1 sentences differ from their closure under logical consequence with respect to weak inconsistency. No conjunction of a proper subset carrying high probability entails a contradiction. Kyburg states in effect that it is legitimate to accept a set of sentences that is inconsistent as long as the conjunction of a sentence and its negation is not in the set. One should avoid "weak" inconsistency. One can live with "strong" inconsistency.

Kyburg's advice is entertainable as long as one is indifferent to the question as to whether one's convictions avoid error. An inquirer whose acceptances are inconsistent in the strong sense often has sufficient logical omniscience to recognize that the set of accepted sentences contains at least one falsity.

Anyone concerned to avoid error in making inductions would want to avoid accepting an inconsistent set of sentences in this strong sense of inconsistency precisely because to accept those sentences is to deliberately import error. Given our lack of logical omniscience, we may fail to be consistent due to inadvertence; but surely we should avoid deliberately importing error as Kyburg recommends.

In response, Kyburg points out (1990a, p. 65), the agent is not in a position to determine which of the accepted sentences is false. It is clear that relative to the initial corpus <u>B</u> the agent does not know which ticket will win. The agent is initially in suspense on this point. After making the inductions, the agent may accept many sentences about the outcome of the lottery; but, according to Kyburg, the agent still is in the dark as to which ticket will win. The inductions to reasonably accepted sentences do not settle the matter. Precisely the same point applies to the case of measurement. If the agent reasonably accepts that at least one report of measurement is false and reasonably accepts each report, we have an inconsistent set of accepted sentences. To be told then that the agent cannot determine which of the set of sentences accepted is false is to concede that the agent cannot determine which of the accepted sentences are true. The agent recognizes this. So accepting these sentences via induction has failed to serve its purpose, which was to fix beliefs as to what the values of the measured magnitudes are with some degree of precision. The agent remains in suspense on this score. Yet, in the very paragraph where Kyburg is admitting that the agent cannot determine which of the elements of the inconsistent corpus are false, he writes:

> To suspend judgment, to say that we don't know the lengths of any of these objects we have measured, is skeptical defeatism at its worst, unless we are prepared to give up acceptance altogether and retreat to the alternative view of scientific knowledge. (p. 65)

But Kyburg himself has admitted that the agent cannot determine which of the elements of the inconsistent corpus are false. This is the very "skeptical defeatism" he rails against.

The trilemma confronting Kyburg is this: High-probability rules for inductive acceptance yield strong inconsistency. Either one respects the concern to avoid error in induction and rejects such rules, one abandons avoidance of error as a concern in inquiry, or one denies that inductive acceptance removes doubts as Kyburg officially claims it does. Kyburg is right to think that inductive expansion adds new information to one's background information. He is wrong to lapse into skeptical defeatism in fending off the demand that a corpus be strongly consistent. There is no acceptable rationale of a high-probability rule as a criterion of inductive acceptance.

The issue at stake here is not the use of the verb *to know* as opposed to the verb *to believe*. Not only does the agent not know which of the tickets will be drawn; he should not have an opinion on the matter other than complete suspension of judgment. This I contend is a conclusion that Kyburg is obliged to concede as long as he maintains that he is sure that at least one ticket will win while not being in position to tell which it is. Kyburg can also say, if he likes, that the agent should accept "Ticket i will not win" for each and every i. Acceptance does not then mean belief in a sense in which the agent acquiring beliefs should be concerned to avoid error. By Kyburg's own testimony, an agent can be in suspense concerning which ticket will win (i.e., neither believe nor disbelieve that ticket i will win for each i) while accepting that ticket i will not win for each.

In the case of the lottery, it seems eminently sensible to conclude that induction does not yield new information. One should refuse to expand altogether no matter how bold one is and should remain in suspense pending the acquisition of new information. If Kyburg has in mind some notion of "acceptance" that is different from coming to full belief as he plainly implies he has, he should explain what it is. Whatever it is, however, induction as ampliative inference is adding new information to a state of full belief. Just as Carnap muddied the terminological waters by calling the assignment of credal probabilities to hypotheses on the basis of evidence "induction," Kyburg promotes unclarity when he declares that induction is inference from evidence to a set of sentences accepted at a certain level of probability.

Induction is, indeed, ampliative inference. But it entails a change in the agent's corpus of evidence or full beliefs to the one reached by inductive expansion. What is accepted via induction is accepted *into* evidence and becomes fully believed. Kyburg rejects this view on the grounds that once an item is admitted into a corpus of full beliefs or evidence and gains probability 1, it becomes incorrigible (Kyburg, 1983, p. 236; 1990a, p. 65).

This claim is simply false. One can have good reasons to give up full beliefs. The account I favor was first sketched (Levi, 1980, chs. 2 and 3) and then elaborated more fully (Levi, 1991, chs. 3 and 4). Kyburg's claim gains whatever shred of cogency it has from the illicit assumption that to have good reason for ceasing to believe some proposition, one needs to have obtained additional evidence and that such new evidence should constitute the sole ground for the contraction.

To rationalize contraction, one does not, in general, appeal to new evidence. Contraction is sometimes coerced by inadvertent expansion into inconsistency so that the only issue confronting the inquirer is what items to give up to escape from contradiction. The choice here does not depend on evidence but on assessments of the value of information to be lost. The other reason for contraction is to give a serious hearing to some explanatorily attractive hypothesis that is certainly false according to the current view. Neither reason justifies giving up some item because new evidence has been acquired. But reasons of this kind can justify giving up full beliefs carrying probability 1.

From the inquirer's point of view, all items in the initial state of full belief are true and are certainly true. Adding the new sentences via inductive expansion incurs risk of error as assessed from that initial view.[15] Once the expansion has been implemented, a new point of view has been adopted and the newly "accepted sentences" carry no risk of error.

Kyburg denies the legitimacy of inductive inference so construed just as the antiinductivists Popper and Carnap do (e.g., in Kyburg, 1990a, p. 67). He even denies that the corpus of evidence from which his pseudoinductive inferences are made are certain – although here his account is far from clear. He declares that we must take into account the uncertainty that infects even our evidential claims. He writes:

This may seem baroque, but if we want to reflect scientific practice, we must allow for the ultimate rejection of even evidential statements. We can accomplish this by a shift of context. If the acceptance of a statement in the evidential corpus is an issue, we shift gears and regard that corpus as the practical corpus, and another corpus of higher level, as the evidential corpus. (p. 67)

There are three features of this passage that deserve consideration.

(a) Kyburg contends that we should not regard our evidential claims as maximally certain if we want to reflect scientific practice. We must allow for the "ultimate rejection" of even evidential statements. Setting aside what "ultimate rejection" can mean for Kyburg, for whom neither acceptance nor rejection is ultimate in any clear sense, I take it he means quite rightly to point out that in science, inquirers must be prepared to change their minds even about what they use as evidence. But conceding and indeed emphasizing this does not preclude insisting that to accept h as evidence at a given time t is to be maximally certain at that time that h is true (Levi, 1980, chs. 1–3; 1984a, ch. 8).

(b) Giving up items from the corpus of evidence is nothing more than contraction. Kyburg suggests that contraction is just a change in context. No doubt when contraction is warranted, the context prior to contraction is relevant to justifying the contraction. And after contraction has occurred, the context of inquiry has itself changed. But appealing to change of context explains nothing. One needs to identify the contextual variables relevant to the warrant for contraction and how changes in such variables might change such warrants. Accounts of legitimate contraction have been proposed elsewhere (e.g., Levi, 1980, 1991). Does Kyburg deny that we need an account of justified contraction – that is, of when an evidential corpus ceases to be an evidential corpus and another corpus becomes the evidential corpus? Why then does he think we need an account of the inverse process of legitimate inductive acceptance? It is not enough to say that we "shift gears."

(c) What can Kyburg mean when he says we shift gears, unless he means that the inquirer fully believes the elements in the evidential corpus as long as it remains the evidential corpus? Of course, when we "shift gears," we move to another corpus as the evidential corpus, as Kyburg himself says. What is the force of saying that one changes the status of a given corpus from being an evidential corpus, to being a "practical" corpus unless it means that the items in the erstwhile evidential corpus cease being certain? The only answer he can give is to abandon the distinction between an evidential corpus and a practical one in favor of a hierarchy of corpora grounded on a system of probability judgments relative to some deeply basic corpus, which, in Kyburgian terms, cannot cease being certain, or, if this can be coherently done, grounded on nothing at all other than logic and mathematics.

In short, Kyburg's acceptance rules are not designed to address the question of inductive expansion of the corpus of full belief, of certainties, or of evidence. What then is the problem for which rules of inductive acceptance are designed?

There is a way to offer an answer suggested by observations made by Kyburg in various places. A Kyburgian corpus is not closed under logical consequence but, as Kyburg noted (1974), a Kyburgian corpus can be partitioned into subsets each of which is a consistent expansion of the initial corpus and, hence, is deductively closed. These sets we call *strands*. They resemble the extensions of an initial corpus or set of factual beliefs generated either by Reiter or Poole defaults.

We have already noted that Reiter extensions (suitably tailored) or Poole extensions can be regarded as representing various expansions of the initial corpus that remain open to the agent *after* expansion. Given an ultimate partition U and a subset U' of U representing the set of elements of U that go unrejected after inductive expansion and, hence, remain serious possibilities not only relative to the initial corpus $\underline{B}$ but to its inductive expansion $\underline{B}^i$ as well, the deductive closure of $\underline{B}$ and any one of the unrejected elements of U' corresponds to an extension of $\underline{B}$.

On the proposals I have favored since the 1960s one should never confuse, as Reiter seems to have done, such an extension with the set of full beliefs obtained by inductive expansion. An extension represents a serious possibility relative to the inductively expanded corpus. Given the set of such extensions, the inductively expanded corpus is the intersection of all such extensions or the set of shared agreements between these extensions. This is the "skeptical" interpretation of default extensions favored by Poole and endorsed by Makinson and Gärdenfors to which reference was made in section 7.2

The idea I am now considering is to think of Kyburg's high-probability criteria for acceptance as principles for accepting hypotheses as serious possibilities relative to the inductively expanded corpus. The set of all such hypotheses fails to be closed under deduction. However, if one partitions them into deductively closed sets, they become Kyburgian strands or extensions of the initial corpus whose intersection is an inductive expansion of $\underline{B}$ and is closed under logical consequence.

Consider the so-called lottery paradox. Each of a million tickets in a fair lottery has a chance of 10^{-6} of being drawn so that the negation of "Ticket i will be drawn" for i ranging from 1 to 1,000,000 carries a very high probability indeed. One Kyburgian strand will consist of the logical consequences of "Ticket 1 will not be drawn." Another will be "Ticket 2 will not be drawn," etc. The intersection of these strands will consist of the background information (e.g., exactly one ticket will be drawn, the lottery is fair) but will represent suspense as to which ticket will win.

Kyburg (1990b) acknowledges explicitly the analogy between his strands and Reiter extensions and points out that a system of Reiter defaults could yield extensions that are consistent closed expansions of the initial corpus by adding "Ticket i will not win" for each i. Kyburg seems to think that because he can generate results analogous to the lottery for Reiter, his account of rational inductive acceptance is one that students of default reasoning ought to take seriously.

Like Reiter, Kyburg resists the skeptical interpretation of his version of default extensions. But unlike Reiter, he does not recommend arbitrarily choosing one of them because of the peremptoriness of choice. Instead, he recommends embracing the union of all default extensions but without closing it under logical consequence.

Kyburg cites the case of the quality inspector who examines the output of a certain manufacturing process to determine whether the items meet certain standards. The chance of an error is low enough (say 1 percent) to pass items. Kyburg points out correctly that this procedure does not require belief that each of the items passing the test meets the standards. But the quality inspector (or, more accurately, the statistician who designed the test, for the quality inspector may be a robot) is designing a strategy for making decisions relative to moral, political, economic, or other "practical" goals and not for making decisions as to what new information to add to a corpus relative to cognitive. The kind of decision-making disciples of the Neyman–Pearson school of statisticians call "inductive behavior" is decision making of the first kind. In general, no inductive inference adding new full beliefs to a corpus of evidence is implemented. The decisions are simply decisions as to what to put up for sale and not what to believe.

It seems clear from this that Kyburg would resist the suggestion I am making to take the meet of the various strands. But if Kyburg is truly interested in inductive inference and not inductive behavior, it seems to me that he should welcome the suggestion as a positive improvement in his proposals; for it not only removes the "paradox" from the lottery. It also preserves the deductive closure and strong consistency of the expansion.

But even if the suggestion is taken seriously, we need to consider whether the skeptical interpretation implied is too skeptical. Perhaps the sound recommendation to suspend judgment in the lottery is the product of a blanket recommendation to suspend judgment in every case. This fear is not, strictly speaking, well founded. If one of the tickets carries probability greater than 0.5 of being drawn, inductive expansion will favor it. It is easy to show that the inductive expansion transformations will be exportable but not importable. $(K*i8)$ will work but not $(K*i7)$.

The trouble with this proposal is that the criterion of expansion is far too weak. It will counsel complete suspension between a set of exclusive and exhaustive alternatives unless exactly one of them carries probability greater than 0.5.

This is bad enough. Kyburg's main objection to the criteria for inductive expansion I favor is their partition sensitivity. But the attempted rescue of Kyburg's idea seems to face the problem of partition sensitivity. For a given language, either one uses the proposal I have made relative to a set of exclusive and exhaustive alternatives consisting of maximally consistent sets. In that case, it is utterly unlikely that any maximally consistent set will carry probability greater than 0.5. Alternatively, one can consider coarser-grained partitions. In that event, the criterion for inductive expansion proposed is partition sensitive as well as deductively cogent.

We have seen in Chapter 6 that it is false to claim that the only serious contender for a criterion of legitimate inductive expansion that takes probability and statistics into account is a high-probability rule that construes probability as a satisficing measure of inductive support. It is important to see this point. Otherwise those who share my view that the failure of logical closure is a real defect in high-probability rules may think that probability-based criteria for inductive expansion are suspect altogether. A retreat may be made to some form of antiinductivism denying legitimacy to any inductive expansion except as a manner of speaking. Or attempts might be pressed to develop criteria for inductive expansion that make no use of probabilistic considerations. These alternatives seem to me no more tenable than advocating Kyburg's high-probability criterion. If we are to take probability as relevant to inductive expansion in general and default reasoning in particular, we had better turn away from high-probability rules to criteria (such as those proposed in Chapter 6) that finesse the central difficulties.

Adams (1975) takes the position that the probability of what he misleadingly calls an "indicative" conditional is the same as the conditional probability of the "consequent" given the "antecedent." Adams is a resolute probabilist. As such, he is committed to an antiinductivist perspective. Rational agents should not fully believe any extralogical and extramathematical truths. The only corpus an agent ought to endorse is the weakest potential corpus of logical, mathematical, and other conceptual truths (if there be such). Relative to this corpus, the agent makes judgments of credal probability meeting the demands of coherence imposed by the calculus of probabilities. These credal probabilities may change over time in response to the impact of sensory inputs.

Those who resist probabilism, like myself, insist that in science and daily life agents do have full beliefs, including belief in the truth of theoretical claims, empirical generalizations, and the testimony of witnesses and of the senses. Like many other probabilists, Adams tacitly concedes that this appears to be the case; but he (again tacitly) insists that this is mere appearance. What appears to be full belief that h in reality is a judgment that the credal probability that h is extremely close to 1.

Adams is not, thereby, proposing a high-probability rule for inductive expansion. There can be no inductive expansion according to Adams; for this requires a change in the corpus of full belief. Such changes must be denied as illegitimate or unintelligible.

What is distinctive of Adams's approach is his treatment of conditionals or what he calls in the misleading terminology currently deployed "indicative" conditionals. According to Adams, the probability of a conditional h > g is equal to $P(g/h)$. Armed with this assumption, Adams is then able to say that h > g is assertible when its probability is near 1 – that is, when $P(g/h)$ is near 1.

Although Adams's version of the high-probability rule is not to be understood as a principle of inductive expansion, he does seem to think that propositions carrying high probability can be used as premises in reasoning as if they were fully believed. He goes out of his way to point out, however, that the soundness of reasoning from premises, all of which carry high probability, is not truth-conditional soundness but a stricter condition of probabilistic soundness. Thus, if an agent judges each sentence of the form "Ticket i will not be drawn" to be highly probable in the case of a million-ticket lottery, the conclusion that truth conditionally follows from these million "premises" that no ticket in the lottery will be drawn is not probabilistically sound; for it will not carry a high probability. Adams writes: "common sense supports the appropriateness of insisting on probabilistic soundness, since it is patent that even truth-conditionally sound inferences from a lot of 'shaky data' can be highly unreliable" (1975, p. 2).

Of course, in the case of the lottery and kindred examples, the "shaky data" are no data at all. "Common sense" does not recommend fully believing or taking for granted as data the million claims of the form that "Ticket i will not win." Common sense recommends complete agnosticism as to the outcome of the lottery.

This remark is made to emphasize how poor a simulation Adams's high-probability criterion of assertibility is to the presystematic view that agents can sometimes reasonably believe fully extralogical propositions. To say this does not decisively refute the probabilistic point of view. Just as Parmenides painted a metaphysical picture of reality strongly at variance with common experience, probabilists can paint a picture of our doxastic condition strongly at odds with common experience. They should not, however, try to paper the abyss over with high-probability simulations of full belief. Such simulations are bad music in the sense according to which Carnap alleged that Heidegger was a bad musician.

Having said this, however, it should be emphasized that the conceptions of probabilistic consistency and entailment that Adams has devised and explored in probing the character of probabilistic soundness do have some uses. Adams himself used the notion of probabilistic entailment to characterize a certain kind of logic of conditionals. Lehmann and Magidor (1992) and Pearl (1992) have noticed that Adams's characterization of probabilistic entailment can be exploited to useful effect in examining the conclusions that may be drawn from assertions of nonmonotonic implication to other such assertions and, as a consequence, in characterizing certain kinds of nonmonotonic implication relations. Thus, Lehmann and Magidor show that if h > g is replaced by h |~ g and h is replaced by T |~ h, then given any probabilistically consistent and finite set S of "assertions" of the form h |~ g, S probabilistically entails the assertion A if and only if S "preferentially" entails A (1992, pp. 26–8). Speaking very loosely, S preferentially entails A if and only if any preferential consequence relation |~ for which all assertions in S hold is one in which A also holds. A nonmonotonic (pseudo) inferential relation |~ @ $\underline{K}$ (in K) is a preferential consequence relation that may be generated from a revision operator * on potential corpora in K satisfying the AGM postulates (with or without (K*5)) minus (K*8). Kraus, Lehmann, and Magidor (1990) offer a semantics for preferential entailment. What Lehmann and Magidor show, in effect, is that Adams's "probabilistic semantics" for conditionals can be reconstrued as a probabilistic semantics for preferential entailment in the case where S is finite and probabilistically consistent.

Thus, someone who either had no sympathy with probabilism or was quite neutral on the issue might find Adams's conception of probabilistic entailment a useful tool for exploring formally entertainable types of nonmonotonic inferential relations. This acknowledgment does not of itself concede that Adams's characterization of conditionals or their assertibility is itself a cogent explication of suppositional reasoning, of which conditional judgment is an expression.

Pearl (1992, pp. 178–9), who is more sympathetic to probabilism than I am, does acknowledge a difficulty due to the lottery paradox. He writes:

Are these paradoxes detrimental to ε semantics, or to nonmonotonic logics in general? I would like to argue that they are not. On the contrary, I view these paradoxes as healthy reminders that in all forms of reasoning we are dealing with simplified abstractions of real-world knowledge, that we may occasionally step beyond the boundaries of one abstraction and that, in such a case, a more refined abstraction should be substituted.

... The lottery paradox represents a situation where ε no longer offers a useful abstraction of reality. Fortunately, however, the consistency norms help us to identify such situations in advance, and alert us (in case our decisions depend critically on making extensive use of the disjunctive axiom) that a finer abstraction should be in order (perhaps full-fledged probability theory). (p. 179)

If ε semantics (which is Pearl's formulation of Adams's account of conditionals as an account of "defaults") are a more inaccurate "abstraction" than probability theory, as Pearl seems to admit, why bother with the inaccurate abstraction when the more accurate abstraction is available? Perhaps it may be more readily programmable. From the philosophical perspective I am taking, that is of marginal concern. Pearl is acknowledging his allegiance to an antiinductivist probabilism. If that is the case, I do not understand why he is bothering with default reasoning at all since it is a species of inductive reasoning.

In spite of the criticisms I have leveled against their views, I remain sympathetic with both Pearl and Kyburg in their insistence on taking probabilistic and statistical reasoning in particular and induction in general seriously. As such they rightly stand opposed to authors like Reiter and Poole who think that probability and statistics may be ignored in ampliative reasoning. My dispute with Pearl and Kyburg is that they fail to take seriously the minimal demands that an account of inductive inference ought to meet. On this score, my sympathies are on the side of those students of nonmonotonic reasoning (including ampliative inference) who take deductive closure requirements seriously. Unfortunately, authors like Gabby, Makinson, Gärdenfors, et al. fail to take the probabilistic and inductive aspects of nonmonotonic reasoning into account as fully as they should and as Kyburg and Pearl try to do.

8.4 Satisficing Support as a Shackle Measure

Kyburgian degrees of belief do have something in common with Shackle degrees of belief under the interpretation summarized in section 6.9. Like Shackle measures, Kyburgian measures are derivable from families of caution-dependent acceptance rules. Unlike Shackle measures, they are not derivable from families of deductively cogent and partition-sensitive rules. Kyburg regards the lack of partition sensitivity as a virtue. In section 6.8, I pointed out that the lack of deductive cogency undermines the utility of a corpus of full belief as a standard for serious possibility. Partition sensitivity, on the other hand, is no obstacle to objectivity. If inquirers differ over the choice of ultimate partition, the dispute is subject to critical control.

In spite of the difference, Kyburgian degrees of belief like Shackle degrees of belief are derivable from an ordered *family* of inductive acceptance rules where the ordering reflects the relative boldness or caution of the rules. The differences between the two kinds of indices concern the properties of the members of such a family. A Kyburgian corpus is not closed under consequence. A family of inductive acceptance rules generating a Shackle measure rule is. Kyburgian inductive acceptance rules do not, therefore, yield inductive expansions of the initial corpus. Acceptance rules generating a Shackle measure do.

Each member of a family of caution-dependent acceptance rules is both ampliative and nonmonotonic. Deductively cogent families of this type determine Shackle measures while recommending inductive expansions of the initial corpus at various levels of boldness.

In section 8.1, it was pointed out that an inquirer may be unprepared to reach a decision as to how to expand relative to $\underline{K}$ because of a failure to fix on a level of boldness or caution. Or an adviser to the inquirer who has done an analysis of the inquirer's predicament may not want to foist on the inquirer a specific level of boldness or caution. In such a case, potential answers to the question under study may be assigned degrees of belief or confidence of acceptance or plausibility. It may then be left to further reflection to fix on a level of caution or boldness or equivalently to fix on a threshold of inductive support to be used in determining how to expand inductively.

Thus, even if someone insists that the conclusions of default reasoning are merely plausible conclusions but not new full beliefs so that no genuine inductive expansion to a new corpus of full belief is implemented, it may be maintained that underlying the assessments of plausibility or degrees of belief is a family of nonmonotonic and ampliative inductive expansion rules. The nonmonotonicity of plausible inference is derivative from the nonmonotonicity of inductive inference.

As I have reconstructed Shackle's measure, the b-value of a hypothesis is determined by the initial corpus $\underline{K}$, an ultimate partition U, a credal probability measure Q, and an informational-value-determining probability N. The b-value then reflects the maximum value of the index q reflecting tradeoffs between informational value and credal probability at which the hypothesis is accepted via induction. An agent might not, at a given stage, be prepared to endorse a specific value for q as reflecting his or her degree of caution or boldness and may, instead, simply indicate the b-value. But even if the inquirer does not actually engage in an inductive expansion, the b-value he assigns reflects the use of a caution-dependent, partition-dependent, and deductively cogent set of criteria for inductive expansion. No induction need take place; but the assignment of b-values to hypotheses will remain grounded on nonmonotonic and ampliative inferences. The particular b-value assigned, for example, to "Tweety flies" will be positive if Tweety is known to be a bird but not if, in addition, Tweety is known to be a penguin.

Thus, even though the conclusions reached in default reasoning may be inductively supported in the satisficing sense to some degree, they are

258 Matters of Degree

inductive expansions of a nondeductive and nonmonotonic variety at some degree of boldness.

I have pointed out (Levi, 1966; 1967; 1984a, ch. 14) that there are other deductively cogent and caution-dependent families of inductive expansion rules alternative to those I have proposed that are entertainable. But whether or not one favors the rules I advocate or rules belonging to some alternative set, they still remain inductive expansion rules with ampliative and nonmonotonic properties.

Those who seek to remain neutral concerning controversies about the appropriate caution-dependent and deductively cogent families of inductive expansion rules to endorse might rest content with taking a measure of degree of belief measure as if it were a primitive and giving it the structure of a formal Shackle b-function. As long as the aim is to defer a verdict on which of the proposed families of inductive expansion rules should be adopted, there can be no objection. However, taking the b-function as primitive without acknowledging the need to ground it in a deductively cogent, caution-, and partition-sensitive family of inductive expansion rules is objectionable.

I do not mean to suggest that the only kind of interpretation of the Shackle formalism is in terms of families of deductively cogent, caution-dependent, and partition-sensitive expansion rules. In section 8.5, I shall discuss just such an alternative interpretation. This interpretation does not allow for understanding the Shackle measures as indexing degrees of belief and disbelief; for they discriminate between propositions the agent already is committed to fully believing – that is, to judging maximally certain. To anticipate, the discriminations focus on vulnerability to being given up in choosing between potential contraction strategies. Although many epistemologists have equated invulnerability to being given up (incorrigibility) with certainty, the perspective I have been advocating insists that the two ideas are distinct (Levi, 1980).

In recent years, many authors have reinvented Shackle's formalism. They have often claimed to apply it as a representation of degrees of belief and disbelief or as a representation of degrees of invulnerability to being given up. For the most part, however, they have failed to provide more explicit explanations of their intentions such as the derivation of degrees of belief from some family of deductively cogent, caution-dependent and partition-sensitive inductive acceptance rules or derivations of assessments of invulnerability to being given up in terms of losses of informational value. As a consequence, claims appropriate to one interpretation are not kept distinct from claims appropriate to another.

Thus, Dubois and Prade (1992, proposition 5, p. 148) remind us of the fact, well known since at least the mid 1960s, that the set of sentences carrying positive degrees of certainty should constitute a deductively closed and consistent theory or corpus.[16] Dubois and Prade seem to think of their measures of certainty or degrees of necessity as degrees of belief. They do not connect them to any additional structure that would indicate whether they are derived from families of deductively cogent, caution-dependent, and partition-sensitive inductive expansion rules if that were their intent. Nor do they derive them from assessments of losses of informational value as befits assessments of entrenchment or incorrigibility. Moreover, they think of assessments of certainty or the associated assessments of degrees of possibility (that Dubois and Prade themselves acknowledge relate to degrees of disbelief or potential surprise) as representations of belief states alternative to using deductively closed theories alone. The result is confusion. The "ordered belief sets" can be construed as representations of belief states together with an assessment of which elements of a corpus are more vulnerable to being given up when the need arises. Or they can be construed as the expansions via induction of states of full belief that would be recommended were one to be maximally bold. These are quite distinct evaluations and ought not to be confused.

Many authors (e.g., Spohn, 1988; Dubois and Prade, 1992) follow Shackle in seeking to find a principled method for updating Shackle-type degrees of belief in the light of new information. If more attention were paid to specifying explicitly the intended applications of Shackle measures and less to describing a formal structure and then appealing to poorly motivated and idiosyncratic intuitions, the effort to obtain such surrogates for Bayesian conditionalization would cease.

Let $\underline{B}$ be an initial corpus of full belief. Let the ultimate partition U consist of four hypotheses BB, GB, BG, and GG specifying whether the color of X's left eye and right eye respectively are brown or green. Suppose the credal probabilities adopted by the agent are 0.4, 0.15, 0.2, 0.25 respectively. Let each element of U carry equal informational value. The rejection rules I have proposed recommend rejecting an element of U if its credal probability is less than $q/4$ where q ranges from 0 to 1. d-values for the elements of U are derivable according to procedures specified in section 6.9 as follows: 0, 0.4, 0.2, 0. If one sought to expand with maximum boldness when q = 1, the result would be the expansion $\underline{B}^i = \underline{K}$ by adding the disjunction of BB and GG.

$\underline{K}$ can be understood to be an "ordered belief set" in the sense of Dubois and Prade.[17] Under the interpretation of degrees of belief in terms of families of inductive expansion rules, the ordering of sentences in $\underline{K}$ with respect to d-value or with respect to b-value is relative to the belief state represented by $\underline{B}$. Relative to $\underline{B}$, $\underline{K}$ is the set of sentences carrying positive b-value. Equivalently, it is the inductive expansion strategy recommended when the agent is maximally bold. Once the expansion is implemented and $\underline{K}$ becomes the state of full belief, all elements of $\underline{K}$ receive the maximum b-value of 1. Thus, whereas BB $\vee$ GG receives b-value of 0.2 relative to $\underline{B}$, it receives b-value 1 relative to $\underline{K}$.

What about BB and GG? Relative to $\underline{K}$, their probabilities are 8/13 and 5/13 respectively. The rejection level is now q/2 so that the d-values are 0 and 3/13. BB is believed to degree 3/13 relative to $\underline{K}$.[18]

Thus, on the interpretation of Shackle measures as degrees of belief and disbelief generated by families of deductively cogent, caution-dependent, and partition-sensitive inductive expansion rules, the belief *state* represented by the ordered belief set $\underline{K}$ is the same as the one represented by $\underline{B}$ without an ordering of the elements of $\underline{B}$ but with an ordering of the potential answers generated by U. And once the expansion strategy relative to maximal boldness is implemented, the degrees of belief are "updated" by updating the credal probabilities and the informational value assessments and then reapplying the family of inductive expansion rules.

Dubois and Prade propose representing belief states by ordered belief sets. This method of representation makes sense when the ordering is with respect to grades of incorrigibility or "entrenchment" in the sense of Gärdenfors (1988). The belief state is a state of full belief. All sentences in the corpus representing such a state are judged certainly true and, hence, judged certainly true to the same degree. Such sentences may be evaluated differently with respect to incorrigibility or entrenchment but not with respect to certainty (Levi, 1980, ch. 3). Degrees of belief (and degrees of disbelief = degrees of potential surprise) should not be confused with degrees of entrenchment or incorrigibility. The rhetoric of Dubois and Prade suggests, however, that they think that the ordering of elements of the corpus is with respect to belief (necessity ordering) or disbelief (inverse of possibility ordering). What they mean remains, therefore, unclear.

The absence of a clear interpretation hampers the discussion of updating by Dubois and Prade and Spohn (1988). According to the account of updating of "degrees of possibility" advanced by Dubois and Prade (1992, secs. 2.2, 2.3), we should expect the update of the d-values assigned to BB and GG when their disjunction is added to $\underline{B}$ to form $\underline{K}$ to leave their d-values unchanged at 0 each. We cannot say why this should be so except by appealing to the formalism Dubois and Prade deploy.

By way of contrast, the interpretation proposed here implies that the d-values assigned to BB and GG should be different when their disjunction is added to $\underline{B}$ to form $\underline{K}$. This recommendation, unlike the one proposed by Dubois and Prade, is based on an explicit indication of the intended applications of judgments of degrees of disbelief. If the agent (relative to $\underline{B}$ or relative to $\underline{K}$ as the case may be) does fix on a level of boldness to use, the assessments of d-values do indicate what conclusions may be legitimately reached at the desired level. Neither Spohn nor Dubois and Prade have provided an account of the application of their proposals as specific as this.

8.5   Degrees of Incorrigibility

Shackle measures, as I have interpreted them, represent degrees of confidence of acceptance and rejection. Like degrees of credal probability, they apply in a significant way to hypotheses that are not in $\underline{K}$ and whose negations are not in $\underline{K}$. All elements of $\underline{K}$ carry 0 d-value and maximum b-value (according to my practice) equal to 1.

There is another interpretation of Shackle measures useful in discussing belief change. It concerns degrees of incorrigibility or degrees of "entrenchment" as Gärdenfors says. Such degrees of incorrigibility ought not to be confused with degrees of certainty or degrees of belief (Levi, 1980). Degrees of incorrigibility apply to items in the current corpus $\underline{K}$. But all items in the belief set or the current state of full belief are maximally certain in the sense that as long as they remain in the current $\underline{K}$, no distinction is made between them with respect to risks to be taken. $\underline{K}$ defines the space of "serious possibilities" in such a manner that there is no serious possibility that any item in $\underline{K}$ is false. From the point of view of the inquirer, there is no risk to be incurred in acting on the assumption that such items are true. There is no risk of error in adding them to the corpus. They are already there and risk is assessed on the assumption that everything in the corpus is true. They are fully believed.

Thus, no discrimination can be made between items in the corpus with respect to expectation determining probability. They all carry probability 1 in the sense of absolute certainty and not "almost" certainty. Moreover, they all carry maximum b-values and their negations maximum d-values. Neither credal probability nor Shackle-like degrees of belief as interpreted here can index degrees of incorrigibility or degrees of entrenchment.

Yet, some items in the corpus are more vulnerable to being given up than others when contraction is required. Truths of logic and mathematics carry maximal incorrigibility. Explanatorily powerful laws tend, on the whole, to be more incorrigible than mere correlations and descriptions of occurrences or processes tend to be more or less incorrigible depending on how much of the explanatory background in the corpus would have to be given up if such claims are abandoned. How in detail to assess degrees of incorrigibility among items taken for granted is a mare's nest, which thankfully we can bypass here. The point I mean to insist on is that degrees of incorrigibility ought not be confused with degrees of certainty. Appraisals of incorrigibility discriminate among items in the current corpus. Appraisals of degrees of certainty apply primarily to items outside the corpus.

Even though I mean to resist discussion of explanatory power, simplicity, and the like as determinants of degrees of incorrigibility, I do want to suggest that whatever these determinants are in detail they are reflections of inquirers' demands for information. When inquirers' are required to have or do have good reason to remove from their corpus of full beliefs some assumption h (i.e., to contract by removing h), they have many contraction strategies at their disposal that meet the requirement. Under such circumstances, it is a mistake to suppose that their choice of a contraction strategy reflects a concern to minimize risk of error. No risk of error can be incurred in removing any item from $\underline{K}$ – that is, no risk of *importing* a false assumption into $\underline{K}$ at that step. How can there be when no assumption true or false is being imported into $\underline{K}$ through the contraction? Nor can one say that in contraction one runs the risk of removing a true assumption rather than a false one. From the vantage point of the inquirer prior to contraction all items in $\underline{K}$ are true. If the inquirer does not think so, $\underline{K}$ is not his or her corpus.

What is true is that contraction incurs a loss of information. It is important to identify incentives an inquirer might have for suffering such a loss of information (see Levi, 1991, ch. 4); but given that, for whatever reason, h is to be removed from $\underline{K}$, the inquirer quite sensibly would want to minimize the loss of valuable information or *informational value*. And such assessments of informational value reflect the inquirer's demands for information including his or her interest in a simple explanatory and predictively powerful scheme. Such demands for information ought to determine a specification of which items in $\underline{K}$ should be removed when h is removed and would, in this sense, reflect an assessment of degrees of incorrigibility or entrenchment.

In section 2.1, I briefly sketched proposals elaborated earlier (Levi, 1991) for assessing losses of informational value in order to characterize an admissible contraction removing h from $\underline{K}$. I assumed that an inquirer's demands for information may be represented by an informational-value-determining probability function M. Given that h is to be removed from $\underline{K}$, the set of potential contraction strategies $C(\underline{K},h)$ removing h from $\underline{K}$ can be constructed by identifying first the set of *saturatable* contractions removing h from $\underline{K}$. These are the contractions removing h such that expanding them by adding ~h yields a maximally consistent corpus in $\underline{K}$ or, more generally, an element of a basic partition V. Several distinct saturatable contractions removing h yield the same maximally consistent corpus when expanded by adding ~h. Among these there will be exactly one that contains all the others and, hence, is the strongest of them. This is the so-called *maxichoice* contraction. Alchourrón, Gärdenfors, and Makinson (1985) restricted the contraction options to intersections of subsets of maxichoice contractions. In my judgment all contraction strategies ought to be considered and these are characterized as intersections of subsets of saturatable contractions (Levi, 1991, ch. 4). These contraction strategies are the ones to be evaluated with respect to informational value.[19]

The undamped informational value of a potential corpus $\underline{J}$ in K is $1 - M(\underline{J})$. Consider the set $C(\underline{K},h)$ of contractions of $\underline{K}$ by removing h. I have argued elsewhere that loss in undamped informational value cannot be the index to minimize in choosing a contraction strategy from this set but rather loss in damped informational value (Levi, 1991). Damped informational value is assessed as follows: The damped informational value of a saturatable contraction is equal to its undamped informational value. The damped informational value of the intersection of a set of saturatable contractions is equal to the damped informational value of the saturatable contraction carrying the least (damped = undamped) informational value (or the highest M-value) in the given set.[20]

To minimize loss of damped informational value is to adopt a contraction that is the intersection of a set of saturatable contractions removing h carrying highest (damped = undamped) informational value (if such contractions exist) and, hence, lowest M-value.[21] Ties in optimality are broken by choosing a contraction of this kind that is the intersection of all such saturatable contractions. The recommended contraction also carries highest damped informational value and, hence, minimizes loss of damped informational value.

Given an evaluation of potential corpora in $C(K, h)$ with respect to damped informational value, it is possible to evaluate sentences in $\underline{L}$ with respect to the degree of incorrigibility of such sentences. The incorrigibility, $in(g/M,\underline{K},h)$, of a sentence g in $\underline{L}$ relative to M, $\underline{K}$, and h is determined by looking at the potential contractions in $C(\underline{K},h)$ that do *not* contain g and determining which of these contractions carries maximum damped informational value. Obviously, the higher this assessment is the more corrigible g will be. If g is in every potential contraction-bearing positive damped informational value, it carries maximum incorrigibility, and if it is in none, it carries minimum incorrigibility. $in(g/M, \underline{K},h)$ is, therefore, defined as follows:

If there is a contraction strategy in $C(\underline{K},h)$ containing g and another that does not, $in(g/M, \underline{K},h)$ is equal to the largest damped informational value assigned a potential contraction strategy in $C(\underline{K},h)$ that does not contain g.
If g is in every member of $C(\underline{K},h)$, $in(g/M, \underline{K},h) = 1$.
If g is not in any member of $C(\underline{K},h)$, $in(g/M, \underline{K},h) = 0$.

The function $in(g/M,\underline{K},h)$ (which I shall write as $in(g)$ when no misunderstanding should arise) is defined for all sentences in $\underline{L}$ and weakly orders them with respect to incorrigibility. Indeed, I have characterized the in-function as a real valued function defined on the unit interval from 0 to 1. But any positive monotone transformation of in-function will do as well. Given our construction of the in-function from M-functions, we can derive the following conditions on in-functions (relative to $\underline{K}$ and h):

(*in*1) If g and g′ are logically equivalent, $in(g) = in(g')$.

(*in*2) $in(f \wedge g) = \min[in(f), in(g)]$.

(*in*3) t is a logical truth if and only if $in(t) = 1$ (= maximum in-value).

(*in*4) $in(g)$ or $in(\sim g)$ is 0.

(*in*5) $in(g) > 0$ if and only if g is in $\underline{K}_h^-$.

Corollary: $in(h) = 0$.

Thus, the measure of incorrigibility has some of the properties of a Shackle measure of degree of belief: (*in*1) corresponds to (b1), (*in*2) to (b2), and (*in*4) to (b4). There are two major differences. (b3) states that all and only elements of $\underline{K}$ carry maximum b-value. (*in*3) states that all and only logical truths carry maximum in-value. Only if $\underline{K}$ coincides with the set of logical truths will these two conditions agree. (b3) also implies that all elements of $\underline{K}$ carry positive b-value. This stands in contrast to (*in*5), which restricts positive in-value to elements of $\underline{K}_h^-$. The difference between these properties of b-functions and in-functions reflects their intended application. Elements of $\underline{K}$ are all maximally certain and maximally believed. They differ in how vulnerable they are to being removed when h is removed from $\underline{K}$. All and only logical truths are maximally incorrigible.

Gärdenfors's measure of degree of entrenchment (Gärdenfors, 1988, secs. 4.6–4.8.) satisfies the following postulates:

(*en*1) If g and g′ are logically equivalent, $en(g) = en(g')$.

(*en*2) $en(f \wedge g) = \min[en(f), en(g)]$.

(*en*3) t is a logical truth if and only if $en(t) = 1$ (= maximum en-value).

(*en*4) $en(g)$ or $en(\sim g)$ is 0.

(*en*5) $en(g) > 0$ if and only if g is in $\underline{K}$.

(*en*1)–(*en*4) parallel (*in*1)–(*in*4). But (*en*5) differs from (*in*5). All elements of $\underline{K}$ receive positive en-value so that there is no automatic clash between (*en*5) and (b3). (The difference between (*en*3) and (b3) remains.)

Although Gärdenfors does not do so, it is possible to derive the en-function from the same M-function that determines the in-function. Instead of considering all saturatable contractions removing h from $\underline{K}$, consider all maximal contractions of $\underline{K}$ of any kind. A maximal contraction of $\underline{K}$ is any contraction of $\underline{K}$ that is distinct from $\underline{K}$ itself and not contained in any other contraction of $\underline{K}$ other than $\underline{K}$ itself. Each one is a corpus or theory having an M-value. Every other contraction of $\underline{K}$ other than $\underline{K}$ itself can be represented as the intersection of such maximal contractions. The damped informational value of a maximal contraction of $\underline{K}$ is the same as its undamped informational value. Assign the intersection of any subset of such maximal contractions the smallest damped informational value assigned a member of the set. The degree of entrenchment of g relative to M and $\underline{K}$ may now be defined as follows:

If g is in $\underline{K}$, $en(g/M,K)$ is the maximum damped informational value assigned a contraction of $\underline{K}$ that does not contain g.

This derivation of en-functions from M-functions differs from derivations of in-functions from M-functions in the following respects:

(i) All potential contractions of $\underline{K}$ are considered and not just those that are contractions removing h. Now a potential contraction of $\underline{K}$ is the intersection of a set of maximal contractions of $\underline{K}$. Suppose we then restrict the set of contractions of $\underline{K}$ to those removing h. Maximal contractions removing h are maxichoice contractions removing h. However, it is a mistake to suppose that the only contractions removing h from $\underline{K}$ are intersections of maxichoice contractions removing h. Given any maxichoice contraction removing h from $\underline{K}$ and given any set of maximal contractions of $\underline{K}$ that do *not*, the intersection of these contractions of $\underline{K}$ is a contraction of $\underline{K}$ removing h that is not itself the intersection of a set of maxichoice contractions. Indeed, it is a saturatable but not maxichoice contraction. The set of potential contractions removing h from $\underline{K}$ is the set of intersections of all subsets of saturatable contractions.

(ii) Consequently, the ordering of maximal contractions of $\underline{K}$ with respect to damped informational value determines the ordering with respect to damped informational value both for contractions of $\underline{K}$ and contractions of $\underline{K}$ by removing h. The latter ordering is the restriction of the former to potential contractions of $\underline{K}$ removing h.

(iii) In assessing damped informational value relative to $\underline{K}$ alone, $\underline{K}$ should be considered a contraction strategy – indeed, the contraction strategy minimizing loss of damped informational value. Hence, minimum en-value should be assigned exclusively to sentences not in $\underline{K}$. In considering contraction strategies from $\underline{K}$ removing h, $\underline{K}$ is not an available contraction strategy. So the minimal loss of informational value is to be incurred by $\underline{K}_h^-$. Hence, any sentence not in this corpus is assigned minimal in-value.

Suppose $\underline{K}$ consists of the consequences of h and g or alternatively of the consequences of the following:

(a) h ∨ g.
(b) h ∨ ~g.
(c) ~h ∨ g.

The set of contraction strategies from $\underline{K}$ includes removing all members of each subset of {(a),(b),(c)}. Suppose that the consequences of {(a),(b)} (= the consequences of {h}) is a corpus carrying lower M-value and, hence, higher informational value than any other maximal contraction. en(h) > en(g). Yet, if we consider the set of contraction strategies removing h from $\underline{K}$, retaining the consequences of {(a),(b)} is not an option. There is no contraction strategy containing h and, since there is at least one containing g – namely {(a), (c)} – incurring positive loss of informational value, in(g) > in(h). This is so even though both measures and orderings are derived from the same M-function.

The optimal contraction $\underline{K}^-_h$ is easily characterized as the corpus containing all positively incorrigible sentences relative to $\underline{K}$ and h. The optimal contraction is not, however, the set of all positively entrenched sentences relative to $\underline{K}$. These are all the members of $\underline{K}$. Gärdenfors (1988) suggested that the optimal contraction consists of the closure under consequence of the intersection of $\underline{K}$ with the set of sentences g in $\underline{L}$ such that en(h) < en(~h ⊃ g). This proposal builds into the definition of optimal contraction satisfaction of the recovery postulate (Levi, 1991, p. 144). No matter what one thinks of recovery, its satisfaction ought not to be built into the definition of $\underline{K}^-_h$. To avoid this infelicity, we may say that $\underline{K}^-_h$ is the set of positively entrenched sentences minus the set of sentences carrying entrenchment at least as small as h. This set coincides with the potential contraction minimizing loss of damped informational value.

The use of one measure or the other is largely a matter of taste. The measure of incorrigibility allows one to say that in removing h from $\underline{K}$ all sentences in $\underline{K}$ carrying 0 or minimal incorrigibility are removed where degrees of incorrigibility are measured by losses of damped informational value relative to a baseline of losses we are constrained to incur by removing h rather than from the baseline of not contracting at all.

Those who favor entrenchment over incorrigibility might argue that entrenchment is preferable because, though it is relative to $\underline{K}$, it is not relative to h. But this too is a matter of taste. In contracting from $\underline{K}$ by removing h, we are already constrained by the requirement to remove h. The incorrigibility ordering relative to M, $\underline{K}$, and h is just the restriction induced by h of the entrenchment ordering relative to M and $\underline{K}$.

Regardless of taste, both incorrigibility and entrenchment are derivable from M together with $\underline{K}$ in the case of entrenchment and $\underline{K}$ and h in the case of incorrigibility. Both incorrigibility and entrenchment are Shackle-like measures satisfying two basic axioms of a Shackle measure of degrees of belief: (b1) and (b2). They differ formally from degrees of belief because Shackle b-measures of degrees of belief in their intended applications assign maximum b-value to all items in the corpus $\underline{K}$ of full belief and may assign positive b-value to sentences not in $\underline{K}$ though consistent with it. Both entrenchment and incorrigibility measures assign maximum b-value exclusively to logical truths. All and only sentences in $\underline{K}_h^-$ receive positive degrees of incorrigibility and all and only sentences in $\underline{K}$ receive positive degrees of entrenchment.

b-values are determined relative to an ultimate partition U and a trade-off represented by the index q of Q-values (determined by a master-expectation-determining probability P and the current corpus $\underline{K}$) and N-values (determined by the informational-value-determining M-function and $\underline{K}$). They reflect trade-offs between risk of error and informational value. Degrees of incorrigibility and entrenchment are determined by the M-function and $\underline{K}$. There is no trade-off between risk of error and new informational value.

Shackle measures of degrees of belief determine dual measures of degrees of disbelief or potential surprise. Measures of entrenchment or incorrigibility have formally parallel duals representing measures of degrees of corrigibility or removability.

Thus, Shackle measures can be interpreted in at least two useful ways: in terms of caution- and partition-dependent deductively cogent inductive expansion rules and in terms of damped informational value of contraction strategies. One interpretation (as an assessment of incorrigibility) plays a role in characterizing optimal contractions and, hence, in all of the Ramsey tests I have considered in this essay. The other interpretation plays a role in characterizing inductively extended expansions and, hence, in inductively extended Ramsey tests. Although the formal structures are the same, the applications are clearly different.

That is all to the good. There is no need to choose one interpretation over the other. They are both useful for their own rather different purposes. However, we should avoid confusing the two applications or treating them as one by focusing on formal similarities while neglecting the differences in the intended interpretations.

This means, for example, that when Dubois and Prade (1992) discuss "ordered belief sets" they should make plain that their measures are *in*-functions or *en*-functions and not b-functions. Every sentence in a corpus representing a given belief state is maximally certain and, hence, carries maximum b-value equal to that carried by logical truths. If this were not so, a belief set or corpus would not represent the state of full belief serving as the standard for serious possibility.

Yet elements of a corpus do differ with respect to their entrenchment or incorrigibility. Sentences counted as equally and maximally certain may, nonetheless, differ with respect to their vulnerability to being given up. Items in a belief set differ in their entrenchment even though they are equally certain.

Dubois and Prade do notice that logical truths are the only items that are maximally entrenched whereas other items may carry maximal certainty even though they are not maximally entrenched. They fail to see this as an indication that measures of degrees of certainty and measures of degrees of entrenchment have different intended applications. To the contrary, they defend their necessity or certainty measures (the analogues of b-measures) as measures of entrenchment in contexts "where some non-tautological sentences must be protected" (p. 147). And they claim that their "ordered belief sets" generalize Gärdenfors's notion of a belief set (p. 148).

Dubois and Prade are by no means alone in conflating degrees of certainty with grades of incorrigibility. Foundationalist epistemologists confound certainty with incorrigibility as a matter of philosophical commitment. Conflating degrees of incorrigibility with degrees of certainty is symptomatic of a lingering commitment to the foundationalist tradition.

Needless to say, once entrenchment or incorrigibility is distinguished from degree of belief as distinct interpretations of the Shackle formalism, it should become even clearer that constructing an algorithm for revising evaluations of either kind with the acquisition of new evidence makes little sense.

9 Normality and Expectation

9.1 Defaults and Normal Worlds

Defaults as I have proposed to understand them formulate a criterion for reducing the space of serious possibilities relative to $\underline{B}$ to a space of serious possibilities relative to $\underline{B}^i$ by identifying which maximal consistent expansions of $\underline{B}$ are maximal consistent expansions of $\underline{B}^i$.

According to the proposals I have offered, the consistent extensions of $\underline{B}$ considered to be maximal are expansions of $\underline{B}$ by adding elements of an ultimate partition U. But they need not be maximally consistent extensions of $\underline{B}$ with respect to the language $\underline{L}$ or, if one likes, logically possible worlds that are seriously or epistemically possible relative to $\underline{B}$. The maximal consistent extensions of $\underline{B}$ relative to U constitute a partition of the set of seriously possible worlds in $\underline{L}$ relative to $\underline{B}$. To each element of U (or the corresponding extension of $\underline{B}$ that is maximal relative to U) there corresponds a unique set of possible worlds. Consequently, we could regard the elements of U as "states" in the sense of Kraus, Lehmann, and Magidor (1990, sec. 3.4) and define a "labeling" function from states to sets of possible worlds. Or, we can take the elements of U to be surrogates for possible worlds and dispense with labeling functions. This is the procedure I shall follow.

I contend that the partition sensitivity thereby introduced reflects the demands for information of the inquirer in addressing the problem under consideration. In the context of the given inquiry, the investigator is not interested in stronger (consistent) information than that represented by elements of U. An inquirer interested in birthrates in Iowa in 1992 simply does not regard information about the dog population in Arkansas in 1863 as a potential answer to his or her question. Of course, the inquirer may at some stage in the inquiry have good reason to want to refine his or her U so as to allow for such potential answers and an account of how such a change should be undertaken ought to be developed (see Levi, 1984a, ch. 7). But in the context of the given inquiry, it is not part of the inquirer's concern to worry about dogs in Arkansas in the nineteenth century.

According to Kraus, Lehmann, and Magidor (and here they reflect a widespread attitude among students of nonmonotonic reasoning), the possible worlds or states in U relative to $\underline{B}^i$ are regarded as the "normal" worlds or states in the set of possible worlds or states relative to $\underline{B}$. Alternatively, $\underline{B}^i$ is the union of all expansions $\underline{B}^+_x$ where x is an unrejected element of U according to the system of defaults adopted.

If $\underline{B}$ is expanded by adding h consistent with $\underline{B}^i$, some elements of U and, perhaps, some normal elements of U relative to $\underline{B}^i$ will be eliminated through ordinary unextended expansion. But some normal elements will survive (since h is consistent with $\underline{B}^i$). Given that the system of defaults remains the same, the surviving normal elements relative to $\underline{B}$ and only these remain normal according to the system of defaults relative to $\underline{B}^+_h$. Hence, the inductive expansion $\underline{B}^{+i}_h$ is the intersection of $\underline{B}^+_x$ for just those x's in U that survive rejection.

According to this way of thinking, when h is consistent with $\underline{B}^i$, $\underline{B}^{+i}_h$ will be identical with $(\underline{B}^i)^+_h$.

What about cases where g is consistent with $\underline{B}$ but inconsistent with $\underline{B}^i$ so that expanding $\underline{B}$ by adding g eliminates all the elements of U identified as normal by the defaults? If there is some way in which the defaults can determine a subset of the elements of U consistent with $\underline{B}^+_g$ that are less abnormal than any others, $\underline{B}^{+i}_g$ is supposed to be the intersection of $\underline{B}^+_{g \wedge x}$ for all x that comprise the least-abnormal elements of U relative to $\underline{B}^+_g$ according to the defaults.

On the view under consideration, defaults determine which elements of U are normal and which are the least-abnormal elements consistent with $\underline{B}^+_g$ for g inconsistent with $\underline{B}^i$. The inductive expansion function thus defined is ampliative but constant or permutable, at least with respect to sentences that are equivalent given $\underline{B}$ to the intersection of some subset of U.

Discriminations between elements of U with respect to degrees of normality or abnormality are based on defaults of the form "A's are normally B's." Shoham (1987), Makinson (1989), and Kraus, Lehmann, and Magidor (1990) offer systematic accounts of nonmonotonic or default reasoning beginning with evaluations of states or worlds with respect to normality. They explore the impact on nonmonotonic reasoning of structural constraints imposed on assessments of degrees of normality.

Kraus, Lehmann, and Magidor (1990, pp. 181–2) introduce the idea of a *cumulative model* consisting of an arbitrary nonempty set of states, a binary relation between members of the set called a "preference relation," that satisfies a smoothness condition. A cumulative model may be interpreted as consisting of an ultimate partition U (the set of states) relative to $\underline{B}$ and a preference relation with respect to normality defined for elements of U.

The smoothness condition guarentees that, for every $U' \subseteq U$ and every $x \in U'$, either x is minimally abnormal (no other member of U' is preferred to x) or there is a minimally abnormal element y of U' strictly less abnormal than x.

Given a cumulative model, $h \mathrel{|\!\sim} g @ \underline{B}$ if and only if every minimally abnormal element in the subset of U that together with $\underline{B}$ entails h also (together with $\underline{B}$) entails g.

Kraus, Lehmann, and Magidor (1990) and Lehmann and Magidor (1992) examine types of nonmonotonic inferential relations that are characterized by cumulative models and various species of cumulative models where the preference relations are subject to additional constraints.

Of particular interest are the *ranked* models (Lehmann and Magidor, 1992, p. 21). A ranked model is a cumulative model where the preference relation yields a strict partial order of the elements of U with respect to abnormality and satisfies a "modularity" condition. Modularity amounts to the existence of a total ordering of the elements of U that preserves the strict partial ordering with respect to abnormality and is such that ranking elements of U that are noncomparable with respect to abnormality as equi-preferred yield an equivalence relation. Lehmann and Magidor show that the nonmonotonic inferential relations characterized by ranked models are *rational* in the sense that they satisfy the following conditions:

(*Closure*) If $h \mathrel{|\!\sim} g_i$ for all g_i in $\underline{A}$ and $\underline{A} \vdash f$, then $h \mathrel{|\!\sim} f$. [Corresponds to (K*1)]
(*Reflexivity*) $h \mathrel{|\!\sim} h$. [Corresponds to (K*2)]
(*Weak Conditionalization*) If $h \mathrel{|\!\sim} g$, then $T \mathrel{|\!\sim} h \supset g$. [Corresponds to (K*3)]
(*Weak Rational Monotony*) If $\not\mathrel{|\!\sim} \sim h$ and $\mathrel{|\!\sim} h \supset g$, then $h \mathrel{|\!\sim} g$. [Corresponds to (K*4)]
(*Left Logical Equivalence*) If $\vdash h \equiv g$ and $h \mathrel{|\!\sim} f$, then $g \mathrel{|\!\sim} f$. [Corresponds to (K*6)]
(*Conditionalization*) If $h \wedge g \mathrel{|\!\sim} f$, then $h \mathrel{|\!\sim} g \supset f$. [Corresponds to (K*7)]
(*Rational Monotony*) If $h \not\mathrel{|\!\sim} \sim g$ and $h \mathrel{|\!\sim} f$, then $h \wedge g \mathrel{|\!\sim} f$. [Corresponds to (K*8)]

These seven conditions contain redundancy but are equivalent to the formulation in Lehmann and Magidor (1992). I list them in this form so as to identify the postulates for *unextended* AGM revision from which these conditions may be derived with the aid of the translation manual offered by Makinson and Gärdenfors (1991) for interpreting unextended, belief-contravening, nonmonotonic pseudoinferential relations in terms of AGM revision.

A rational inferential relation is consistency preserving if and only if it satisfies the following condition as well:

(*Consistency Preservation*) If h |~ ~T, then h ⊢ ~T. [Corresponds to (K*5)]

In the subsequent discussion, we shall consider consistency preserving rational inferential relations. In this case, h |~ g @ $\underline{B}$ if and only if g is in $\underline{B}_h^*$ where * is AGM revision.[1]

For convenient reference, I mention the following two important conditions that rational inferential relations also satisfy:

(*Cut*) If h |~ g and h ∧ g |~ f, then h |~ f.
(*Cautious Monotony*) If h |~ g and h |~ f, then h ∧ g |~ f.

The fact that rational, consistency-preserving inductively unextended pseudoinferential relations can be interpreted in terms of AGM revision seems to be insufficient for the purpose of understanding the inferential relations studied by Kraus, Lehmann, and Magidor and Lehmann and Magidor. These relations are intended to be inductively extended. In section 5.4, a manual for interpreting nonmonotonic pseudoinferential relations in terms of expansive transformations was provided that includes unextended expansion, unextended AGM revision (the case covered by Makinson and Gärdenfors), unextended Ramsey revision, and their inductively extended counterparts. If we replace the unextended pseudoinferential relation |~ in the postulates by the inductively extended pseudoinferential relation ‖~, we cannot derive the postulates from the Makinson–Gärdenfors manual and the conditions on the corresponding AGM revision.

However, the postulates for consistency-preserving, rational, inductively extended nonmonotonic inferential relations can be derived from the characterization of pseudoinferential relations in terms of expansive transformations given in section 5.4 provided (i) the expansive transformation used is inductively extended AGM revision *i and (ii) the underlying inductive expansion function i is permutable.

Consider, for example, the inductively extended version of weak conditionalization. The Gärdenfors–Makinson translation manual could be used to derive this principle from (K*i3) of section 5.7. But (K*i3) is not a valid condition on inductively extended AGM revision (or, for that matter, Ramsey revision) because of the ampliativity of inductive expansion.

The manual in section 5.4 provides a way of deriving weak conditionalization from inductively extended AGM revision. The condition from which weak conditionalization is derived is (K*i3m). But (K*i3m) is not a valid requirement on all inductively extended AGM revisions.

It is, however, a constraint on inductively extended AGM (and Ramsey) revisions based on permutable inductive expansion functions. (K*i3m) states that $\underline{B}_h^{*i}$ is a subset of $\underline{B}_h^{+i}$. If i is permutable, this condition becomes equivalent to the claim that $(\underline{B}^i)_h^*$ is contained in $(\underline{B}^i)_h^+$. This is a substitution instance of (K*3) imposing a constraint on *unextended* AGM revision. This substitution instance of (K*3) and permutability suffice to yield (K*i3m). In the case of inductively extended AGM revision, permutability is necessary and sufficient for the joint satisfaction of (K*i7) and (K*i8) and, hence, of (K*i3m) and (K*i4m). (See the main thesis of section 5.8.) The translation manual of section 5.4 then suffices to yield the postulates for consistency preserving and rational inductively extended pseudoinferential relations. The kind of inductively extended revision used in the translation is inductively extended AGM revision. Indeed, neither inductively extended expansion nor inductively extended Ramsey revision will yield all requirements on consistency preserving inductively extended rational inferential relations.

We may, therefore, claim that among the expansive transformations all and only inductively extended AGM revisions based on permutable inductive expansion functions generate inductively extended rational pseudoinferential relations that seem to capture the intended domain of application involved in default reasoning. (The special case where the inductive expansion function is the identity function covers the case where the AGM revision is unextended. This too yields rational inferential relations as Gärdenfors and Makinson show because the identity function is permutable.)[2]

Why should permutability be endorsed? I do not think it should be. I have urged that inductive expansion functions should be derivable using deductively cogent, caution-dependent, and partition-sensitive inductive expansion rules of the sort discussed in Chapter 6. If the suggestion is taken together with the necessity of the assumption of permutability for the rationality of inductively extended pseudoinferential relations, the only way to achieve the condition of rationality is to use bookkeeping rules with maximum boldness.

I have argued in section 6.10 that, where U is finite, bookkeeping with maximum boldness yields excessively strong conclusions from a presystematic perspective. In the appendix to Chapter 6, I argue, in effect, that in some cases where U is infinite, bookkeeping with maximum boldness can lead to inconsistency in inductive expansion and, hence, to violations of consistency preservation.[3]

The main philosophical point advanced in Lehmann and Magidor (1992) is that rational inferential relations are to be favored over the alternative inferential relations rehearsed in Kraus, Lehmann, and Magidor (1990). Bookkeeping with maximum boldness, however, leads to unacceptable, overly strong results and, hence, argues against using such inductively extended relations.

On the other hand, if one lets q be less than 1 with bookkeeping or does not use bookkeeping at all, not only does rationality fail, but none of the weaker nonmonotonic inductively extended inference relations surveyed by Kraus, Lehmann, and Magidor are inductively extended inference relations. Closure, reflexivity, consistency preservation, and left logical equivalence hold for all inductively extended inference relations. When bookkeeping is not used, all of the other conditions can fail. When bookkeeping is used, Cut may be added to the conditions from the list that are satisfied; but nothing else may be added. When q is less than 1, inductively extended inference relations are not "cumulative" (they do not satisfy cautious monotony). Only when q = 1 are such relations cumulative in which case the relations are rational as well.

Thus, when nonmonotonic inference is understood to be inductive and ampliative expansion is grounded in caution-dependent, deductively cogent, and partition-sensitive inductive expansion rules, *none* of the kinds of inference relations studied in Kraus, Lehmann, and Magidor (1990) and Lehmann and Magidor (1992) are satisfactory.

Clearly something has gone awry. The ideas found in Kraus, Lehmann, and Magidor (1990) and Lehmann and Magidor (1992) for classifying nonmonotonic inference relations are intriguingly elegant. Moreover, their work does illuminate the *belief-contravening* nonmonotonic pseudoinference relations deployed in suppositional reasoning. In such applications, the preference relation can be derived from an ordering of saturatable contractions of some corpus $\underline{B}$ which generates an entrenchment ordering. This, however, is an ordering with respect to entrenchment and not with respect to normality or expectedness.

However, as I have been insisting since Chapter 6, default reasoning is ampliative inductive reasoning and not belief-contravening suppositional reasoning. What has gone wrong, I believe, is that Kraus, Lehmann, and Magidor, who quite clearly are thinking of ampliative inductive reasoning have, nonetheless, structured it along the lines of belief-contravening suppositional reasoning. The pivotal step at which the slide from belief-contravention to inductive reasoning takes place is to be found in the way normality and the ordering with respect to normality are understood.

To explain this, consider what an ordering with respect to normality of the elements of U might be.

The inductive expansion rules I have explored provide us with two kinds of candidate orderings to consider:

1. Orderings of elements of U with respect to expected epistemic utility modified by equating elements of U for which the expected utility $Q(x) - qN(x)$ is nonnegative as maximally preferred.

2. Orderings with respect to degrees of potential surprise (d-values), with respect to degrees of Shackle-type belief (b-values) or with respect to the function $q(x) = 1 - d(x)$ corresponding to Zadeh's (1978) notion of possibility.

Where bookkeeping with $q = 1$ is used and U is finite (I shall focus on finite cases for the most part), the elements of U that survive rejection are those for which $Q(x) - N(x)$ or $Q(x)/N(x)$ is a maximum.[4] When bookkeeping with maximum boldness is used, therefore, maximum $Q(x)/N(x)$ represents maximum normality. When N assigns equal value to all elements of U, the most normal states are those carrying maximum probability.

When bookkeeping with $q = 1$, the elements of U carrying maximum Q/N–values are also those that carry 0 surprise or d-value and, in this sense, are least abnormal.

Thus, both considerations of expected utility and surprise single out the same set of normal states. The question arises whether ordering elements of U with respect to Q/N or with respect to d or both are acceptable orderings with respect to normality (abnormality).

Although the maximally normal states coincide with the minimally surprising, the normality order is not precisely the reverse of the surprise order. Example 1 of section 7.1 illustrates the point. If the probability distribution given in Table 7.1 is compared with the d-function for bookkeeping given by Table 7.3, ~r~f~b is the most normal element of U according to both. rf~b is the second-most-normal according probability. ~rf~b and ~r~fb are third-most-normal according to probability. The ranking according to the d-function ranks rf~b together with the third-most-normal elements with respect to probability as second most normal. Only one of these orderings can serve to represent normality or abnormality in degree.

To decide which of these two rankings is a preference ranking with respect to normality or abnormality, we need to consider what happens when the most normal element of U is eliminated by expanding the background theory $\underline{B}$ and then forming the inductive expansion via bookkeeping with $q = 1$. That is to say, form $\underline{B}_a^+$ where a = ~(~r~f~b). The inductive expansion rule applied to this case where the N-function over the truncated U remains uniform is clearly going to yield rf~b as the most normal among the surviving elements of U. By reiterating the procedure, we easily obtain the result that the ordering with respect to $Q(h)/N(h)$, which, in this case, is the ordering with respect to $Q(h)$, yields the desired results. This conforms with the claims made about prototypicality in section 6.2.

Consider, however, what happens when the index of boldness q is less than 1 and bookkeeping is applied. Suppose, in particular, that $q = .44$. Table 7.3 implies that the four elements of U including the top ranked in d-value and the three second ranked go unrejected. Moreover, if the boldness is increased, only the top ranked in d-value survives unrejected. Similarly the ordering with respect to Q or Q/N specified in Table 7.1 specifies that the most probable, second-most-probable, and the two third-most-probable elements of U go unrejected. Neither ordering is an ordering with respect to normality since the most normal elements have different rankings in both systems of "preferences."

Still we might construct new orderings by equating all unrejected elements of U when q = 0.44 with respect to normality and otherwise preserving the ordering with respect to d-value (or q-value) and the ordering with respect to Q/N. But the orderings constructed by these lopping-off techniques cannot serve as the preference orderings needed to obtain Kraus, Lehmann, and Magidor ranked models that agree with the inductive expansion rules. Suppose, for example, that the disjunction of rf~b and ~r~f~b is added to $\underline{B}$ by ordinary expansion and then an inductive expansion is made via bookkeeping. rf~b will be eliminated. It should, due to this consideration, be less normal than ~r~f~b. But in making the inductive expansion $\underline{B}^i$, these were ranked as equally normal.

There is, in this case, no preference for normality relation defined on the elements of this finite U with the inductive expansion rule as specified that can yield a cumulative model and, hence, any other kind of model of the sort canvassed by Kraus, Lehmann, and Magidor including the ranked models of Lehmann and Magidor. The lopping-off technique used to obtain a most normal set of states ignores differences that become relevant as additional information is added to $\underline{B}$.

The need to use the lopping-off technique is even more urgent when bookkeeping is not used. It is then required even when q = 1. However, the distinctions suppressed by lopping-off become relevant when new information of the right sort is added.

The upshot is that, except for the case where bookkeeping is used with q = 1, there is no preference relation over the elements of U that can capture the preference relation required by the Kraus, Lehmann, Magidor modeling of nonmonotonic inferential relations.

According to the view I have been advocating, the "preference" ranking that is relevant to inductive expansion is not an ordering of elements of U or of states alone but of all inductive expansion strategies. These are representable by subsets of U that survive rejection. The preference represents an evaluation of expected (epistemic) utility as explained in Chapter 6. It is, to be sure, assumed that rejecting all but one element of U is an inductive expansion strategy so that the overall preference for inductive expansion strategies does, indeed, induce a preference over elements of U; but a most preferred inductive expansion strategy may and, in general, will exist that is different from any of these. Indeed, when the Q and N distributions are numerically determinate and U is finite, any inductive expansion strategy consisting of a subset of expansions optimal among those adding exactly one optimal element of U must be at least as good and possibly better than an optimal element of U. Even when bookkeeping is deployed, this will be so. Preferences among elements of U are not being maximized. Preferences among expansion strategies are.[5]

As pointed out earlier, however, in the case where we are considering contracting from a corpus $\underline{K}$ by removing ~h and then expanding by adding h, it is possible to characterize the process relative to an ordering of "maximal" contractions in the sense of saturatable contractions. Such saturatable contractions may serve as the "states" over which preferences are defined in cumulative models in general and ranked models in particular. I surmise that even though Kraus, Lehmann, and Magidor (1990) and Lehmann and Magidor (1992) clearly are interested in ampliative reasoning, the preference relations they study seem better suited to representing entrenchment or incorrigibility than ordering with respect to normality.

This argument depends, of course, on the idea that one eliminates all but the most normal worlds or states consonant with the information being expanded by induction. That is the idea utilized by Kraus, Lehmann, and Magidor. We could, however, view things differently. We could think of grades of normality as something to be *satisficed*. Then either the d-function or q-function ordering can be utilized as an assessment of normality. But the results obtained will not yield a rational nonmonotonic inferential relation or, indeed, even a cumulative one except when bookkeeping with q = 1 is deployed.

The idea of satisficing is closely related to the notion of adding new information (that need not be restricted to elements of the ultimate partition but includes disjunctions of subsets of such elements) to an initial corpus if its Shackle degree of belief is high enough. In that setting, it seems more comfortable to speak of degrees of belief or, perhaps, degrees of disbelief (potential surprise). Moreover, it offers an alternative way to broach the question of obtaining rational inferential relations.

In Chapter 8, it was argued that measures of entrenchment and incorrigibility are, like measures of potential surprise and degrees of belief, Shackle-like. Perhaps there is a way of tying these features together that can yield a defense of rational nonmonotonic implication relations for inductive expansion.

The perspective just mentioned is developed by Gärdenfors and Makinson (1993). Their ideas will be explored in section 9.2.

9.2   Expectations and Inductive Expansion

Makinson and Gärdenfors (1991) propose a method of translating nonmonotonic implications into AGM revisions. The prescription is to replace expressions of the form x |~ y @ $\underline{K}$ by expressions of the form y ∈ $\underline{K}_x^*$ where * is an AGM revision. I have been using their manual throughout this book and have adapted it to accommodate revisions according to Ramsey and inductively extended revisions in order to explore the properties of nonmonotonic inferential relations both when they are unextended pseudoinferential relations representing belief-contravening suppositional reasoning and inductively extended strictly inferential relations capable of characterizing genuine inductive inference.

The aim of the 1993 essay by Gärdenfors and Makinson appears to be to explicate the idea of default reasoning that sometimes is used to constrain nonmonotonic implication relations by exploring the recommendations made by default reasoners concerning what is to count as legitimate nonmonotonic reasoning.

Gärdenfors and Makinson begin by drawing a distinction between "firm beliefs" that an agent has at a time and "expectations."

The fundamental idea is that when we reason, we make use of not only the information that we firmly believe, but also the expectations that guide our beliefs without being quite part of them. Such expectations may take diverse forms, of which the two most salient are *expectation sets* and *expectation relations*. (p. 2)

Expectation sets are informally characterized as

propositions, drawn from the same language as our firm beliefs and indeed not differing from them in kind but at most in the ways in which we are prepared to use them. So understood, expectations include not only our firm beliefs as limiting case, but also propositions that are regarded as plausible enough to be used as a basis for inference so long as they do not give rise to inconsistency. (p. 2)

Thus, given a corpus $\underline{B}$ of firm beliefs, the expectation set $\underline{K}$ is taken to the set of propositions that are "plausible enough" relative to $\underline{B}$ "to be used as a basis for inference so long as they do not give rise to inconsistency." $\underline{K}$ contains $\underline{B}$ (whose elements are maximally "plausible" relative to $\underline{B}$) and both may be taken to be closed under logical consequence.

Given the informal characterization of expectation sets, it would appear that $\underline{K}$ has the formal properties of an inductive expansion of $\underline{B}$ – that is, $\underline{K} = \underline{B}^i$ for some inductive expansion rule that, informally speaking, appeals to some satisficing measure of plausibility.

One qualification must be made to this observation. The expectation set $\underline{K}$ behaves exactly like the corpus $\underline{B}$ of full beliefs to the extent that the elements of $\underline{K}$ may be used as a "basis for inference." $\underline{K}$ can be used in this way, however, as long as it does not "give rise to inconsistency."

This remark is rather puzzling. The corpus $\underline{B}$ of firm beliefs cannot be used as a basis for reasoning any more than the expectation set $\underline{K}$ can when it "gives rise to inconsistency." Adding x inconsistent with $\underline{B}$ to $\underline{B}$ requires that either elements of $\underline{B}$ be removed, x be removed, or both to secure a consistent basis for inference. If one's basis for reasoning has changed from $\underline{B}$ to $\underline{K}$, precisely the same observation may be made.

In my judgment, the distinction between firm beliefs and expectations would have been drawn in a more useful way by saying that, when $\underline{B}$ is the corpus of firm (= full) beliefs, a proposition x is an expectation in $\underline{K}$ when it is plausible enough to add it to the corpus of firm beliefs via an inductive expansion. Relative to $\underline{B}$, x is not a firm or full belief. But once inductive expansion is implemented, $\underline{K}$ and not $\underline{B}$ becomes the corpus of firm beliefs.

There is another way to draw the distinction that is rather different. It is clear that plausibility is alleged to come in degrees – a point to which I shall return shortly. Instead of saying that $\underline{K}$ is the corpus of propositions plausible enough relative to $\underline{B}$, it is the corpus that results if one considers all propositions that are plausible to some positive degree where propositions carrying 0 degree of plausibility are understood to be those that cannot be plausible enough to be added via induction to $\underline{B}$. On this reading, the corpus of expectations $\underline{K}$ is the corpus of *positive* expectations.

Notice that the corpus of positive expectations is the strongest or largest corpus the inquirer could legitimately adopt relative to $\underline{B}$ and the evaluation of hypotheses with respect to plausibility by setting a standard of what it is to be plausible enough. It appears tantamount to the corpus the inquirer would adopt were he or she maximally bold in making inductions. But the inquirer might not be so bold as that. He or she might have standards of plausibility that are more stringent. Even so, prior to inductive expansion from $\underline{B}$, the inquirer might usefully consider the range of inductive expansions at various levels of boldness and the corpus of positive expectations would constitute an explicit demarcation between inductive expansions that are starters from those that are nonstarters.

Thus, an important difference between the interpretation of $\underline{K}$ as the expectation set and the corpus of positive expectations according to my proposed usage is that the former construal presupposes a recommendation that the inquirer be maximally bold in making inductive expansions and an interpretation that has no such implication.

Under *both* interpretations, the elements of $\underline{K}$ are *not* firm or full beliefs as long as inductive expansion from $\underline{B}$ with maximum boldness is not implemented. The inquirer may want to collect more data or other information x through experiment, observation, and testimony from others prior to making an induction. The result would be an expansion of $\underline{B}$ but not via induction. The new corpus of full belief would be $\underline{B}_x^+$ and induction would be based on that corpus. Or the inquirer might suppose for the sake of the argument that x is true and seek to determine the inductively extended expansion $\underline{B}_x^{+i}$ with maximum boldness or conceivably at other levels of boldness.[6]

Adopting the suppositional version of this idea, we can define the inductively extended nonmonotonic inferential relation x $\|\sim$ y @ $\underline{B}$ as holding if and only if y is a logical consequence of $\underline{B}_x^{+i}$ where the transformation +i is understood to operate at some specific level of boldness. According to the approach of Gärdenfors and Makinson, the level of boldness is taken to be the maximum allowable and unless otherwise specified, I shall subsequently understand $\underline{B}^i$ to be the corpus adopted at this level of boldness.

Gärdenfors and Makinson formulate their approach differently. They understand x $\|\sim$ y @ $\underline{B}$ to hold if and only if y is a logical consequence of K_x^* $= (\underline{B}^i)_x^*$ where $*$ is unextended AGM revision. That is to say, they tacitly take for granted that x $\|\sim$ y @ $\underline{B}$ if and only if x $\mid\sim$ y @ $\underline{K}$.

This understanding is more than a mere definition. It embodies the substantive assumption that when x is consistent with $\underline{B}$, $\underline{B}_x^{+i} = (\underline{B}^i)_x^*$. This holds in general only if inductive expansion is permutable. Permutability, as we have seen, does presuppose maximum boldness in inductive expansion from $\underline{B}_x^+$.

$\underline{B}^i = \underline{K}$ is the corpus of expectations or the corpus of positive expectations relative to the firm beliefs $\underline{B}$. $\underline{K}$ is then the inductive expansion of $\underline{B}$ to be recommended if the inquirer were maximally bold and added all hypotheses to $\underline{B}$ that carry some positive degree of plausibility.

If we are not going to insist that inductive expansion should be implemented only with maximum boldness where all that is required is that what is added should be plausible to some positive degree, we should not rest content with a distinction between what is positively plausible and what is not. We need an account of grades of plausibility that requires that sentences are added to the firm beliefs when they are "plausible enough" according the standard of boldness that is adopted.

Gärdenfors and Makinson (1993, sec. 3.1) introduce the notion of an expectation ordering over sentences in $\underline{L}$ that satisfies the following three conditions:

(E1) The weak ordering relation $\le$ with respect to expectation is transitive.

(E2) If x $\vdash$ y, x $\le$ y.

(E3) For every x and y, x $\le$ x $\wedge$ y or y $\le$ x $\wedge$ y.

(E2) implies the reflexivity of ≤. (E3) implies its completeness so that ≤ weakly orders the sentences in <u>L</u>. Because the ordering with respect to expectation is relative to the corpus of firm or full beliefs <u>B</u>, (E2) should be strengthened to say that if <u>B</u>, x ⊢ y, x ≤ y. This insures that not only do logical truths carry maximal expectation but so do sentences in <u>B</u>. It is also easy to see that all sentences inconsistent with <u>B</u> carry minimal expectation.

Gärdenfors and Makinson (1993, sec. 3.2) show that this ordering with respect to expectations relative to <u>B</u> can be represented by an order preserving "belief scale" which can by convention be taken to be real numbers in the closed interval from 0 to 1. The belief function from sentences to real numbers thereby generated is a Shackle measure of degree of belief. Unlike the b-functions, as I have interpreted them, the "belief valuations" of Gärdenfors and Makinson are not derived from a family of deductively cogent, caution-dependent, and partition-sensitive inductive expansion rules but from a primitive expectation ordering satisfying (E1) – (E3) relativized to <u>B</u>. But a b-function derived from the expectation ordering relative to <u>B</u> can be readily understood as a satisficing measure of inductive support. This conforms to the idea that b-values are indices of plausibility, as Gärdenfors and Makinson put it in their informal presentation, and that the expectation set consists of those sentences that are plausible enough. Nothing is lost of significance to this discussion by restricting attention to derivations of b-functions from expectation orderings over sentences in <u>L</u> that are equivalent, given <u>B</u>, to assertions that exactly one of a given subset of elements of an ultimate partition U is true.

One can specify a family of deductively cogent, caution-dependent, and partition-sensitive inductive expansion rules by requiring a threshold value in the unit interval such that all and only sentences in <u>L</u> whose b-value is above that level are in the inductive expansion of <u>B</u>. Because there are, in general, many (infinitely many) b-functions defined on the real interval between 0 and 1 that preserve the expectation ordering, one must be careful in specifying the threshold levels of acceptance relative to the particular b-function one has adopted and to adjust threshold levels when the representation is changed from one b-function to another order preserving one. That is because the b-function is obtained from the expectation ordering and is then used to define a family of inductive expansion rules rather than deriving the b-function from an independently specified family of caution-dependent, deductively cogent, and partition-sensitive expansion rules.

Suppose, however, that the inductive expansion principle is used to generate a family of caution-dependent, deductively cogent, and partition-sensitive expansion rules. Such a family of inductive expansion rules recommends maximizing $Q(x) - qM(x)$. Here the degree of boldness q is the ratio $(1 - \alpha)/\alpha$ where α takes values from 0.5 to 1. $b(h) = d(\sim h) = 1 - q(\sim h)$

and the threshold b-value is equal to $1 - q$ where q is the index of boldness. In this way, the index of boldness has been standardized by the derivation from the expected epistemic utility function to be maximized. Some of the arbitrariness in the way b-function is derived from an expectation ordering has been removed.

My reservation with the Gärdenfors–Makinson procedure depends on more than this relatively minor technical point. The more serious complaint is that it fails to provide an interpretation of the expectation ordering just as Shackle failed to provide an interpretation of his measures of potential surprise and the corresponding d-functions. The formal structure of the ordering is specified; but, as was noted, there are several possible applications of b-functions and of formal structures like expectation orderings. As Gärdenfors and Makinson themselves acknowledge, expectation orderings are to be distinguished from entrenchment orderings (1993, sec. 4.) However, even there, their distinction rests primarily on structural differences in the two orderings rather than on a specification of the differences in interpretation required by the contrast in the intended applications. (See sections 8.4–8.5 for an elaboration of the difference.)

Nonetheless, whether or not expectation orderings that are not derived from families of inductive expansion rules are taken as primitive, it is true that an inductive expansion of $\underline{B}$ relative to a given threshold of b-value (that increases with decreasing boldness or increasing caution) can be defined as consisting of all sentences whose b-value is above the threshold.

Gärdenfors and Makinson recommend a standard threshold when inductive expansion is from $\underline{B}_x^+$ for x consistent with $\underline{B}$. They do so by specifying necessary and sufficient conditions for $x \parallel\sim y @ \underline{B}$, which holds if and only if y is in $\underline{B}_x^{+i}$. (1993, sec. 3.1, definition 3.1). I present it here in a version taking into account explicitly the relativization to the corpus of firm beliefs and x consistent with $\underline{B}$.

$(C|\sim) \ x \parallel\sim y @ \underline{B}$ iff $y \in Cn(B \cup \{x\} \cup \{z: \sim x < z\})$.

$(C|\sim)$ implies that $\underline{B}_x^{+i}$ should consist of the logical consequences of $\underline{B}$, x, and all those sentences whose expectations are greater than that of $\sim$x. Since the sentences of $\underline{B}$ perforce carry maximum expectation and $\sim$x is not implied by $\underline{B}$, this is tantamount to the logical consequences of $\underline{B}$ and the set of sentences whose expectations are greater than $\sim$x.

In the special case where x is a consequence of $\underline{B}$, $(C|\sim)$ determines the inductive expansion $\underline{B}^i$ of $\underline{B}$. It consists of all sentences carrying greater expectation relative to $\underline{B}$ than $\sim$T – that is, the minimum expectation. Stated in terms of b-values, $\underline{B}^i$ is the set of sentences carrying positive b-value relative to $\underline{B}$. In this case, the threshold b-value is set at $0 = 1 - q$. $q = 1$ and this is tantamount to maximum boldness.

It is easy to see that $(C|\sim)$ also gives $\underline{B}^{+i}_x$ for x consistent with $\underline{B}$ at maximum boldness. Expanding $\underline{B}$ is to take the consequences of $\underline{B}$ and x. Inductively expanding that corpus $\underline{B}^+_x$ invokes an expectation ordering relative to $\underline{B}^+_x$. $(C|\sim)$ implies that this ordering is the same as the ordering relative to $\underline{B}$ except that those sentences whose expectations are not greater than ~x are set at the minimum 0 b-value. The corpus of positive expectations relative to $\underline{B}^+_x$ is then the set specified by $(C|\sim)$.[7]

But there is another way to present the results of using $(C|\sim)$ equivalent to this one. This is the one Gärdenfors and Makinson explicitly deploy. The consequences of the union of $\underline{B}$ and the set of sentences carrying greater expectation than ~x is a subset of $\underline{K} = \underline{B}^i$. It is a contraction of $\underline{K}$ removing ~x from $\underline{K}$ if it is present.

In determining this contraction, the expectation ordering is being used as if it were an entrenchment ordering of the elements of $\underline{K}$ rather than an expectation ordering relative to $\underline{B}$. I have argued at length in Chapter 8 against confusing degrees of belief with degrees of entrenchment and incorrigibility. I am not sure whether Gärdenfors and Makinson endorse the conflation. But on interpretations of their view according to which the expansion from $\underline{B}$ to $\underline{K}$ is not actually implemented, adding x to $\underline{K}$ is not a genuine change in full belief. The contraction removing ~x from $\underline{K}$ is not a genuine contraction and the entrenchment ordering used is not the entrenchment ordering reflecting assessments of losses of damped informational value relative to $\underline{K}$. Since the target is to explicate $\underline{B}^{+i}_x$ in terms of $\underline{K}^*_x$, the revision transformation is a pseudorevision (section 7.6) and the entrenchment ordering is a pseudo–entrenchment ordering coinciding in the relevant range of expectation ordering relative to $\underline{B}$. On this interpretation, $(C|\sim)$ states that $\underline{B}^{+i}_x$ may be equivalently represented as $(\underline{K}^-_{\sim x})^+_x = \underline{K}^*_x = (\underline{B}^i)^*_x$.

Thus, $(C|\sim)$ is a very powerful assumption. It implies that (i) the inductive expansion function recommended favors maximum boldness, (ii) the inductive expansion function is permutable, and (iii) the expectation ordering relative to $\underline{B}^+_x$ sets the sentences whose expectation relative to $\underline{B}$ is no greater than that of ~x at the minimum 0 value, their negations at the maximum value of 1, and otherwise agrees with the expectation ordering relative to $\underline{B}$.

Thus far no effort has been made to derive the expectation ordering from a family of deductively cogent, caution-dependent, partition-sensitive inductive expansion rules. I shall once more consider the inductive expansion principle of section 6.6 to obtain such a family.

Because the Gärdenfors–Makinson type inductive expansion rule must be permutable as (ii) states, when expectation orderings are derived in the manner indicated, not only must q = 1 as condition (i) requires but bookkeeping must be used. In section 9.1, we have also seen that grades of normality understood as grades of probability or grades of Q/N over elements of U (but not in general disjunctions of such elements) will be preserved for elements of the truncated partition U_x relative to $\underline{\mathbf{B}}_x^+$ where these orderings of elements of U determine optimal expansions of $\underline{\mathbf{B}}$ and $\underline{\mathbf{B}}_x^+$ respectively.

Consider now the ordering of sentences equivalent given $\underline{\mathbf{B}}$ to disjunctions of elements of U and the ordering of sentences equivalent given $\underline{\mathbf{B}}_x^+$ to disjunctions of elements of the truncated U_x with respect to expectation. To fix ideas, return to example 1 in section 7.1 and consider Table 7.3 specifying d-values for elements of U when bookkeeping is used and each element of U carries equal N-value. As already pointed out in section 9.1, the ordering of elements of U with respect to Q/N (i.e., in this case, the ordering of elements of U with respect to probability) is *not* the same as the ordering of elements of U with respect to d-value relative to $\underline{\mathbf{B}}$ (in the reverse direction). Abnormality is not the same as surprise. A fortiori normality is not the same as expectation. Nonetheless, the expectation ordering relative to $\underline{\mathbf{B}}_x^+$ is related to the expectation ordering relative to $\underline{\mathbf{B}}$ in the manner specified according to condition (iii).

This elaborate bit of reconstruction is designed to explain how the Gärdenfors–Makinson approach can be made to square with the interpretation of expectation orderings derived by use of the inductive expansion principle. It also illustrates the point already emphasized in the previous chapter that, although maximizing indices of inductive support that lurk behind talk of grades of normality and satisficing indices that lurk behind talk of grades of expectation and surprise are distinct kinds of appraisal, these appraisals, when understood appropriately, nonetheless yield equivalent results in inductive inference.

The ultimate aim of the Gärdenfors–Makinson discussion of expectation seems to be to assimilate nonmonotonic ampliative inference within the framework of AGM revision. I have shown that this effort can be made to square with the use of families of deductively cogent, caution-dependent and partition-sensitive inductive expansion rules (such as those based on the inductive expansion principle) provided that inductive expansion is permutable so that bookkeeping with maximum boldness is deployed and that the AGM revision of $\underline{\mathbf{B}}^i = \underline{\mathbf{K}}$ by x when x is inconsistent with $\underline{\mathbf{K}}$ is a pseudorevision. The device of using pseudorevisions finesses the charge of confusing entrenchment with degrees of belief. But even if we allow this, the assumption of the permutability of inductive expansion remains to be addressed. This assumption demands that bookkeeping with maximum boldness be deployed. It cannot be repeated too often that bookkeeping with maximum boldness is a very questionable and rash procedure. To the extent that it is unacceptable, the attempt at assimilating ampliative inference to a formal facsimile of AGM revision along the lines suggested by Gärdenfors and Makinson is untenable as well. This is essentially the same conclusion reached in section 9.1 in discussing the ideas of Kraus, Lehmann, and Magidor, and in examining the ideas of Reiter and Poole in Chapter 7.

9.3   Probability-Based Induction

There are families of deductively cogent, caution-dependent, and partition-sensitive inductive expansion rules that are not derivable from the inductive expansion principle. The feature of this principle philosophically attractive to me is that it is grounded in a point of view that understands inductive expansion as decision making designed to optimize certain epistemic goals. Niiniluoto (1986) seeks to do the same relative to a different conception of these goals than I favor. Which conception is preferable or whether some alternative is even better is controversial. Nothing I have said here argues in favor of one or the other. Both approaches are consonant with the view of Peirce and Dewey that inquiry aimed at fixing belief involves adapting means to ends. Both views do so without endorsing the vulgar pragmatism of those who insist that the ends are moral, political, economic, personal, religious, or aesthetic ends.

Skeptics may wonder whether any sensible account of epistemic goals can be offered. In order to address those skeptics without arguing the case for epistemic values, I have sought to identify some features of inductive inference that do not depend on any particular view of the cognitive aims of inquiry or, indeed, on the assumption that there are any. Many of the major points emphasized are sufficiently robust to be relevant to the views of those who are inductivists – that is, allow for inductive expansion according to deductively cogent, caution-dependent, and partition-sensitive families of inductive expansion rules.

Doing so is enough, in particular, for default reasoners interested in inductively extended nonmonotonic inferential relations. It seems to me that the conclusion that rational inferential relations and the associated inductively extended AGM revision operations must demand bookkeeping with maximum boldness casts considerable doubt on the legitimacy of such nonmonotonic inferential relations as representations of legitimate ampliative inference. Default reasoners recommend boldness to the point of rashness.

There is another feature of the previous discussion that merits some emphasis. The inductive expansion principle generates families of inductive expansion rules that are *probability* based. These rules cannot be formulated exclusively in terms of credal probability functions (Q-functions). But the credal probability judgments represented play a central role in the formulation. Probability-based rules have the salient virtue of yielding nonmonotonic inferential relations satisfying left logical equivalence, consistency preservation, closure, and reflexivity. High-probability rules (which obviously qualify as probability based) do not pass this test except when the threshold of acceptance is at probability 1 or, if probability functions take values in the nonstandard reals, when the threshold is $1 - \varepsilon$ for some infinitesimal value ε.

High-probability rules setting the threshold of acceptance at $1 - \varepsilon$ are unsatisfactory for other reasons. In particular, they recommend caution to the point of timidity. If a fair coin is tossed a thousand times, no hypothesis as to the relative frequency of heads passes the threshold. Nor does a disjunction of any subset of such hypotheses except the disjunct of all of them. Yet, it does seem reasonable to expand one's corpus of full beliefs by predicting that the relative frequency will be in some small interval around 0.5 where the size of the interval depends on the degree of boldness. At least inductivists (including, so it seems, default reasoners) ought to say so.

But there are other probability-based rules including one that is fairly simple and computable. Consider a Q-function defined over U relative to $\underline{B}$ and expand according to the following principle:

The Ratio Inductive Expansion Principle:

> Let â be an element of U carrying maximum value for $Q(x)/N(x)$ among elements of U. Let q be a real number between 0 and 1. Reject an element x of U if and only if $Q(x)/N(x) < q\ Q(\hat{a})/N(\hat{a})$.

The ratio inductive expansion principle generates a family of probability-based, caution-dependent, deductively closed, partition-sensitive inductive expansion rules relative to fixed $\underline{B}$, U, Q, N, and values for q ranging from 0 to 1 – just as the inductive expansion principle does. Unlike the latter principle, I do not know how to derive it from a view of the cognitive aims of inquiry that seems attractive. On the other hand, it does not founder on the difficulties of high-probability rules.

The ratio inductive expansion principle has several interesting features:

(1) Bookkeeping with the rule at a fixed value for q is equivalent to applying the rule without any iterations.

(2) b-values (expectation values) and d-values can be derived from such rules.

(3) Most crucially, the results of using such rules approximate the use of rules derived from using the inductive expansion principle relative to the same $\underline{B}, U, Q, N$, and q in the following sense:

Approximation Thesis:

> Let b(x) be the b-function derived from the inductive expansion principle with bookkeeping relative to <u>B</u>, U,Q,N and let b′(x) be the b-function derived from the ratio inductive expansion principle with bookkeeping relative to the same factors. If b(h) > b(g), then b′(h) > b′(g).

The approximation thesis may be defended as follows. Let x and y be elements of U. If d(x) < d(y), q(x) > q(y). For each z in U, q(z) – the maximum q-value at which x fails to be rejected via bookkeeping using the inductive expansion principle – is correlated uniquely with k(z) = $Q(z)/N(z)$ so that k(x) > k(y) if q(x) > q(y). Hence, $Q(x)/N(x) > Q(y)/N(y)$. If k(x) and k(y) are divided by $Q($â$)/N($â$)$, this comparison must be preserved. Hence, it is clear that the maximum q′ level q′(x) at which x fails to be rejected relative to the ratio inductive expansion principle is higher than the corresponding value q′(y) for y. Hence, d′(y) > d′(x). Let h and g be any pair of disjunctions of elements of U. Let b(h) > b(g). Then d(~h) > d(~g). Let x be the minimally surprising element of U in the subset of U whose truth given <u>B</u> entails ~h and let y be the same for ~g. d(x) > d(y). Hence, d′(x) > d′(y). d′(~h) > d(~g). Consequently, b′(h) > b′(g), as was to be shown.

The converse to the approximation thesis does not hold. A counter-example is found in example 1 of section 7.1. The ratio inductive expansion principle assigns a higher b′-value to the negation of the disjunction of ~r~f~b and rf~b than to negation of the disjunction of ~r~f~b and ~rf~b. This is because (given the equal M-values assigned all elements of U), the ratio of the probability of rf~b to ~r~f~b is greater than the ratio of the probability of ~rf~b to ~r~f~b. Yet the d-values for rf~b and ~r~f~b are equal as are the d-values of the two disjunctions resulting from disjoining one of these with ~r~f~b. Hence, the b-values of these two disjunctions are equal.

Thus, we have a precise and useful sense in which the results of using the ratio inductive expansion principle yield an expectation ordering almost agreeing with the expectation ordering obtained with the aid of the inductive expansion principle. Even if the ratio inductive expansion principle cannot be rationalized as optimizing with respect to appropriate cognitive goals, it approximates a principle that does so by taking into account the contextual features deployed in using the inductive expansion principle. In particular, it is probability based. Finally, computations for applying rules using the ratio inductive expansion principle seem easier than for rules using the inductive expansion principle – a point that ought to please students of artificial intelligence.

When q = 1, the nonmonotonic inferential relations derived from using the ratio inductive expansion principle are permutable (i.e., ampliative but constant). The inductively extended versions of (K*1) – (K*8) as specified in section 9.1 all hold. And, indeed, the agreement between the inductive expansion principle and the ratio inductive expansion principle in verdicts rendered is exact.

When q < 1, we have the same story as can be told using the inductive expansion principle with bookkeeping when q < 1. Reflexivity, left logical equivalence, closure, consistency, and cut are satisfied. But nothing else.

10   Agents and Automata

Rational agents and inquirers are sometimes capable of evaluating their beliefs, values, and actions with respect to their coherence or consistency – that is, with respect to prescriptive standards of rationality. But this is not always so. Often the predicaments faced are too complex or the time available before a judgment is to be made too short or the cost of deliberation too great for even the most intelligent and well balanced agents to engage in such self-criticism. And even when these considerations present no obstacle, emotional difficulties and indolence may impede this type of activity.

Because agents sometimes can and sometimes cannot fulfill the demands of reason by themselves, they need help. Education in logic and mathematics can contribute. Practical training in various types of deliberation is often useful. Whether current forms of psychotherapy are of value is a matter of dispute but their chief importance to us is their claim or promise to do so. The same is true of reading good literature.

We also use prosthetic devices when they become available. We write notes to ourselves when we don't trust our memories. We consult handbooks as stores of information and resources for how to make exact or approximate calculations. And more recently we have become fond of using the products of the burgeoning technologies that furnish us with calculators and other automata that enhance our capacity to engage in self-criticism.

I make these banal observations to emphasize a simple but important point. The reason for designing "expert systems" of the sort that seem to have occasioned much of the preoccupation with nonmonotonic reasoning appears to have been to facilitate the design of automata that serve as prosthetic devices to rational agents, enhancing their capacities to calculate, deliberate, and inquire.

The point is important because it is sometimes thought that the aim is to contribute to modeling the human or inquiring agent – that is, to contribute to designing *l'homme machine*.

In my judgment, this activity is misdirected. I do not wish here to enter the contentious arena where the feasibility of simulating the human mind

289

mechanically is debated. Whatever the prospects might be in the future, the chief social value right now of the work of computer scientists and students of artificial intelligence derives from their contributing to the design of automata that enhance the capacities of inquiring and deliberating agents to be rational. This is quite a different activity than constructing machines that think like we do.

The activity is an important one. The results have revolutionized the lives of everyone living in societies where the new technologies are available – not only by offering us a variety of games for our amusement but by enhancing our capacities to do the calculations needed to solve hard problems.

In designing automata, it is important to understand what ideal standards of rationality are supposed to be. For otherwise the automata that are designed might ingeniously enhance our irrationality. This, I believe, is relevant to studies of nonmonotonic and default reasoning that have been the subject of the preceding three chapters. I have not pretended here to contribute to these studies in a way that improves the technologies used in the design of programs for expert systems. I have been concerned with elaborating a network of standards for rational suppositional and inductive reasoning that seems to provide at least part of the prescriptive standards relative to which the proper functioning of expert systems may be assessed.

Thus, I have downplayed questions of computational complexity as irrelevant precisely because I have not been worried about the problem of implementation. I do not dispute the fact that our capacity to meet worthy standards of rationality is bounded. That is precisely why we need prosthetic devices to enhance our capacities to push the boundaries back a little further. Even if the devices developed fail to meet the standards of rationality perfectly, it is desirable that they help us do better than we do unaided by them in performing the tasks they are designed to perform. The themes developed in this book are concerned with identifying standards for what it is to do better.

Thus, the theses advance here are frankly prescriptive and, insofar as they are contentious, they conflict with other prescriptions concerning what it is to do better.

I have argued (in Levi, 1991, and in this book) that the AGM scheme is inadequate as an account of rational belief change. In the first four chapters of this book, I acknowledged that the AGM approach fares better as an account of suppositional reasoning for the sake of the argument. Even here, however, I contend that it should be amended by making Ramsey revision rather than AGM revision the key transformation.

In Chapters 5 and 6, I argued that the account of suppositional reasoning, of conditionals as expressions of such reasoning and the logic of conditionals and supposition reasoning elaborated in Chapters 1–4 should be extended to take into account the results of inductive inference. In doing so, I came down on the side of inductivists in opposition to probabilists and Popperians. I reviewed an approach to inductive inference I had developed beginning in the 1960s that is decision theoretic in approach and seeks to formulate a sensible account of how considerations of probability and such "informational" values as explanatory power, simplicity, and the like interact in theory choice, statistical estimation, and ordinary induction from observations. I have shown how the ampliativity of such inductive inferences combined with the assumption that induction represents a shift from one belief state representable by a deductively closed corpus or theory to another implies the nonmonotonicity of the resulting inferences. This nonmonotonicity is contrasted with the belief-controvening variety exhibited in the suppositional reasoning discussed in the first four chapters. The general conditions where they do and do not come together have been identified.

Chapters 7–9 have sought to support the claims that the sort of "default" reasoning studied by students of artificial intelligence is, indeed, inductive reasoning, to argue that the motivations that appear to lurk behind the study of default reasoning tend to favor reasoning under the conditions where induction and AGM inference come together, and to insist that these conditions limit rather severely the legitimacy of the applications of default reasoning understood as inductive inference.

The main conclusion I mean to draw from the argument of these three chapters is, therefore, that the demands being imposed by Poole, by Lehmann and Magidor, and by Makinson and Gärdenfors on default reasoning restrict the range of legitimate applicability of default reasoning too severely. It is arguable that the procedure proposed by Reiter, when amended as was suggested in Chapter 7, reflects better the prescriptive standards that may be imposed on inductive expansion. The claim is not based on the computational complexity of these demands but on the dubiety of taking them to be prescriptive standards for inductive inference, except in extremely circumscribed circumstances. If anything, this conclusion sustains a modified version of the weaker demands imposed by Reiter as these modifications are explained in Chapter 7.

The issues I have raised are no doubt contentious. But just as I think cognitive psychologists and economists who explore the extent to which human agents do or do not conform to standards of rationality ought to consider more carefully what those standards are or ought to be, anyone engaged in designing automata that are intended to enhance the capacity of human agents to be rational agents in their deliberations and inquiry should do the same. Those who would relegate such matters to philosophers endorse a division of labor in these matters that is both undesirable and untenable.

Notes

Chapter 1

1 Just as judgments of subjective or personal probability are sometimes grounded in full belief or knowledge of objective probability, judgments of serious possibility are often grounded in assumptions about objective possibility (i.e., ability). But such objective backing is not always available. Yet, coherent and rational agents cannot avoid making judgments of serious possibility either explicitly or implicitly. See Levi, 1980.

2 For a more accurate characterization of the structure of a set of potential answers, see Levi, 1967; 1980; 1984a, ch. 5; or 1991. A brief sketch will also be found in section 6.1.

3 Strictly speaking, a minimal revision by adding h in the sense intended by Gärdenfors's h is an expansion when h is consistent with $\underline{K}$ and is a transformation of the sort just sketched otherwise.

4 In general, contraction by removing h from $\underline{K}$ when $\underline{K}$ is consistent is a transformation removing h from $\underline{K}$ to yield a deductively closed corpus contained in $\underline{K}$. Admissible contractions remove h with minimal loss of informational value. For further discussion, see section 2.1 and Levi, 1991, ch. 4. Contraction from the inconsistent belief state requires special treatment. See Levi, 1991, sec. 4.8.

5 Most commentators interpret Ramsey as having offered a unitary derivability-from-laws account of conditionals of the sort Hansson (1994) calls a "derivability theory." Chisholm (1946) went so far as to combine a passage from Ramsey in the text of his paper that supports the derivability-from-laws interpretation with an earlier passage from a footnote that I and many others are fond of citing as the source of "Ramsey tests." The impression is left that Ramsey is a precursor of Chisholm (1946) and Goodman (1947). As asserted in the text, Ramsey's test for open conditionals does not require that the consequent be derivable with the aid of laws from the supposition. To illustrate, let X be certain that a die has been tossed and has either landed with an even number of spots up (e) or with an ace showing (a). Let X be in suspense as to whether e or a is true. According to Ramsey's test, the open conditional (*) "Had the die landed with an odd number of spots up, it would have shown an ace" is acceptable even though X cannot derive this from laws. Consider now the case where X is not in

suspense but is sure that e is true. (*) is, from X's point of view, counter-factual. The suppositional interpretation requires that X should remove e from his set of full beliefs. In doing so, he must give up at least one of the sentences e ∨ a or e ∨ ~a. In particular, whether X retains e ∨ a and, hence, finds (*) acceptable or does not retain it and finds (*) unacceptable depends on the loss of informational value incurred. Derivability from laws is relevant to reckoning such losses but is not required to warrent the ac-ceptability of conditionals. J. Mackie (1962) took the view that even belief contravening conditionals expressed "telescoped" or "condensed" arguments from suppositions to conclusions where the suppressed premis-es are the beliefs of the agents modified in order to make consistent room for the supposition. According to Mackie, when items need to be removed from the agent's beliefs, beliefs poorly supported by the evidence are more vulnerable to being given up than well-supported beliefs in such modifica-tions. If one insists, as I do, that all items in an agent's body of full belief are maximally certain and, hence, equally well supported, Mackie's criterion for vulnerability to being given up is useless. I contended (Levi, 1984, ch.12) that what is vulnerable to removal is what is less valuable as information. Even so, Mackie's approach is suppositional in the sense under discussion here because Mackie accounts for reasoning from belief-contravening supposition by invoking a notion of corrigibility graded according to a standard of epistemic value.

Mackie (1973) shifted to what he called a "suppositional" view that did not expect the agent to supply the suppressed premises of the condensed argument on demand (p. 94). To say "if h then g" is to assert g within the scope of the supposition that h (Mackie, 1973, p. 93). I, for my part, am wedded to a distinction between commitment and performance (Levi, 1991, ch. 2) that allows one to say that the agent is committed to certain premises of the condensed argument being within the "scope" of the presupposition, even if the agent cannot fulfill the commitment by supplying the premises explicitly. According to this view, the distinction between the condensed argument and presupposition view of belief, contravening conditionals becomes invisible. Counter to Hansson's (1994) interpretation of Mackie, both stand in opposition to derivability interpretations.

J. Collins (1991) offers an account of conditionals as expressions of suppositional reasoning according to which the transformation of the agent's belief state demanded by supposition is by imaging. Imaging transformations are parasitic on specifications of truth conditions for conditionals along lines suggested by Stalnaker (1968) and Lewis (1973) (see note 6 and section 3.8 for further discussion). According to imaging, such reasoning does not proceed in the open case as it does according to the requirements of Ramsey. Thus, (*) would not be acceptable even in the open case. Imaging is not suppositional as I understand that idea. Hansson claims that derivability theories come in "ontic" and "epistemic" versions in the sense of Lindstrom and Rabinowicz (1994). No account of conditionals that endorses the Ramsey test for open conditionals (as mine does) is either ontic or a derivability theory. The ontic accounts offered by Stalnaker (1968) and Lewis (1973) and the epistemic imaging proposal of Collins based on Lewis's approach are not derivability theories even though they flout the Ramsey test for open conditionals. Gärdenfors (1978, 1988) offers what appears to be a suppositional interpretation of conditionals in the epistemic sense suggested by Ramsey and Mackie. But, as we shall see in Chapters 2 and 3, the status of Gärdenfors's theory as a suppositional interpretation is open to question. It deviates from Ramsey's approach to the open case, allows conditionals to be objects of belief as ontic theories do, and seeks to support a version of conditional logic resembling Lewis's favorite VC.

6   R. Stalnaker (1968, p. 102) is apparently responsible for the suggestion that the Ramsey test recommends making a "minimal revision" of $\underline{K}$ to insure that p is in the result. Minimal revisions are transformations of $\underline{K}$ both in cases where neither p nor ~p is in $\underline{K}$, when p is in $\underline{K}$ and when ~p is in $\underline{K}$. Several types of transformations fill this description. Stalnaker interpreted "minimal revision" as a transformation by what I shall call "imaging." The term *imaging* was introduced by D. Lewis (1973) to characterize a form of updating probability judgments alternative to so-called conditionalization. P. Gärdenfors (1988, ch. 5) discussed and generalized Lewis's idea and showed how it might be related to changes in states of full belief. J. Collins (1991) has engaged in a systematic exploration of the relations between minimal revision accounts based on generalized imaging transformations and those based on nonprobabilistic versions of conditionalization. Studies of this kind are also found in Katsuno and Mendelzon (1992). Imaging transformations will be discussed briefly later on. However, it is worth mentioning here that such transformations do not guarantee satisfaction of the core condition of adequacy for being a Ramsay test. Other interpretations of minimal revision do. Ramsey provided a clear view concerning the acceptability of conditionals in belief-contravening and belief-conforming cases. However, his name and authority should not be associated with a view that rejects the conditions of acceptability he clearly intended to offer for the open case. The wide popularity of the imaging interpretation of minimal revision and the persistent effort to associate this interpretation with Ramsey's ideas are graphic evidence for the desirability of setting the record straight.

7   There is a third class of "if" sentences identified by Dudman ("generalizations") that are not expressions of suppositional reasoning. I suggest interpreting such sentences as attributions of dispositional properties. "If Fannie was contradicted, she cried" is understood as claiming that Fannie had the disposition to cry upon being contradicted. See Levi, 1984a, ch. 12, and 1980, for discussion of dispositions and ability.

8   This point has been disputed by advocates of imaging who have invented a decision theory according to which their account of conditionals can be construed as capturing suppositional reasoning of the kind relevant to

practical deliberation. J. Collins, who is one of the more sophisticated apologists for this approach (see Collins, 1991), even calls this theory "suppositional" decision theory rather than "causal" decision theory as is the custom in order to distance the approach from excessively close conceptual ties to causation. In Levi, 1948a, ch. 15; 1982; 1983;1985a; and 1985b, I argued that no precedents in presystematic practice support the recommendations of causal or suppositional decision theory where they conflict with other accounts of decision making under risk or uncertainty, some recommendations of the theory are clearly absurd, and the theory is itself committed to metaphysical assumptions concerning which principles of rational choice ought to be neutral. I shall not reiterate the case for these views here.

Chapter 2

1 To say that h > g is acceptable according to $\underline{K}$ is to claim that anyone whose corpus is $\underline{K}$ should be prepared to rule out ~g as impossible in a fantasy where a corpus that is a transformation $T(\underline{K})$ of $\underline{K}$ is used as the standard for serious possibility. Whether such a conditional judgment of serious possibility is a full belief that h > g is true is a question to which we shall return in Chapter 3.

2 In the appendix to Chapter 6, I argue that when there are infinitely many maxichoice contractions, no more than finitely many distinct M-values should be assigned to them. This guarantees the existence of a minimum M-value in $C(\underline{K},h)$. The assumption that minimum exists addresses a problem similar to that addressed by D. Lewis's limit assumption for spheres of accessibility (Lewis, 1973, sec. 1.4).

3 The measure of damped informational value defined over potential contractions of $\underline{K}$ removing h induces a measure of incorrigibility (Levi, 1980, 1991) and also of entrenchment (Gärdenfors, 1988) over sentences in $\underline{K}$. The relation between entrenchment and incorrigibility will be discussed further in section 8.5.

4 The notions of an admissible contraction removing h from $\underline{K}$ and of a Ramsey contraction removing h from $\underline{K}$ are defined only for consistent $\underline{K}$. Contraction from inconsistent corpora is an important topic in accounts of the conditions under which sincere belief changes are justified. See my earlier discussion (Levi, 1991, sec. 4.8). But, as explained in section 1.1, there is no need to consider such contraction in discussion of reasoning from suppositions made for the sake of the argument or of the acceptability of conditionals. Moreover, when an inquirer does inadvertently expand into inconsistency by consulting the testimony of the senses or reliable or expert witnesses, retreat from inconsistency requires determining whether to remove the new information in a way that returns (subject to some qualifications) to the initial belief state, revising the initial belief state by retaining the new information and removing items from the initial state or doing both. The contraction from inconsistency is not uniquely determined

by the inconsistent belief state, the assessment of informational value and specification of some item to be removed as it is in the case of admissible contraction.

5   Conditional credal probability judgments carry another significance. The agent is prepared at the time $\underline{K}$ is the corpus to adjust his credal state so that the conditional probabilities according to Q(./h) become the agent's unconditional probabilities on expanding by adding h to $\underline{K}$. In effect, the agent endorses a rule for adopting credal states relative to various alternations of his or her state of full belief that has this property of converting conditional to unconditional probabilities under the right conditions. However, the agent is not irrational if he or she reneges on this *confirmational commitment* on adding h to $\underline{K}$. The agent might change his or her confirmational commitment (Levi, 1974, 1980). Here too the condition h is consistent with $\underline{K}$. Sense can be made of belief-contravening conditional probability judgments by appealing both to the agent's confirmational commitment at the given time and to the criteria (such as informational-based Ramsey tests) for belief-contravening conditional judgments of serious possibility given the agent's state of full belief. The agent's state of full belief and confirmational commitment are two separate contextual parameters involved in assessing belief-contravening conditional credal probability judgments. Only the state of full belief is implicated in belief-contravening judgments of serious possibility.

Authors who have been attracted by the idea of embedding accounts of conditionals in accounts of conditional probability *change* the customary conception of conditional probability so that the condition h can be belief contravening – that is, inconsistent with $\underline{K}$. In doing so, conditional probability loses its force as an index for determining fair betting rates in called-off bets, and belief-contravening conditional judgments are no longer seen as functions of two separate contextual parameters – the confirmational commitment and state of full belief. Far from being a useful tool in a fine-grained account of change in full and probabilistic belief, appeal to such Popper functions ignores important distinctions.

Enthusiasts for the Popper function approach compound confusion by conflating inconsistency with $\underline{K}$ with carrying 0 credal probability relative to $\underline{K}$ (see Levi, 1980, sec. 10.3).

6   Notice that the saturatable contractions that give up the claim that the coin is tossed must under the circumstances give up the claim that it landed heads. But this claim factors into the two disjunctions: "either the coin is tossed or it lands heads" and "either the coin is not tossed or it lands heads." The first of these disjunctions must be given up because of lawlike background information that insists that tossing is a necessary condition for the coin landing at all. The question arises whether the second disjunction may be given up as well. Whether or not it is given up will not alter the saturatable character of the contraction; but only when it is retained is the contraction maxichoice. It is possible to endorse a probability measure M for determining undamped informational value that accords the two

contractions equal M value and, hence, undamped informational value. But then the intersection of the two contractions (which is the same as the second one) will have the same damped informational value. In the magisterial essay by Alchourrón, Gärdenfors, and Makinson (1985), potential contractions removing h from $\underline{K}$ are restricted to intersections of maxichoice contractions. There is no justification for this restriction. Maxichoice contractions are always stronger than saturable contractions that are contained in them. That does not imply that they carry more informational value or content but only that they do not carry less. In any case, intersections of maxichoice contractions are weaker than maxichoice contractions. If this meant that they carried less informational value, we could not be minimizing loss of informational value in contraction unless the contractions are maxichoice (see Levi, 1991, sec. 4, and the discussion in section 2.2 of this volume.)

7   The ordering of maxichoice contractions that lurks at the basis of the AGM approach to contraction can be formally represented as a revealed preference defined in terms of choice functions satisfying certain "choice consistency" conditions. (See Levi, 1986a, for a discussion of revealed preference and references to the literature.) But identifying such choice functions defined over the set of maxichoice contractions will not afford a decision theoretic rationale for the AGM approach to belief contraction as Lindström (1990) and Rott (1993) seem to think. To do that, the choice function must be defined over all potential contractions of the given corpus or at least all potential contractions removing some specific proposition so that a "revealed" preference is given that indicates when the meet of a set of saturable contractions does come out optimal. Neither Rott nor Lindström does this. In particular, each fails to show when a choice function *defined over all potential contractions* guarantees that an optimal contraction satisfies recovery – that is, must be the meet of maxichoice contractions.

8   D. Makinson (1987) has pointed out correctly that AGM revision yields the same results whether or not the recovery postulate is satisfied. Hence, so does RTGL. This point has been invoked to downplay the importance of doubts about recovery to either the account of belief change proposed by Alchourrón, Gärdenfors, and Makinson or the approach to conditionals advanced by Gärdenfors. I have indicated the points of agreement and disagreement between my view of belief change and that offered by Alchourrón, Gärdenfors, and Makinson elsewhere (Levi, 1991) and have argued that abandoning recovery does make a significant difference to accounts of belief change. However, the considerations I adduce do not touch the question as to whether the issue of recovery makes any difference to conditions of acceptability of conditionals according to the Ramsey test. It does not when RTGL is the criterion of acceptability. But it does when RTR is used. This should count as an additional reason for adopting the latter criterion of acceptability; for it is preferable to favor a criterion that does not prejudice in advance the significance of a controversial issue over a criterion that does.

9   F. Jackson (1990) has offered a forceful defense of the classification of (1')
    with (3') rather than with (2'). I shall offer at most a sketch of a rebuttal
    here. He begins by assuming that sentences like (2') are to be interpreted
    according to the "natural approach" requiring that two conditions
    (Support) and (Conditional noncontradiction) are satisfied (pp. 136–7). He
    claims that (1') like (3') violates either (Support) or (Conditional noncon-
    tradiction) or that (Conditional noncontradiction) is obeyed in a context
    dependent way. If Jackson were right, (2), like (2'), should carry a truth
    value. The argument of Chapter 3 seeks to defend the view that neither
    (2) nor (2') carries truth values. (2) is a judgment of conditional possibility
    that, like judgments of conditional probability, lack truth values. If this is
    right, (Support) is not applicable to (2) any more than it is to (1) or, for that
    matter, to (3). It should not be applicable to (1'), (2'), or (3') either. I
    have already conceded, in effect, that consensus conditionals violate
    (Conditional noncontradiction) even when the context is held fixed and,
    hence, agree that (3') and its kin do so. A person who thought that Oswald
    did not kill Kennedy can affirm both (3) and (3') and might very well intend
    to convey that message by uttering both in tandem. I also think that (2') and
    its grammatical cousins obey this principle when context is held fixed (but
    not when it is allowed to vary). One cannot jointly affirm (2) and (2') coher-
    ently. In this respect, I argue that (1') and its grammatical cousins resemble
    (2'). Whatever the merits of a more elaborate version of my counter to
    Jackson's argument might be, one cannot refute Dudman's classification
    by begging contentious semantical issues.

Chapter 3

1   If the "first-order" credal probability distributions are treated as truth-
    value-bearing hypotheses, I assume (I conjecture fairly noncontroversial-
    ly) that the conditional credal probability that h conditional on the credal
    probability the h = r is itself equal to r. ("r" is here a standard numeral.)
    The calculus of probability then insures the result stated in the text. It is
    worth emphasizing that when x is distinct from 0 or 1, the first-order
    credal probability is uniquely determined. The agent should be certain (on
    the assumption we are making for the sake of the argument that first-
    order credal probability judgments are truth value bearing) of the truth
    of this probability distribution p_i. But it gets probability 0 according to
    the second-order credal probability distribution over p_1 and p_2. That is as
    it should be if we are seeking to suspend judgment between these two
    alternatives. The demand that these two alternatives be in suspense and
    that the inquirer should be certain that p_i cannot be consistently satisfied.
    Yet, they both should be satisfied if first-order credal probability
    distributions represent truth-value-bearing hypotheses. The difficulty
    raised is a version of Savage's 1954 worry about "unsure probabilities."

2   If credal states are allowed to be indeterminate (Levi, 1974, 1980), the
    permissible credal probabilities for p_i will be values x belonging to some

subinterval of the unit interval, and the permissible credal probabilities for p_2 will be values of $1 - x$ for permissible values of x. Then the credal state for h and ~h would be some convex subset of the convex hull of p_1 and p_2. If this set of permissible probabilities does not contain p_1, the hypothesis represented by this set should be taken as incompatible with p_1. Shifting to this credal state clearly betrays the concern to avoid error. To consider both p_1 and p_2 to be serious possibilities, the permissible values of x would have to range over the closed interval from 0 to 1. But then all the probabilities between p_1 and p_2 should be considered as truth-value-bearing serious possibilities and should be "supported" by the second-order credal state for the probability assignments for h. Subintervals of positive Lebesgue measure of the interval of values for x from 0.1 to 0.9 should get positive credal probability. They do not. To be sure, the higher-order probabilities can be modified so that all of the first-order probabilities are supported. In that case, we are prevented from suspending judgment between *just* p_1 and p_2. In the indeterminate case, as in the case where higher-order credal probabilities are numerically determinate as discussed in note 1, incoherence threatens.

It should be emphasized that the reservations raised here and in note 1 concerning the legitimacy of making higher-order probabilties are directed at the second-order credal probability judgments of agent X at time t concerning X's first-order credal probability judgments at time t. They do not concerning X's views of Y's judgments at t or of X's judgments at t′. Nor do they concern credal probability judgments concerning statistical, objective, or physical probabilities.

3   This and the virtually equivalent argument in section 3.1 dealing with Gärdenfors's restriction of the Ramsey test to the positive part is a variant of H. Rott's simplified argument for P. Gärdenfors's (1986) trivialization result (Rott, 1989, pp. 93–5). Unlike Gardenfors and Rott, however, this argument presupposes only weak nontriviality (Rott, 1989, p. 93). That is to say, there is a potential corpus $\underline{K}$ in K that lacks both h and ~h for some sentence h in $\underline{L}^1$. Rott is prepared to abandon the claim that the revision of $\underline{K}$ by adding h when neither h nor ~h is in $\underline{K}$ is identical with the expansion of $\underline{K}$. He still claims to be deploying a Ramsey test for conditionals because h > g is acceptable if and only if g is in the revision of $\underline{K}$ by adding h. However, revision is understood in a manner conflicting with a core presupposition of Ramsey tests. He complains about the view taken in Levi, 1988, and reiterated here that every potential corpus $\underline{K}$ is uniquely determined by one and only one $\underline{K}^1$ (given the revision operator and the Ramsey test associated with it) but confuses it with my rejection of the reduction condition, which he is anxious to retain. I would have thought that anyone favoring an epistemic approach to the acceptability of conditionals would agree that the set of acceptable conditionals is uniquely determined by the set of nonmodal judgments (together with the appropriate test for acceptability). This in itself does not imply rejection of the reduction condition. However, if one takes the

position that the only properties of beliefs that ought to matter to an inquirer are properties of the nonmodal part of his or her belief state, then the rejection of the reduction condition is quite natural. Rott has misunderstood my contention that modal and conditional judgments lack truth values. One can, if one likes, construct a truth semantics for conditionals. But insofar as we are concerned to avoid risk of error, we are concerned to avoid risk of error in the nonmodal part of the corpus. Rott complains that I fail to offer a systematic way to recognize the distinction between objective modalities and subjective ones (Rott, 1989, p. 106). I am not sure what he is asking for, but it seems to me that a clear distinction has been drawn by indicating whether a sentence can be assigned a subjective or credal probability or, more generally, represent the so-called object of a propositional attitude. Only statements of objective modality can be such objects. Insofar as Rott wants an analysis of dispositions, abilities, and other objective modalities such as statistical probabilities, he should check Levi, 1984, ch. 12, and 1980. S. O. Hannson (1992) also acknowledges a distinction between the nonmodal part of a corpus and the modal part and restricts application of the Ramsey test to the nonmodal part. To this extent, he shares my 1988 view. Rott, Hansson, and I share similar understanding of the options to be faced in the light of the "trivialization" results of Gärdenfors and others. We make different choices.

4   Levi (1988) and Rott (1989) show that by using both the positive and negative RTGL, weak triviality is threatened. It is possible to allow the Ramsey test to live with (K*r4) if one abandons consistent expandability; but then application of the Ramsey test in the core cases yields incoherent verdicts and the weakened version of (K*5) is violated. However, the requirement that K be closed under expansions need not be threatened by Gärdenfors's reasoning. On the other hand, one could follow Rott (1989) and give up closure under expansion as well as consistent expandability and introduce a new operation on corpora that serves as the surrogate for expansion when neither h nor ~h are in $\underline{K}$. This approach conflicts with the Ramsey test criterion as severely as Gärdenfors's approach does.

5   Insisting that for every consistent $\underline{K}$ either h > g or h > ~g is in $\mathbf{RT}(\underline{K}^1)$ does not help. On that assumption, the positive Ramsey test implies that the expansion of $\underline{K}^1$ by adding h (by adding ~h) is maximally consistent in $\underline{L}^1$. This result is surely unacceptable.

6   Rott (1989 , pp. 104–5) speculates that the reason I withold truth values from modalized judgments is because conditionals appear to be only about subjective changes of belief. Conditionals are not claims "about" anything. If they were claims about something, they would be true or false. His contention that there is a genuine difference between genuine changes in belief state and changes for the sake of the argument is more interesting. Consider the following sentence:

(1) If Gorbachev had died shortly after his assumption of power, the world's peace would be in greater danger now (i.e., 1989) than it actually is.

Rott seems to want to argue that (1) must clearly carry a truth value whether or not one wants to provide a Ramsey test criterion for its acceptability. I am not sure I understand his argument. I shall use a different example where attention can be focused on what seems to me to be the relevant issue suggested by the example.

The agent knows that there are two main driving routes from New York to Boston: one through Hartford and the other through Providence. The agent also knows that driving time is longer on the Providence route than on the Hartford route. Let h = "Jones traveled from New York to Boston via Hartford" and h′ = "Jones traveled from New York to Boston via Providence." Neither h nor h′ (nor their negations) are in the agent's corpus <u>K</u>.

Scenario 1: The agent fully believes that Jones has not as yet taken the trip. The agent should then be in a position to say:

(2) If Jones drives via Hartford, it will take him less time than if he drives via Providence.

Scenario 2: The agent knows that Jones has already taken the trip, though he or she remains in ignorance of the route Jones took. The agent could still say:

(2′) Had Jones driven via Hartford, it would have taken him less time that it would have had he driven via Providence.

In these two cases, the agent does not know the route Jones took.

Scenario 3: The agent knows that Jones drove via Providence. Then the agent might have said:

(3) Had Jones driven via Hartford, it would have taken him less time than it actually did.

Scenario 4: Suppose the agent, beginning in the situation depicted in scenario 3, becomes convinced for good reason that Jones drove via Hartford. According to Rott, the agent would not then judge that the time it took him to drive is less than the time it actually took him to drive. At least, that is what I think his objection is. He concludes from this that a genuine change of view will fail to yield the kind of verdict that is clearly given in scenario 3 when h is a supposition for the sake of the argument.

It is clear that in scenario 4, the agent will not make the incoherent judgment. But the judgment is just as incoherent as the product of revision by adding h. The use of sentence 3 to manifest a conditional attitude of serious possibility cannot be read as registering the acceptance of the incoherent judgment in either scenario. What scenario 3 (and 1, 2, and 2′) does is convey two conditional judgments of serious possibility: one conditional on h and one conditional on h′. In addition, they make a comparison of the consequences of the transformed belief states involved

– in particular, concerning driving times to Boston according to the two routes. In scenario 4, the revision of the corpus by adding h is not a fantasy supposition for the sake of the argument. In scenario 3 it is. Yet in both scenarios, the revision of the corpus by adding h is compared with the revision of the corpus in scenario 3 by adding h'. And the comparison leads to the same verdict. Of course, scenario 3 is no longer appropriate to express it. One might have in scenario 4 "Had Jones traveled via Providence it would have taken him longer than it actually did."

The examples considered here and Rott's example about Gorbachev concern what I like to call "cross fantasy" comparisons where comparisons are made of the contents of belief states transformed by different suppositions. The conceptual feasibility of making such comparisons is very important but does not, in my judgment, support a realist construal of conditionals. There is more to be said, on both sides, regarding this issue. My remarks are targeted at Rott's worry. Space does not permit a more elaborate discussion.

7   Darwiche and Pearl (1994) and Freund and Lehmann (1994) have recently addressed the question of iterated revision of belief states. Both pairs of authors have independently noted that an account of such revision can be furnished by supplementing the AGM postulates for revision with additional postulates. Of particular interest is the Freund and Lehmann postulate ($K*9$) that states that if ~h is in the intersection of $\underline{K}$ and L', $\underline{K}^*_h = \underline{L}^*_h$. They observe that their postulate amounts to keeping the method of belief revision fixed over iterated change. They acknowledge alternative approaches found in the work of Boutilier (1993), Boutilier and Goldszmidt (1993), and Williams (1994) that do not obey this requirement. Pearl and Darwiche also criticize Boutiliers's proposal. ($K*9$) appears to be precisely the condition that is satisfied when AGM revision is defined by deriving * from the informational-value-determining M-function and a basic partition V where the M-function and V are held fixed over iterated revisions. In genuine belief change, I contended in section 2.1 that rationality does not require that V remain fixed. On the other hand, M should remain constant unless some genuine global change in demands for information is justified. In suppositional reasoning, one should keep both M and V fixed, for the iterated conditional judgments being made are relative to the initial corpus and the evaluations of informational value made in the setting where that is the inquirer's belief state. To allow the M-function or V to vary in iterated suppositional reason without an explicit stipulation to that effect in the "if clause" seems to go against the intent of suppositional reasoning. To this extent, I favor the attitude of Freund and Lehmann. I might add that the importance of the basic partition V and the M-function was recognized in Levi, 1991, pp. 123–4, 144, in the context of genuine belief change. The details presented in this book slightly modify my previous approach and adapt it to suppositional reasoning.

8 Gärdenfors claims that this analysis is equivalent to Pollock's analysis of "even if" conditionals. He appeals tacitly to the recovery condition on contractions to show this. But the appeal to recovery is not necessary. g must be in $\underline{K}$ in order for the "even if" conditional to be acceptable whether or not recovery holds.

9 As Collins appreciates, once this step has been taken, one can, if one likes, jettison the realistic construal of conditionals although the approach proposed does make serious use of the truth theoretic semantics.

10 The term *imaging* was introduced by D. Lewis (1976) as an alternative principle for changing credal probability states when new information is added to the corpus to the method of conditionalization. Gärdenfors (1982; 1988, sec. 5.3) generalized this idea. However, it is apparent that imaging as understood by Lewis and Gärdenfors may be modified so as to become an account of one way of changing states of full belief. Ramsey did not represent the way belief states are revised in testing conditionals as changes in states of credal probability. J. Collins has shown how to dispense with probabilities in characterizing imaging tests. In my judgment, not only is it unnecessary to use probabilities, it is undesirable to do so. The same point applies mutatis mutandis to imaging tests.

11 If Stalnaker's semantics for conditionals is used, suspension of judgment obtains between the first two alternatives. The results are, however, the same in all respects pertinent to my argument.

12 According to Lewis's favorite elaboration of the structure of the similarity relations between possible worlds, T $>$h cannot be equated with $\square$ h but rather with h itself. $\square$h is equated instead with ~h $>$ ~T. It is easy to see that in a state of suspense between h and ~h, the inquirer might also be in suspense between. $\Diamond$h (= ~[h $>$ ~T]) and ~$\Diamond$h (= $\square$~h = h $>$ ~T). There is, to be sure, a relation between the judgment that h is objectively possible and that it is seriously possible. Let h be judged objectively possible according to $\underline{K}$ so that the claim that h is objectively possible is in $\underline{K}$. Then the judgment that h is seriously possible is acceptable according to $\underline{K}$. That is to say, ~h is not in $\underline{K}$. The converse, however, does not hold as the discussion in the text indicates.

As I argue in the text, I do not see that such a notion of objective de dicto possibility has any useful application to date. But that does not mean that there is no use for a concept of objective possibility. The concept of an ability does play a role in scientific inquiry as is familiar in the statistical notion of a sample space. A die has the ability to land with one of the six faces showing up on a toss. These abilities are a property of the die. Abilities are dual properties to surefire dispositions. To say that the die has the ability to show one spot on a toss is equivalent to saying that it lacks the surefire disposition to fail to show one spot on a toss. If the inquirer knows that the die has the ability to show a one spot on a toss and that the die is tossed on some specific occasion, then sometimes the inquirer is obliged to judge it seriously possible that the die will show a one spot on that occasion. But this will not always be so. If the inquirer knows, for example, that the toss is by Morgenbesser and that the die has the surefire disposition to land with two spots showing on a toss by Morgenbesser, he should judge it not seriously possible that the die will land showing a one spot.

It may be thought that in this scenario the agent cannot consistently fully believe that the die (i) has the ability to show a one spot on a toss, (ii) has an ability to show two spots on a toss by Morgenbesser, and (iii) was tossed by Morgenbessr. This, I conjecture, is the view that Lewis would take if he were to speak in the idiom of abilities. (Though not explicitly discussed there, it appears to be an implication of Lewis, 1973, sec. 1.9.) Consider then a gas that is prepared by setting it at a given energy level. The particles in the gas are then, according to classical statistical mechanics, located at some point in a definite phase space and hence are to travel in the trajectory determined by the deterministic laws of classical mechanics for particles in that state. The gas is able to be at one of those points (to be in one of those states) and travel in the associated trajectory on having its energy level set in the given way. If we were in a position not only to know this but also the trajectory on which it traveled, we would be obliged to give up our conviction that it has the ability to be at any one of the other points in phase space associated with the given energy level. Thus, the assumption that the underlying mechanical system is deterministic obliges us to withdraw the attribution of the objective ability if we know too much about the details of the experiment and implies that the ability is too epistemically implicated to count as an objective feature of the gas. On this understanding, there can be no objective abilities when the "underlying" theory is deterministic.

This is, of course, precisely the view taken by J. Bernoulli (1713) and extended by him to reach the conclusion that there can be no objective probabilities given an underlying determinism. I have argued (Levi, 1990) that one can allow for objective probability or chance while remaining neutral with respect to underlying determinism by allowing for the consistency of claims like (i), (ii), and (iii). There is, of course, a cost to doing so. One needs to provide an account of "direct inference" from knowledge of objective abilities to judgments of serious possibility and (as Venn, Peirce, Fisher, and Reichenbach, among others, acknowledged) from knowledge of chances to judgments of serious possibility that specifies the kind of knowledge of the experiment that does and does not interfere with making the judgments. My point in mentioning this matter here is that objective abilities that do not have the requisite neutrality with respect to underlying determinism lack useful application in science. This is arguably so even if one assumes that the underlying physics is quantum mechanical.

13 D. Lewis originally introduced imaging as an alternative to temporal credal conditionalization as a constraint on changing credal probability judgments (Lewis, 1976, p. 311) Gärdenfors (1988, pp.108–18) extended

and elaborated the idea as "generalized imaging" in this setting. However, imaging can be redescribed as a mode of revising potential corpora or states of full belief without reference to probability judgments. Consider the set **K** of possible worlds in which all sentences in $\underline{K}$ are true. The imaging revision of $\underline{K}$ by adding h consists of the corpus or theory of sentences that are true in all possible worlds that are "nearest" or "most similar" worlds in which h is true to some world in **K**. h > g is acceptable relative to $\underline{K}$ if and only if g is in the imaging revision of $\underline{K}$ by adding h. This characterization is equivalent to the one given in the text. By construing imaging revisions as minimal revisions, many authors consider imaging tests of acceptability to be generalizations of Ramsey tests (Collins, 1991; Katsuno and Mendelzon, 1992). But, as explained in the text, such tests differ from the Ramsey test as Ramsey conceived it for the open conditional. John Collins has been exploring the merits of belief revisional accounts of conditionals based on a conception of revision through imaging.

14   For a list of postulates for imaging comparable (with slight adjustments for formulation) with postulates for Gärdenfors revision, see Katsuno and Mendelzon, 1992, p. 189, with discussion on pp. 189–91.

Chapter 4

1   Attention will be restricted to languages that at a minimum contain all so-called truth functional connectives. In addition, the interpretations are required to assign truth values to truth functional compound sentences as a function of their constituents in the customary manner. Indeed, to be useful for representing states of full belief, the language will also be expected to contain at least first-order logic and a suitable system of constraints on the notion of an interpretation for predicates, functors, and quantifiers.

2   Throughout this discussion, I shall suppose that logical consequence satisfies a compactness requirement. I do this for the sake of convenience; but I do not want to be committed to the view that for purposes of regulating the rational coherence of states of full belief, compactness must be satisfied by the consequence relation holding between belief states.

3   The ideas put forth in this discussion are the outcome of conversations with H. Arló Costa. He shares in credit for any insights that may emerge but should not be blamed for defects. The fruits of our discussions of conceptions of positive and negative validity are presented in Arló Costa and Levi, 1995.

4   Strictly speaking, the pair $\langle K,M \rangle$ should be the triple $\langle K,M,V \rangle$ to take account of the relativity to the basic partition of section 2.1 Because including it makes no material difference to this discussion, I leave it out of account.

5   The reason is that the intersection of a corpus in richer language ML with the nonmodal language $\underline{L}^1$ must be a nonmodal corpus $\underline{K}^1$ such that

$\text{QRT}(\underline{K}^1)$ is contained in the intersection of the corpus in ML with the language $\underline{L}^2$ capable of expressing noniterated conditional judgments. The consistent expandability assumption, (K*4w), and (K*5) all make demands on what may count as an expansion. Inconsistency can arise through expansion, therefore, without adding information incompatible with what is already in the corpus.

6   The expansion of $\underline{K}$ by adding h is the smallest deductively closed set in the language containing all the consequences of $\underline{K}$ and h and, in addition, meeting the requirements for being a corpus representing a potential belief state. In the example discussed in the text, the expansion of $\underline{K}^2$ by adding the negation of the given substitution instance of (A6) is not the deductive closure of $\underline{K}^2$ and the negated instance. The reason is that this deductive closure is not a potential corpus. Yet, the expansion in question is well defined. The smallest potential corpus containing the logical consequences of $\underline{K}^2$ and the negation of the substitution instance of (A6) is the inconsistent corpus. Consistent expandability fails even though neither the sentence added nor its negation is in the initial corpus and even though closure under expansion is satisfied.

Gärdenfors (1988, p. 49–51) introduces a set of axioms for expansion and defines an expansion transformation as a function from potential corpora to potential corpora that for given $\underline{K}$ yields the smallest potential corpus satisfying the axioms. It is easy to show that the expansion of $\underline{K}$ by adding h must contain all the logical consequences of $\underline{K}$ and h. Gärdenfors claims, however, that he can demonstrate that the expansion of $\underline{K}$ by adding h is identical with the set of all the logical consequence of $\underline{K}$ and h. This will be true in those cases where the set of potential corpora consist of all deductively closed theories in the language in question. The potential corpora in $\underline{L}^2$ are, however, a proper subset of the deductively closed theories in that language thanks to the constraints imposed by QRT.

7   H. Arló Costa (1990, pp. 560–1) has shown that the strong version of (K*5) used by Gärdenfors that takes the consistency of $\underline{K}$ to be sufficient for the consistency of $\underline{K}_h^*$ (the success postulate) conflicts with the quasi Ramsey test in the face of the reduction condition. The point I am making is that (K*5) as stated here (weak success) conflicts with the quasi Ramsey test as long as (K*4w) is retained as a constraint on *. Arló Costa also contends that weak success is not required for deriving VC and, in particular, for showing the validity of (A7). The postulate LA he uses instead cannot establish the positive validity of (A7) without assuming consistent expandability (Lemma 3, p. 563). However, a modified version of LA equivalent to PosVal does show (A7) to be positively valid. And PosVal is consistent with (K*4w). The spirit, if not the letter, of Arló Costa's observation is correct.

I shall say that the set of potential corpora is closed under *strict expansion* if and only if it is closed under expansion and for every potential $\underline{K}$,

$\underline{K}_h^+ = Cn(\underline{K} \cup \{h\})$. Given (K*3), closure under strict expansion implies PosVal. The converse, however, does not hold. PosVal is the weakest condition insuring (A7).

8   Suppose ~h is in $\underline{K}_h^*$ for some consistent potential corpus $\underline{K}$. $\underline{K}_h^*$ is either inconsistent or is not a potential corpus. By PosVal, ~h is positively valid. It is a member of every consistent potential corpus $\underline{K}$ in $\langle K,* \rangle$ and is, of course, in every inconsistent $\underline{K}$. Hence, whether $\underline{K}_g^*$ is consistent or not, ~h is a member. For every consistent $\underline{K}$ and every g, g > ~h is accepted relative to $\underline{K}$. (A7) is positively valid in brsw[2]'s satisfying PosVal. Conversely if (A7) is positively valid in a brsgw, PosVal must be satisfied by that system.

9   Suppose the negation of (A7) consisting of h > ~h and ~(g > ~h) is in $\underline{K}$. ~h must then be in $\underline{K}$ in virtue of (K*3). g cannot be in $\underline{K}$. Otherwise (K*4w) would yield g > ~h in $\underline{K}$ and we would have inconsistency. But then $\underline{K}_g^*$ must remove not only ~h but also h > ~h. This means that the revision of $\underline{K}_g^*$ by adding h is consistent. However, if this is so, by (K*2), h is in a consistent corpus. Hence, ~h cannot be negatively valid as the supposition of our argument and NegVal require. So the supposition must be rejected and (A7) is shown to be negatively valid. This argument does not, however, support the positive validity of (A7). As before, the converse is trivial.

10 Katsuno and Mendelzon, however, use (K*5) in a version that clashes with (K*4w) when a quasi Ramsey test and reduction condition are introduced. Katsuno and Mendelzon are not concerned, however, with providing a semantics or conditions of acceptability for conditionals, even though they appear to be aware of the possibility of doing so.

11 According to Lewis's semantics, T > h is true at a world if and only if h is true at that world. Hence, h and T > h share the same truth value in every possible world so that their equivalence is valid. Their equivalence is not positively valid when conditionals are Ramsey test conditionals. (T > h) $\supset$ (T $\supset$ h) but the converse fails as our discussion of (A6) reveals. Hence, T > h is to be interpreted as $\sim\Diamond\sim$h – that is, as claiming that ~h is not seriously possible. But T > h cannot be construed as claiming that ~h is impossible in some nonepistemic sense that entails the truth of h but is not entailed by it. In Lewis's system, $\sim\Diamond\sim$h = $\Box$h is defined by ~h>~T.

12 Gärdenfors (1988, p. 150, lemma 7.6) claims that satisfaction of (K*6) is necessary and sufficient for the validity of (A8). It is surely necessary, but it is not sufficient. Gärdenfors points out (pp. 57–8) that (K*7) and (K*8) imply (K*<7, 8>), which he labels 3.13. This, in turn secures (K*6) but not the other way around (p. 55).

13 Assume (K*7), (K*r8), and (K*$\langle 7, 8\rangle$) are satisfied. If ~g is not in $\underline{K}_h^*$, then either g is or is not in $\underline{K}$. In the latter case, the conditions for applying (K*r8) obtain and this implies that (K*8) applies as well. If g is in $\underline{K}$, (K*8) cannot be obtained from (K*r8). But in that case, h $\wedge$ g is in $\underline{K}_h^*$ and h is in $\underline{K}_{h\wedge g}^*$. (K*$\langle 7,8\rangle$) then secures the identity of the two revisions and the satisfaction of (K*8).

14   Gärdenfors shows that (K*7) is equivalent to claiming that the intersec-
     tion of $\underline{K}^*_h$ and $\underline{K}^*_g$ is contained in $\underline{K}^*_{hvg}$ (1988, pp. 57, 211–2, condition
     3.14). Let h > f and g > f be both acceptable according to $\underline{K}$. Then f must
     be in the intersection of the corpora $\underline{K}^*_h$ and $\underline{K}^*_g$. (K*7) then implies that
     f is in $\underline{K}^*_{hvg}$ and, hence, that h ∨ g > f is acceptable relative to $\underline{K}$. This claim
     holds for Ramsey revisions as well as Gärdenfors revisions. Hence, (A9)
     is positively valid for both brsg's and brsr's.

15   Gärdenfors (1988, p. 151) does not use negative RTGL and, as a conse-
     quence, he cannot say that satisfaction of (K*8) is necessary and sufficient
     for the validity of (A10). I, however, am using negative RTGL.

16   In the counterinstances to modus ponens furnished by examples 1–3,
     h > (g > f) is acceptable. So is h. g > f is unacceptable. What about
     h ∧ g > f? In both cases 1 and 2, this conditional must be acceptable
     in conformity with the exportation principle: (exportation) h ∧ g > f ⊃
     [h > (g > f)].

     That exportation must sometimes fail even for RTGL is clear. Even
     with (K*8) in place, we cannot guarantee that $\underline{K}^*_{h\wedge g}$ is a subset of $(\underline{K}^*_h)^*_g$.
     The interesting point is that the counterinstances to (A5itt) that demon-
     strate the failure of negative validity of (A5) (when recovery is violated
     and RTR is used) arise in contexts where exportation must be satisfied.

17   W. Rabinowicz (in correspondence) has raised an objection to the use of
     RTR that may be discussed usefully in this connection.

     Suppose h and f are in $\underline{K}$ but that f is not in $\underline{K}^{*r}_h$. According to RTR,
     h > f is not acceptable at K.

     Suppose, in particular, that h is "coin a is tossed once" and f is "coin a
     lands heads," and it is known that coin a is tossed and lands heads. RTR
     yields the plausible verdict that h > f is not acceptable at $\underline{K}$.

     Rabinowicz objects to the plausibility claim. He invites us to consider
     an arbitrary proposition g such that neither g nor ~g is in $\underline{K}$ – for exam-
     ple, the next president of the United States will be a Republican. h ∧ g > f
     is acceptable at $\underline{K}$ according to RTR.

     Rabinowicz thinks it odd that h > f is unacceptable even though for an
     irrelevant g h ∧ g > f is acceptable.

     Observe, however, that g > (h>f) is not acceptable at $\underline{K}$ according to RTR
     precisely because h > f is unacceptable not only at $\underline{K}$ but at $\underline{K}^*_g = \underline{K}^+_g$. (We
     have a clear violation of exportation even though (A5) is not violated.)

     Furthermore, g > f is acceptable at $\underline{K}$ whether or not one uses RTR or
     RTGL and whether or not one endorses recovery. Moreover, h > (g > f)
     is acceptable at $\underline{K}$ (in conformity with exportation).

     Rabinowicz should find this result worrisome if he finds his objection
     to RTR compelling. The outcome of the future election has "nothing to
     do" with the outcome of the toss. To avoid this conclusion, however, it
     would be necessary to abandon (K*r4) and preservation. These are
     conditions satisfied by both RTGL and RTR. Conformity with these
     conditions is mandatory if the condition of adequacy for Ramsey test

conditionals is to be obeyed. Hence, if Rabinowicz's objection is at all compelling, it should not be directed against RTR but against the demand that the condition of adequacy for open conditionals be satisfied.

Rabinowicz's objection, therefore, is raised from the perspective of those who think that conditionals assert or express some kind of causal or other dependency between g and f. It is not an objection according to those who endorse a characterization of conditionals as judgments of serious possibility conditional on suppositions.

Chapter 5

1　One of the main themes of Levi, 1991, is that the approach to belief change initiated earlier (in Levi, 1967, 1980) has as one of its ambitions to give an account of the conditions under which changes in states of full belief are justified, whereas the proposals of Alchourrón, Gärdenors, and Makinson, (1985) are restricted for the most part to formally characterizing the transformations of belief state without regard for the conditions under which the changes represented by such transformations are justified. In this essay, I focus on supposition where changes are stipulated for the sake of the argument. Hence, my concern here coincides more closely with the preoccupations of Alchourrón, Gärdenfors, and Makinson. Even so, when suppositional reasoning involves generalized expansion, the question of justifying changes in belief state cannot be neglected.

2　Under usual circumstances, the "potential answer" asserting that the die lands showing n spots carries the same informational value for each n from 1 to 6. In such cases, the appropriate response to a question about the outcome of the toss is to declare oneself in suspense as to what will happen. This seems to be what presystematic judgment supports. It is also what the approach to inductive expansion I favor recommends. This approach has been discussed previously (Levi, 1967,1980, 1984a, 1986a) and will be discussed at length in Chapter 6.

3　H. E. Kyburg (1961) recognized the dilemma very clearly a long time ago. He has exercised as much ingenuity as anyone in devising ways and means for living at ease with the dilemma rather than giving up the high-probability criterion.

4　C. S. Peirce initially claimed that there were two kinds of ampliative inference: inductive and abductive; but he correctly recognized in later years that the types of ampliative inference he was calling "abductive" should be classified among the inductive inferences.

5　The term *inductively extended Ramsey test* as well as the concept represented by the term were introduced and discussed by Fuhrmann and Levi (1994), where some of the distinctive feature of conditionals meeting the requirements of such tests were sketched.

6　Of course, not all of this information is explicitly listed in the premises in customary linguistic representations of such inference. Nor is it the case that the agent could do so on command. The corpus represents the agent's doxastic commitment – that is, the information the agent is committed to judge true with certainty and take for granted in inquiry at the moment. In practice, the agent cites only a few of the premises. But the agent is committed to the judgment that the uncited part of the corpus or evidence contains nothing that vitiates the inference to the conclusion.

7   Many authors, like R. Carnap and K. Popper, are antiinductivists. They deny that ampliative inductive inference is ever legitimate. Of course, Carnap and Popper differed from one another; but not over the issue of antiinductivism. They disagreed over probabilism. Probabilists mute their antiinductivism by trying to recast inductivist problems as questions about assigning probabilities. Popper criticized this project and, although his own positive proposals seem to me questionable, Popper was right to do so. It seems to me that antiinductivists should be candid about their antiinductivism. They should not waste so much paper writing about probabilistic or inductive support. The problem of characterizing indices of inductive support makes sense from an inductivist perspective. Inductive support in the maximizing sense is an index ordering potential answers to a given question on the basis of the amount of support so that one can recommend choosing the answer with the maximum support. Inductive support in the satisficing sense is an index such that the inquirer should accept as part of his new corpus all and only claims whose satisficing support is high enough. Both conceptions of inductive support presuppose that inductive inference is sometimes legitimate. Antiinductivists cannot accept this. And given this stance, it becomes unclear why they are interested in inductive support. Popper, to his credit, took his measures of corroboration not to be indices of inductive support but as indices of testworthiness. To be sure, Popperians can still be asked what the point of evaluating hypotheses with respect to testworthiness is if that appraisal is not ancillary to an inquiry that will eventuate in an ampiative inference.

Are there forms of ampliative inference that are not inductive? This may turn out to be a verbal dispute whose resolution depends on the usefulness of distinguishing between direct inference from information about statistical probabilities and the implementation of experiments to predictions of their outcomes, inverse inferences from the outcomes to conclusions about statistical probabilities, theory choice, and the like. I am not concerned with such matters here. Inferences of all of these varieties are sometimes legitimate according to the inductivist stance I am taking.

8   Observe that the nonmonotonicity of variably ampliative functions is proved for expansions of $\underline{B}$ by adding sentences that are consistent with $E(\underline{B})$ and not merely $\underline{B}$. In the case of ampliative but constant E-functions, monotonocity is never violated as long as expansions of $\underline{B}$ are restricted to inputs consistent with $E(\underline{B})$.

9   Since no undercutting prohibits undercutting in the case where $\underline{K}_h^+$ does not contain ~f, it must also be the case that h is consistent with $\underline{K}$. In such cases, AGM revision of $\underline{K}$ by adding h is identical with expansion of $\underline{K}$ by adding h. No undercutting is not equivalent to (K*8) nor can it be derived from

(K*8). The claim that expansion is equivalent to AGM revision when h is consistent with $\underline{K}$ presupposes both (K*8) and (K*3). And, as we have seen, we can reach no undercutting without (K*8) as long as we have (K*3). Observe, that if we change the requirement that ~f not be a member of $\underline{K}_h^+$ to ~f not being a member of $\underline{K}_h^*$ in the no undercutting principle, (K*8) contains an injunction against undercutting in such situations. No undercutting prohibits undercutting in contexts where undermining is allowed.

Chapter 6

1   Credal states need not be representable by single probability distributions. Credal probability judgment may be credally indeterminate. In such cases, it is representable by nonempty convex sets of probability distributions (see Levi, 1974, 1980).

2   The assessment of informational value may be represented by a convex set of M-functions that is not a singleton. Informational value like credal probability may go indeterminate. For simplicity of exposition this important consideration shall be neglected in this discussion.

3   Because Cut is satisfied by unextended * and *r, it is sufficient to consider cases where inductively extended revision is inductively extended expansion when neither h nor ~h is in $\underline{K}$. For essentially similar reasons, we focus on cases where f, g, and their negations are consistent with $\underline{K}$ as are the conjunctions constructible from h, f, and their negations and the conjunctions constructible from these with h and its negation.

   Suppose that f is in $\underline{K}_h^{+i}$ and g is in $\underline{K}_{h \wedge f}^{*i}$ for some h, f, and g in $\underline{L}$. This means that relative to $\underline{K}_h^*$, $f \wedge g$, $f \wedge \sim g$, $\sim f \wedge g$, and $\sim f \wedge \sim g$ must be consistent potential answers generated by the ultimate partition U_h. For the purpose of this discussion, we may think of these four sentences as the elements of the ultimate partition U_h relative to $\underline{K}_h^*$. By inductive expansion at least the third and fourth elements are rejected at the given level of boldness q. The ultimate partition $U_{h \wedge f}$ relative to $\underline{K}_{h \wedge f}^*$ may be simplified for the purpose of this discussion to the first two elements of U_h. Suppose the first two elements of U_h carry credal probabilities greater than 0.25 and the N-function assigns equal N-value of 0.25 to each of the four elements of U_h. Equal N-values of 0.5 are therefore assigned to the two elements of $U_{h \wedge f}$. Let $f \wedge g$ carry greater Q value than $f \wedge \sim g$ in U_h. Relative to $\underline{K}_h^*$ both go unrejected for all values of q. However, relative to $\underline{K}_{h \wedge f}^*$, $f \wedge \sim g = \sim g$ must carry Q-value less then 0.5. Hence, for q = 1 and some positive values less than 1, this conjunction is rejected so that g is in the inductive expansion for some range of values of q including q = 1. g is in $\underline{K}_{h \wedge f}^{*i}$ but not in $\underline{K}_h^{*i}$, even though f is. Cut is violated.

4   One might allow ampliative inductive expansion by distinguishing between hypotheses carrying probability 1 relative to $\underline{K}$ whose negations are inconsistent with $\underline{K}$ and those whose negations are consistent with $\underline{K}$. The former hypotheses are "certain," the latter "almost certain." Instead

of using the standard reals, this distinction is represented by replacing the standard reals as values of probabilities with nonstandard reals. The difference is stylistic. Either way, setting the threshold k at 1 in the standard reals (which is tantamount to allowing the acceptance of every hypothesis carrying probability differing from 1 by an infinitesimal amount) will indeed yield a theory but one that brings with it the following unsatisfactory consequence. Suppose we consider all values of the binomial parameter p for some process to be unknown and our concern is to estimate the true value relative to <u>K</u>. The credal probability distribution is continuous assigning 0 credal probability (probability differing by an infinitesimal amount from 0) to all point values of p. The criterion will recommend declaring that each one of the values of p is not the true value, even though we also know that exactly one of them is. If the consequence relation is compact, there will be no inconsistency in the theory formally. But I, for one, would not regard such a theory as coherent. For each standard real value for p between 0 and 1, the inquirer is initially almost certain that p is not the uniquely true value while being certain that there is such a uniquely true value among the standard values between 0 and 1. In this initial state, there is no incoherence. The agent who is almost certain that 0.5 is not the true value is in suspense as to whether it is or is not the true value. However, if the agent contemplates inductively expanding so that the new belief state converts these almost certainties to certainties, the agent should recognize beforehand that he or she is absolutely certain to import error into his or her belief state. And should the expansion strategy be implemented, the agent will become committed to being certain that no one of the standard values of p is the true one while being simultaneously committed to being certain that exactly one of them is. This is clearly an objectionable position.

One might seek to avoid the problem by suggesting that the potential answers generated by the partition consisting of the noncountably many standard real values of p should be restricted only to those elements in the Boolean algebra corresponding to principle filters or ultrafilters. That would rule out becoming certain to each of the values of p that it is not the true one. But then it becomes obscure what precisely the recommendation of the high-probability principle is when k =1.

The problem under consideration is, of course, the one illustrated by the "lottery paradox" made famous by H. E. Kyburg (1961). I shall return to this topic in section 8.3.

5   Earlier, (Levi, 1967, p. 94), I wrote that an agent might "detach" two distinct ultimate partitions and inductively "accept" conclusions relative to each. Kyburg (1983, p. 244) argued that I had admitted that the set of inductively accepted sentences could fail to be deductively closed so that my view was no different than his in this respect. I (Levi, 1967) drew a distinction between mere acceptance and acceptance as evidence. An inquirer merely accepted a potential answer to a question if the question was "seriously raised" and the ultimate partition relative to that question

was such that the conclusion was warranted. An agent could recognize different questions as serious at a given time and merely accept h relative to one of them and ~h relative to the other. But mere acceptance was not regarded as sufficient for acceptance into evidence. For that, a sentence would have to be accepted relative to all seriously raised questions. So Kyburg was right to say that the set of sentences merely accepted might be inconsistent. But mere acceptance is itself just simply the result of a bit of suppositional reasoning. The inquirer asks: "Suppose the only question of concern to me were this one. What would the best answer be?" The question of inductive expansion concerns what to accept into evidence. The only reason for being concerned with mere acceptance is that examining this topic clarifies the question of inductive expansion. By way of contrast, Kyburg's account of acceptance fails to explain why anyone should care about the topic.

I subsequently came to feel, however, that the perspective I had on the question of simultaneous detachment of ultimate partitions was at best misleading. I argued (Levi, 1984a, ch. 7; 1980) that when an inquirer confronts simultaneously two serious questions with different ultimate partitions, the inquirer should shift to the coarsest common refinement.

In any case, I have never endorsed the idea that one should deliberately expand the corpus of full belief into an inconsistent one or favored masking such expansion into inconsistency by refusing to close the set of inductively accepted sentences under logical consequence.

6   Among the many rediscoverers of Shackle's measures are: D. Lewis (1973), L. J. Cohen (1977), W. Spohn, (1988), L. Zadeh, (1978), Dubois and Prade, (1992), and Gärdenfors (1988). G. Shafer (1976) recognized Shackle's measure as a special type of "support function" – the so called "consonant" support functions. The first proposal to interpret Shackle measures in the manner now to be sketched is to be found in Levi, 1966 and 1967. See also Levi, 1984a, Chapter 14.

7   I pointed out that Shackle measures are interpretable as indices of "weight of argument" in 1967a, Chapters 8 and 9, where the weight of argument determines the extent to which the evidence justifies the agent in stopping inquiry and expanding by adding a given answer into the corpus of settled assumptions. Thus, Shackle measures were construed as what I now call "satisficing" measures (Levi, 1986b) of evidential support. The "satisficing" construal is also explicit in 1984a, Chapter 14, p. 221. See also section 8.4.

8   I derived (Levi, 1966; 1967; 1984a, ch. 14) Shackle measures from deductively cogent and caution-dependent rules for inductive expansion or acceptance. I did not explicitly mention partition sensitivity although the rules I favor are partition sensitive. I did explore rival views whose acceptance rules were not partition sensitive but where cogency breaks down.

9   The q index equals the ratio $(1 - a)/a$ where a is the "weight" assigned the value Q and $1 - a$ is the weight assigned the value of M. Because a can

range between 0 and 1, q can go from 0 to infinity. However, because no error should be assigned higher epistemic utility than any correct answer, a can be no less than 0.5 and q is restricted to the interval from 0 to 1. Observe, however, the weighted average being used assumes that the utility of avoiding error and the utility of error are 1 and 0 respectively and that the informational value is $1 - M$, which is also restricted to values between 0 and 1. However, each of these utility functions can be subjected to a positive affine transformation independent of the other. There will be a weight b such that the weighted average of the transformed utility of avoiding error and utility of information functions is a positive affine transformation of the original weighted average. $q' = (1 - b)/b$ will equal cq where c is the ratio $a/a*$ of positive constants by means of which the utility of avoiding error and the utility of new informational value are transformed. q' will range from 0 to c. q' can serve just as well as an index of boldness as q. It represents the weighting of the utility of avoiding error and of acquiring new informational value just as well as the "standardized" index I deploy does.

10   Condition (d2) must be extended to cover where the e_i's constitute an infinite set. The proper generalization is not obvious. If all the elements in the refinement of h by the e_i's carry positive d-value but the infimum is 0, are we prepared to set the degree of disbelief that h at 0?

      If the family of deductively cogent, caution-dependent, and partition-sensitive inductive expansion rules is of the kind I have favored where the index of caution or boldness is representable by a real value parameter taking all values within a finite interval (including end points), then there is a largest value of q at which the e_i's in the set all fail to be rejected. That will be least upper bound of the values of q associated with an element in the set. So the d-value assigned the claim that at least one of the e_i's is true is equal to the greatest lower bound of the d-values assigned an element of this set. A similar result applies mutatis mutandis to b-values assigned the joint truth of infinitely many hypotheses.

11   In my judgment (Levi, 1979; 1984a, pp. 235–41), Cohen's interpretation of the Shackle formalism in terms of his conception of inductive support is inadequate. Unlike other reinventors of the Shackle measures, however, Cohen does not think that a formal system together with an appeal to intuition is a surrogate for a clear account of the intended application.

12   The inductive expansion principle was formulated for finite U in section 6.6. The bookkeeping procedure has also been described for finite U. With the aid of bookkeeping the criteria for inductive expansion may be extended to categories of infinite ultimate partitions of general interest. An explanation will be given in the appendix. The idea of bookkeeping was initially proposed in Levi, 1967, pp. 151–2. Repeating the procedure until a fixed point is reached was not, however, considered. Reiterating to a fixed point was first discussed in Levi, 1976, pp. 41–8. It was further

elaborated in Levi, 1980, pp. 53–6, where the sentences added to $\underline{K}$ due to bookkeeping to a fixed point were said to be stably accepted.

13   According to the proposed inductive expansion rules, h in U is rejected if and only if $Q(h)/M(h) < q$. Let U^n be the subset of U surviving rejection at nth iteration. Let $\Sigma^n Q$ and $\Sigma^n M$ be the total Q-value and M-value respectively of elements of U^n. Consider now the n + 1st iteration. h* in U^n is rejected if and only $Q(h^*)/M(h^*) < q\Sigma^n Q/\Sigma^n M$. Every unrejected element h* is such that $Q(h^*)/M(h^*) \geq q\Sigma^n Q/\Sigma^n M$. We perform an induction on the first application of the rules with fixed q at which a fixed point is reached. If n = 1 and q = 1, $Q(h)/M(h) = \Sigma^1 Q/\Sigma^1 M = 1$ for every h* in $U^1 = U$; for otherwise $Q(h^*)/M(h^*) < 1$ for some h* in U^1 and every such h* is rejected in contradiction with the assumption that a fixed point is reached. It follows that for every h* in U^1, $Q(h^*)/M(h^*)$ is a maximum among elements of U. Let n be the first iteration at which the fixed point is reached when q = 1. As the induction step we assume that all and only the elements of U^n carry maximal Q/M-values among elements of U when q = 1. Then for a case where q = 1 and the n + 1st iteration is the first at which a fixed point is reached, $U^{n+1} \subset U^n$ when q = 1 but $U^{n+2} = U^{n+1}$. Then every unrejected element h* of U^n at the n + 1st iteration is such that $Q(h^*)/M(h^*) = \Sigma^{n+1} Q/\Sigma^{n+1} M$, whereas every rejected element h** of U^n is such that $Q(h^{**})/M(h^{**}) < \Sigma^n Q/\Sigma^n M$. Since, U^n is not a fixed point, $Q(h^*)/M(h^*) = \Sigma^{n+1} Q/\Sigma^{n+1} M$ must be greater than $\Sigma^n Q/\Sigma^n M$. So $Q(h^*)/M(h^*) > Q(h^{**})/M(h^{**})$. This proves the claim in the text. This result was asserted in Levi, 1976, p. 44. A version of the argument reported here is given in Levi, 1980, p. 54.

14   Consider a finite ultimate partition U relative to $\underline{K}$. Let bookkeeping to a fixed point with constant q be the inductive expansion ruled deployed to define $E(\underline{K}) = \underline{K}^i$. Let f be in $\underline{K}^{+i}_h$ and g in $\underline{K}^{+i}_{h \wedge f}$. Since f is in $\underline{K}^{+i}_h$, at some iteration in the bookkeeping process, enough elements of the ultimate partition U must have been rejected to entail f. The resulting set of unrejected elements of U is U^{hi}. The corpus $\underline{K}^+_{h \wedge f}$ leads to rejecting either the very same subset of U or a smaller subset so that the resulting set of unrejected elements of U, $U^{h \wedge f}$, is a superset of U^{hi}. If $U^{hi} = U^{h \wedge f}$, then bookkeeping to a fixed point or stably accepted conclusion must give the same result whether we start with $\underline{K}^+_h$ or with $\underline{K}^+_{h \wedge f}$. If $U^{h \wedge f}$ is a proper superset of U^{hi}, then, starting with $\underline{K}^+_{h \wedge f}$, finitely many iterations of the bookkeeping method with fixed q must lead from $U^{h \wedge f}$ to U^{hi} and thereby return us to the first case. $\underline{K}^{+i}_{h \wedge f} = \underline{K}^{+i}_h$ and the Cut condition is guaranteed no matter what the value of q is. (K*i7) will fail, in general, unless q = 1. So will (K*i8) even for *i.

Hilpinen (1968, p. 102) pointed out that the inductive expansion rules I proposed (Levi, 1967) violate a condition (CA4) essentially equivalent to Cut. Because he thinks an inductive acceptance rule should satisfy Cut and because bookkeeping iteration yields no new information when Cut is satisfied, he thinks bookkeeping iteration is pointless or illegitimate.

I refuse to be impaled on the horns of this dilemma. Inductive expansion need not satisfy Cut. However, iteration can be legitimate and, when it is carried out to a fixed point, Cut is satisfied by the inductive expansion rule that is thereby generated.

Chapter 7

1 Observation 4 of Makinson (1994) furnishes a proof of the failure of (K*7) as a proof of the failure of "distribution."

2 Makinson and Gärdenfors (1991) read Poole's approach as follows. Poole does not take the set of facts to be closed under deduction, so Makinson and Gärdenfors consider the special case where the set of facts is a singleton x. They take a maximally consistent set D of instances of Δ to be a maxichoice contraction removing $\sim$x from an initial corpus $\underline{B}$ consisting of all instances of Δ. A Poole extension is the set of deductive consequences of such a maxichoice contraction and x. Taking the deductive closure of the total set Δ of default instances to be the initial corpus, one can then say that x $\mathrel{\mid\!\sim}$ y according to Poole defaults if and only if y $\in \Delta_x^*$ in the special case where AGM revision is a full meet revision in the sense of Alchourrón and Makinson (1982). In my judgment, this reading is misleading at least when Poole systems are used for the purpose of prediction. On the intended applications of default reasoning in "prediction," the initial information in the belief set or corpus $\underline{B}$ consists of the logical consequences of the "facts." New information is added via default by an ampliative inductive expansion of some expansion (say, by adding x consistent with $\underline{B}$) of the facts. If one begins with an initial corpus $\underline{B}$ and adds new information x consistent with $\underline{B}$, Poole's theory yields an account of what the inductively extended expansion $\underline{B}_x^{+i}$ should be. It is to add the intersection of all Poole extensions of $\underline{B}_x^+$. $\underline{B}^i$ is simply the intersection of the Poole extensions of $\underline{B}$. Since this inductive expansion is closed under logical consequence, it does not matter whether the set of default sentences initially given is closed under consequence or not. The Makinson–Gärdenfors interpretation is equivalent to mine under the assumption that $\Delta = \underline{B}^i$. This will be so if (i) $\underline{B}$ is the set of logical truths, (ii) a maxichoice contraction removing $\sim$x from Δ is a maxichoice contraction from the logical closure of Δ, and (iii) inductive expansion is permutable. Then we might look at Pool's defaults as prescribing how the inductive expansion $\underline{B}^i$ $(= \Delta)$ should be revised by adding x and claiming (correctly) that this revision will be a full meet revision. I do not claim the hermeneutical insight required to ascertain whether this conforms to Poole's intent; but whether or not these conditions are met, the reading I favor does make sense of the intention to apply Poole defaults to the so-called task of prediction, which seems clearly to be a case of ampliative reasoning without claiming that the set Δ itself is closed under logical consequence. Of the three assumptions needed to secure the equivalence of the Makinson-Gärdenfors reading of Poole and mine, condition (i) is specific to the special case under discussion and may

readily be extended to an arbitrary consistent corpus $\underline{B}$. If permutability of inductive expansion (iii) is adopted, (ii) will follow as a matter of course. Makinson (1994) has established (in effect) that Poole systems satisfy $(K*i7)$. The question is whether $(K*i8)$ holds. This reduces to the question of the exportability of inductive expansion. Let us suppose that $\underline{B}^i$ does not contain ~g. This means that ~g is not in $\underline{B}$ and that Δ does not entail both g $\supset$ w and g $\supset$ ~w for any w. If one assumes that neither w nor ~w is in $\underline{B}$, w will be in $(\underline{B}^i)^+_g$ if and only if g $\supset$ w is in Δ. The same is true for $\underline{B}^{+i}_g$. In virtue of this circumstance, $(\underline{B}^i)^+_g$ must be a subset of $(\underline{B}^+_g)^i$ so that inductive expansion is exportable. Inductive expansion according to Poole is ampliative but constant.

3   Let the defaults be g; **M**h/h, **M**g/g, and h; **M**~g/~g. Let the initial corpus $\underline{B}$ contain neither h nor ~h and neither g nor ~g. The only Reiter extension of $\underline{B}$ contains $Cn(\{h,g\})$. Hence, h is in $\underline{B}^i$ and so is g. Consider, however, the operation of the defaults on $\underline{B}^+_h$. One extension, as before, includes $Cn(\{h,~g\})$. But there is a second extension including $Cn(h,~g)$. The intersection of the two extensions cannot contain g as a member. Hence, even though h is in $\underline{B}^i$, $\underline{B}^i$ is not a subset of $\underline{B}^{+i}_h$. Inductively extended cautious monotony fails and with it $(K*i8)$.

4   Notice that if "Tweety is a penguin" were part of the background, the supposition that Tweety is a bird would not license the conclusion that Tweety flies – not even by default.

5   This foundationalist view is not to be confused with the claim that when new information obtained by routine expansion conflicts with information already in the belief state, relief from conflict is to be sought by removing either the new information or some element of the background or both. This latter view (which, I conjecture, foundationalists would also endorse) does not claim that the decision as to whether to give up the new information, old background information, or both depends on the reasons on the basis of which the new and old information were originally obtained. Foundationalists endorse this latter requirement.

6   The default assumption is constituted out of the default restriction and the revision assumption. Gärdenfors endorses the revision assumption in virtually all of his writing on belief change. Nonetheless, Gärdenfors disclaims any commitment to foundationalism. This he can consistently do as long as he does not endorse the default restriction as well. In the face of new information that Tweety is a penguin, an inquirer might give up the claim that no penguins fly rather than the claim that Tweety flies depending on the relative "entrenchements" of these claims. There is no principled requirement in Gärdenfors's approach or in the theory of Alchourrón, Gärdenfors, and Makinson that requires giving up the claim that Tweety flies because it had been obtained previously by default reasoning. Their theory would have to be supplemented not only by the revision assumption (that Gärdenfors clearly endorses) but also the default restriction that stipulates that what needs to be brought into question are the results of previous inductive expansions.

Chapter 8

1   Of course, in the future, the agent may have good reason to alter
    conclusions previously reached in this way. But any extralogical or extra-
    mathematical belief is corrigible in this sense – although not necessarily
    in a way satisfying the default assumption.

2   Probabilism ought not to be confused with Bayesianism. Strict Bayesians
    hold that rational agents are in a state of all belief and are in a credal
    state representable by a unique probability distribution over the
    serious possibilities relative to the state of full belief. They endorse the
    maximization of expected utility as the fundamental maxim for rational
    decision making. And they endorse confirmational conditionalization as
    a constraint on rules for changing credal state with change in state of full
    belief (these rules are called "confirmational commitments" in Levi, 1974
    and 1980). Strict Bayesians who are also probabilists deny the legitimacy
    of inductive expansion . Some probabilist strict Bayesians (e.g., Carnap,
    1960) allow for expansion through observation. More radical proba-
    bilists (e.g., Jeffrey, 1965) deny the legitimacy of any kind of expansion.
    Previously (Levi, 1974, 1980), I proposed deviating from strict
    Bayesianism by allowing that rational agents adopt credal states
    recognizing more than one probability function to be permissible. Still I
    endorse a generalization of the principle of expected utility maximization
    and confirmational conditionalization. And I am an inductivist.

3   Ramsey (1990, p. 82) thought that probability judgment is ampliative
    even though the best arguments against this position are those I am now
    cribbing from him.

4   Carnap (1960) did, however, acknowledge his antiinductivist opposition
    to allowing inductive expansion to be legitimate. On this point, his view
    and Popper's are in agreement. Moreover, both authors agree that their
    opposition is motivated by a desire to avoid Hume's problem.

5   L. J. Cohen (1977) is one of the reinventors of the Shackle formalism. He
    has also interpreted Shackle measures as satisficing measures of belief
    worthiness. He calls his measures degrees of "Baconian probability." He
    appeals to a somewhat different measure of what he calls "inductive
    support" of lawlike generalizations to derive assessments of Baconian
    probability for singular propositions. Although he does not seem to
    recognize this, his procedure can be reformulated as appealing to a family
    of deductively cogent, caution-dependent, and partition-sensitive rules.
    The family differs from the kind I favor. I discuss his proposals in Levi,
    1984a, pp. 235–241.

6   Holding $\underline{K}$ fixed, the expansion strategy may be represented by the single
    sentence d. Strictly speaking, however, inductive support applies to the set
    of sentences that is the output of an expansion strategy and only indirectly
    to a sentence accepted as strongest via induction. Inductive support in
    the satisficing sense applies directly to sentences and determines the

expansion strategy as the set of sentences carrying support above a given threshold.

7   Observe that adding logical truths or sentences in the initial corpus $\underline{K}$ does not, in general, carry maximum evidential support in the maximizing sense.

8   Tversky and Kahneman speak of experimental subjects being subject to cognitive illusions, being in error, and committing fallacies in making probability judgments. They do not mean to suggest that assigning a probability to a proposition can be in error. The errors they discuss are failures to conform to the constraints imposed by probability theory – that is, the calculus of probability – on the coherence or consistency of probability judgments. But as they recognize, a failure to conform to the calculus of probability need not be a mistake or fallacy. It might mean that the judgments made by the experimental subjects are not probability judgments. Tversky and Kahneman address this point by identifying an experimental standard for judging that experimental subjects have committed a fallacy. They say that if such a subject is prepared "after suitable explanation" to concede that they made a nontrivial error, that the error was conceptual and they should have known the correct answer or a procedure to find it, they are to be judged as having committed a fallacy. If the experimental subjects reject these propositions, they have misunderstood the question rather than committed a fallacy.

The difficulty with this proposal is to identify what constitutes a "suitable explanation." It is not enough to point out why a failure to conform to the calculus of probability has occurred. One should present alternative explanations relative to which no mistake is made and ascertain whether the experimental subjects favor one interpretation of their intent or the other. It is far from clear that Tversky and Kahneman have done this.

9   In public comment on a talk I gave that covered this matter, Tversky claimed that he and Kahneman had anticipated objections like the one I am raising on pp. 311–12 of their 1983 paper. There is no foundation whatsoever for this claim.

Kahneman and Tversky are responding to the suggestion that experimental subjects in the Linda problem are responding in conformity with Gricean conversational rules, which urge speakers to convey all the relevant error-free information according to their beliefs. Tversky and Kahneman seek to extend this idea to the case of an uncertain speaker and consider the maxim: Maximize the "expected value" of the message to be conveyed, which is described as the information value of the message if true weighted by the probability that it is true. They assert (without argument) that it is unlikely that their experimental subjects interpret the request to rank statements by their probability as a request to rank them by their expected informational value. They also point to the presence of conjunction fallacies in numerical estimates and choices of bets (p. 312).

The passage does not anticipate my worry. Kahneman and Tversky are concerned with an objection to their experiment based on a view as to what experimental subjects will say in response to a question given their beliefs. My objection is based on a view as to how experimental subjects approach forming new beliefs and not new utterances.

It is true that the objection I am advancing denies that the experimental subjects are concerned to maximize probability and that is a feature it shares in common with the view considered by Tversky and Kahneman. But Tversky and Kahneman consider and dismiss without argument a proposed index being maximized (a product of probability and informational value) quite different from anything I consider and inappropriate to the evaluation of potential answers that are candidates for belief. So not only is their first objection directed to the Gricean conversational theory they reconstructed, it objects to maximizing a nonprobabilistic index, which, I agree, is not an appropriate candidate as an index to be maximized and, moreover, they do so without argument.

The second objection they raise concerns the fact that conjunction fallacies are committed in estimation and betting rate problems. I concede that and did so in the lecture to which Tversky responded as well as (Levi, 1985c, pp. 336–9); here I first discussed the conjunction "fallacy" and discussed this matter on the basis of the 1983 paper. But I denied in 1985, in the talk to which Tversky replied and deny here that the fact that such fallacies are committed in the estimation and betting rate problems implies that they are committed in the Linda and medical diagnosis problems.

10 Proposals of this sort have been favorites of authors with probabilist predilections from the time of H. Jeffreys and R. Carnap and remain of interest to so-called Bayesians to the present time – see, for example, the comments of Horwich (1982), Howison and Urbach (1989), and the sometime Bayesian, Earman (1992). I first raised the objection to this view in a paper published in 1967 (see, Levi, 1984a, ch. 5). More recently, Popper and Miller (1983, 1984) attacked this probabilist conception of support as a measure of inductive support because h can be factored into an ampliative part $h \vee \sim x$ and an explicative part $h \vee x$ where the ampliative factor can never receive positive probabilistic support. Jeffrey (1984) among others responded by pointing to other factorizations that do allow the ampliative factor to carry positive support. The debate was then turned into a controversy over who had the right factorization. My own view is that this debate rests on faulty assumptions. Any ampliative factor is equivalent given the evidence to h itself and, hence, all ampliative factors should receive the same inductive support. The difference between a posterior and a prior fails to meet this requirement. So Popper and Miller have the right conclusion; but their argument is insufficient to establish the claim. (See Levi, 1984b, ch. 5, 1986b.)

11 Keep in mind that the conditional credal probability $P(x/h)$ determines the fair betting rate for bets on x that are called off when h is false. This is

the definition given by Ramsey (1990, ch. 4, p. 76), de Finetti (1964), and Shimony (1955), among others. As de Finetti (1972) appreciated, this allows conditional credal probability to be well defined when h carried credal probability 0, provided it remains a serious possibility. I have provided (Levi, 1980. sec. 5.6; 1989) additional discussion of this point. One way of representing the difference is by assigning possible propositions carrying 0 probability in a representation in the standard reals a positive infinitesimal probability. But one can proceed in the usual way with probability functions taking values in the standard reals and say that propositions whose probabilities are 0 but seriously possible are distinguishable for important practical purposes from propositions judged impossible. Whether one regards probability as taking values in the standard reals or in the nonstandard reals is a matter of style and not of substance.

12   The value of $P(x/h)$ when $P(h) = 0$ cannot be derived from the values of the unconditional probabilities $P(h)$ and $P(x \wedge h)$ both of which are 0. The conditional probability must be specified independently either in terms of called-off bets as described in the previous footnote or in some other way.

13   To my knowledge, Kyburg initiated the use of the term *corpus* as a set of "accepted" sentences. A long time ago, I adapted Kyburg's term to my own purposes. His corpus is not deductively closed. Because I want a corpus to represent a state of full belief used for a standard of serious possibility, corpora, according to my usage, are.

14   Kyburg himself does not introduce what I am calling "Kyburgian degrees of belief"; but the notion is easily defined within his framework and provides one way of approaching the characterization of corpora at the various levels of acceptance that he does deploy. I mentioned the possibility of introducing measures of this sort previously (Levi, 1984a, ch. 14).

15   Avoidance of error should not be a concern of someone determining what to accept in Kyburg's sense as we have seen. It is not clear, however, whether Kyburg's sense of acceptance is the same as that of van Fraassen (1980), according to whom when one accepts a theory one should not be concerned to avoid error except among the phenomena. Van Fraasen's use of "acceptance" compounds confusion. Although a person who accepts a theory apparently is to conduct inquiries and deliberations as if the person fully believed the theory, the person does not believe the theory but only accepts it. This is because the agent was not concerned with the truth of the theory in forming his or her opinion. Van Fraassen's distinction does not, however, mark a difference in the attitude one acquires after changing one's view but a difference in the goals one had prior to the change. It would have been much clearer to claim that the agent does come to believe the theory and, indeed, comes to believe that theory is true but, prior to fixing this belief, was not concerned to avoid error in coming to this conclusion. Confusion is compounded by van Frassen's contention that accepting a theory is adopting a weaker and,

hence, less risky conclusion than believing it. This is a flat mistake. The claim believed and the claim accepted are the same. Neither is weaker than the other. There is no difference in probability of error. The only sense in which risk of error is reduced is by the adoption of cognitive goals according to which avoidance of error in deciding which theory to believe (= accept) does not matter as long as it does not impact on one's views concerning the phenomena. As in all risky decision making, one can reduce risk by changing probabilities or by changing utilities. Adopting weaker answers is reducing risk in the first way. Van Fraassen says he is doing this. But it is clear that he is reducing risk in the second way – that is, by changing utilities.

16 Dubois and Prade are prone to use the modal phraseology of Zadeh's possibility theory. b-functions are necessity measures. The function π such that $\pi(h) = 1 - b(\sim h)$ represents the degree of possibility that h. $d(h)$, the degree of disbelief or potential surprise that h, is equal to $b(\sim h)$, and, hence, to $1 - \pi(h)$. $\pi(h)$ is precisely $q(h)$ defined in section 6.9 as the maximum q-value at which h fails to be rejected. (See also Levi, 1967, pp. 131ff.) In discussing their views, I make the required translation from talk of possibility to surprise or degree of belief.

17 See Dubois and Prade, 1992, p. 148. Dubois and Prade characterize "epistemic states" by means of an ordering of sentences in $\underline{L}$ in terms of "qualitative necessity," which, it turns out, is representable by a Shackle b-function. The set of sentences carrying maximum b-value belongs to the "kernel" of the "ordered belief set" $\underline{K}$ consisting of all sentences carrying positive b-value. No interpretation is given of the ordering, except it is intended to compare sentences with respect to "levels of certainty" (p. 144).

18 I have outlined (Levi, 1967) the procedures for updating in this fashion.

19 Alchourrón, Gärdenfors, and Makinson (1985) consider only intersections (meets) of subsets of maxichoice contractions and show that contractions characterized this way satisfy the recovery postulate. Thus, the merits of recovery stand or fall with the merits of restricting the potential contractions removing h from $\underline{K}$ before determining which contraction strategy removing h from $\underline{K}$ to choose. In my judgment, one should begin by considering all contractions removing h from $\underline{K}$ and identify which strategy is best among these. If recovery is to be obeyed, one must show that a contraction from $\underline{K}$ removing h that is not an intersection of subsets of maxichoice contractions is always inferior to some intersection of subsets of maxichoice contractions. I have yet to see a convincing argument for this view and there are substantial presystematic precedents for supposing that recovery should not be obeyed (see Levi, 1991). See also the remarks in Lindström (1990) and Rott (1993) discussed in Chapter 2, note 7.

20 A saturatable contraction from $\underline{K}$ by removing h that carries maximum M-value and, hence, minimum undamped informational value must exist in $S(\underline{K},h)$. Expanding such a contraction by adding $\sim h$ yields a maximally consistent set. The ordering of saturatable contractions with

respected to M-value and, hence, with respect to damped (= undamped) informational value must agree with the ordering of the worlds determined by saturatable contractions with respect to M-value. Consider the set of all such "worlds" or maximally consistent sets. The total M-value of this set must be finite. Therefore, for any given world carrying positive M-value, at most finitely many other worlds carry M-value as great as it does. Hence, there must be a world in the set that carries maximum M-value. The M-value of a saturatable contraction removing h that yields that world when ~h is added must carry maximum M-value (and, hence, minimum damped informational value) among the set $S(\underline{K},h)$ of such saturatable contractions.

21   The existence of a minimum M-value can be guaranteed by requiring, as I proposed in the appendix to Chapter 6, that in a countably infinite set of alternatives, M should be bounded away from 0. See also Chapter 9, notes 3 and 4.

Chapter 9

1   Closure, left logical equivalence, and reflexivity characterize what Lindström (1990) calls an *inference relation*. Gärdenfors and Makinson (1993) consider these three conditions plus weak conditionalization, weak rational monotony, and consistency preservation to be the *basic* postulates for nonmonotonic inference. When these are supplemented by conditionalization and rational monotony, the postulates constitute the *extended* set of postulates for nonmonotonic inference.

2   In spite of the views of Arló Costa (1994) and Freund and Lehmann (1994), this means that (K*3) and (K*4) *as conditions on unextended AGM revision* have an essential role to play in illuminating the nonmonotonic features of rational inferential relations. They clearly do in the case of *inductively unextended* rational pseudoinferential relations where all nonmonotonicity is belief contravening. They also do for inductively extended rational pseudoinferential relations. All consistency preserving rational pseudoinferential relations are derivable via the translation manual of section 5.4 from permutable inductive expansions of unextended revisions obeying *all* of the AGM conditions. Since the identity inductive expansion function is permutable, the case where the pseudoinferential relations are unextended is just a special case. Of course, when the inductive expansion function is not only permutable but ampliative as well, *i does not obey (K*3). And $\underline{B}^{*i}_T$ is not, in general, identical with $\underline{B}$. Arló Costa is right about these points. But these facts are symptoms of the ampliativity of the inductive expansion function i. Ampliativity is extremely important to nonmonotonicity as the proof of the ampliativity thesis of section 5.6 shows. Moreover, *i is a composition of * and i and * obeys both (K*3) and (K*4). Finally, when i is permutable, (K*i3m) and (K*i4m) are equivalent to substitution instances of (K*3) and (K*4) construed as regulating unextended AGM revision.

3   According to Lehmann and Magidor (1992), ranked preference relations
    that generate rational inferential relations must be smooth and modular.
    But the preference relation need not be well founded so that infinitely
    ascending chains (in terms of increasing preference) are precluded. Lemma
    2.7 provides an example of such a mathematical possibility. The example
    specifies a countable set of states. According to the interpretation I am
    proposing for inductive expansion, such states should be understood as
    elements of a countably infinite ultimate partition. The order yielded by the
    preference relation consists of an infinitely ascending chain bounded by a
    most preferred element. In the case where we are concerned with inductive
    expansion, this situation can arise only if the informational determining N-
    function assigns 0 value to the maximal element rendering it equal in
    informational value to a contradiction. In the appendix to Chapter 6, I
    contend that demands for information of this kind ought to be ruled out of
    consideration. I do not deny the mathematical possibility noticed by
    Lehmann and Magidor but only the dubiety of the demand for information
    that it represents in this context of application. It should also be observed
    that a similar restriction ought to apply to demands for information that
    yield corresponding preference relations for the purpose of modeling AGM
    revision. When the set of states or elements of U is noncountably infinite, I
    have allowed for the possibility of infinitely ascending chains and indeed
    for the failure of the smoothness condition. But in those cases, the models
    fail to be ranked or, indeed, cumulative.

4   A more qualified story needs to be told when U is infinite. In the appendix
    to Chapter 6, the topic is discussed for three main types of examples. The
    procedure developed shows how to obtain results for such cases by appeal-
    ing to the inductive expansion principle for the finite case.

5   Lindström (1990) and Rott (1993) exploit the methods of revealed prefer-
    ence theory as explained, for example, in Herzberger (1973) and Sen (1982)
    to derive "preference" relations for use in constructing models for nonmo-
    notonic inferential relations but it is misleading to claim that these are
    choice theoretic approaches. In the case of choosing how to contract from
    a corpus in, let us say, AGM revision, the set of potential contraction
    strategies (the analogue of the set of options) is not identical with the set of
    maxichoice contractions in the sense of AGM or the set of saturatable
    contractions in the sense of Levi,1991. Preferences over the latter elicited
    by appeal to choice functions does not yield the relevance preference
    ranking. The same is true for inductive expansion. Deriving a preference
    over U by utilizing choice functions defined for subsets of U does not yield
    a preference over all the inductive expansion strategies. The relevant
    domain in this context is all elements of the power set of U (including the
    empty set or expansion into inconsistency). It is the latter preference that is
    relevant according to a throughgoing decision theoretic approach.

6   Trouble would arise if x is inconsistent with $\underline{B}$. One would first have to
    revise $\underline{B}$ before inductive expansion takes place. In the previous discussion,

attention has been focused on inductively extended nonmonotonic inference of the form x ‖~y where such inferential relations are derived from inductive expansion rules applied to consistent expansions of an initial corpus <u>B</u>. Cases where x is inconsistent have been disallowed by treating <u>B</u> *as if* it were a set of logical truths and understanding the reflexivity, left logical equivalence, and closure conditions as requiring that x ‖~ ~T. Revisions of <u>B</u> by adding x could be considered if the need arises; but doing so would require replacing <u>B</u> with another weaker corpus <u>B</u>′ consistent with all x's that we wish to consider as left-hand terms in the relation x ‖~y. Because our interest here is in consistent inductively extended expansions of <u>B</u>, we can rest content with focusing on cases where x is consistent with <u>B</u> and the inductively extended nonmonotonic inferential relations generated relative to <u>B</u>. Although Gärdenfors and Makinson do not put the matter in this way, they do by implication presuppose a corpus <u>B</u> of "firm" beliefs such that it is treated as maximally incorrigible in the given context of application. This need not mean, however, that if x inconsistent with <u>B</u> should be added, it would be inappropriate to revise <u>B</u> by x but that, for the context-bound purposes at hand, we do not intend to consider such cases. The pseudoincorrigibility of <u>B</u> is postulated as a technical convenience in order that conditions on nonmonotonic inferential relations may be conveniently formulated for situations where x inconsistent with <u>B</u> on the left-hand side of x ‖~y is not seriously contemplated.

7   According to Gärdenfors and Makinson (1993, sec. 3.3), defaults of the type F's are normally G's are expressed by saying that for all b, $Fb \supset \sim Gb$ is less expected than $Fb \supset Gb$. That is to say that $Fb \land Gb$ is less surprising than $Fb \land \sim Gb$. $(C \mid\sim)$ then states that relative to $\underline{B}^+_{Fb}$, Gb is positively expected.

Bibliography

Adams, E. (1975). *The Logic of Conditionals*. Dordrecht: Reidel.

Alchourrón, C., Gärdenfors, P., and Makinson, P. (1985). "On the Theory of Logic Change: Partial Meet Functions for Contraction and Revision." *Journal of Symbolic Logic* 50: 510–30.

Alchourrón, C., and Makinson, D. (1982). "The Logic of Theory Change: Contraction Functions and Their Associated Revision Functions." *Theoria* 48: 14–37.

Arló Costa, H. (1990). "Conditionals and Monotonic Belief Revisions: The Success Postulate." *Studia Logica* 49: 557–66.

Arló Costa, H. (1994). "Epistemic Conditionals, Snakes and Stars." In *Conditionals and Artificial Intelligence*, edited by G. Crocco, L. Fariñas del Cerro, and H. Herzig. Oxford: Oxford University Press.

Arló Costa, H., and Levi, I. (1995). "Two Notions of Epistemic Validity." Unpublished manuscript.

Arló Costa, H., and Shapiro, S. (1992). "Maps between Nonmonotonic and Conditional Logic." In *Principles of Knowledge Representation and Reasoning: Proceedings of the Third International Conference*, edited by B. Nebel, C. Rich, and W. Swartout, pp. 553–64. San Mateo, Calif.: Morgan Kaufman.

Bernoulli, J. (1713). *Ars Conjectandi*. Translated by Bing Sung. Cambridge, Mass.: Harvard University, Department of Statistics, Technical Report 2.

Boutilier, C. (1993). "Revision Sequences and Nested Conditionals." In *Proceedings of the 13th International Conference on AI (IJCAI)*, pp. 519–25. San Mateo, Calif.: Morgan Kaufman.

Boutilier, C., and Goldszmidt, M. (1993). "Revision by Conditional Beliefs." In *Proceedings of the 11th National Conference on Artificial Intelligence (AAAI)*, pp. 649–654. San Mateo, Calif.: Morgan Kaufman.

Brewka, G. (1991). "Cumulative Default Logic, in Defense of Nonmonotonic Inference Rules." *Artificial Intelligence* 50: 183–205.

Carnap, R. (1936). "Testability and Meaning." *Philosophy of Science* 3: 419–71; 4: 1–40.

Carnap, R. (1960). "The Aim of Inductive Logic." In *Logic, Methodology and Philosophy of Science*, edited. by E. Nagel, P. Suppes, and A. Tarski, Stanford, Calif.: pp. 302–318. Stanford University Press.

Chisholm, R. (1946). "The Contrary-to-Fact Conditional." *Mind* 55: 289–307.

Cohen, L. J. (1977). *The Probable and the Provable*, Oxford: Clarendon Press.

Collins, J. (1991). "Belief Revision." Ph.D. dissertation, Princeton University.

Darwiche, A., and Pearl, J. (1994). "On the Logic of Iterated Belief Revision." In *Proceedings of the Fifth Conference on Theoretical Aspects of Reasoning about Knowledge (TARK, 1994)*, pp. 5–23. San Mateo, Calif.: Morgan Kaufman.

de Finetti, B.(1964). "Foresight: Its Logical Laws, Its Subjective Sources." In *Studies in Subjective Probability*, edited by H. Kyburg and H. Smokler, pp. 93–158. New York: Wiley.

de Finetti, B. (1972). *Probability, Induction and Statistics*. New York: Wiley.

Dempster, A. P. (1967). "Upper and Lower Probabilities Induced by a Multivalued Mapping." *Annals of Mathematical Statistics* 38: 325–39.

Doyle, J. (1979). "A Truth Maintenance System." *Artificial Intelligence* 12: 231–72.

Doyle, J. (1992). "Reason Maintenance and Belief Revision, Foundations vs. Coherence Theories." In *Belief Revision*, edited by P. Gärdenfors, pp. 29–51. Cambridge: Cambridge University Press.

Dubois, D., and Prade, H. (1992). "Belief Change and Possibility Theory." In *Belief Revision*, edited by P. Gärdenfors, pp. 142–82. Cambridge: Cambridge University Press.

Dudman, V. H. (1983). "Tense and Time in English Verb Clusters of the Primary Pattern." *Australasian Journal of Linguistics* 3: 25–44.

Dudman, V. H. (1984). "Conditional Interpretations of If-sentences." *Australasian Journal of Linguistics* 4: 143–204.

Dudman, V. H. (1985). "Towards a Theory of Predication in English." *Australasian Journal of Linguistics* 5: pp.143–93.

Earman, J. (1986). *A Primer of Determinism*. Dordrecht: Reidel.

Earman, J. (1992). *Bayes or Bust*. Cambridge, Mass.: MIT Press.

Freund, M., and Lehmann, D. (1994). "Belief Revision and Rational Inference." *Leibniz Center Technical Report 94–16*. Jerusalem: Hebrew University.

Fuhrmann, A. (1993). "Observations on Validity and Conditionals in Belief Revision Systems." *Journal of Applied Non-Classical Logics* 3: 225–38.

Fuhrmann, A., and Levi, I. (1994). "Undercutting and the Ramsey Test for Conditionals." *Synthese* 101: 157–69.

Gabbay, D. (1985). "Theoretical Foundations for Nonmonotonic Reasoning in Expert Systems" In *Logic and Models of Concurrent Systems*, edited by K. R. Apt. Berlin: Springer Verlag.

Gärdenfors, P. (1978). "Conditionals and Changes of Belief." In *The Logic and Epistemology of Scientific Change*, edited by I. Niiniluoto and R. Tuomela, pp. 381–404. Acta Philosophical Fennica 30. Amsterdam: North Holland.

Gärdenfors, P. (1982). "Imaging and Conditionalization." *Journal of Philosophy* 79: 747–60.

Gärdenfors, P. (1986). "Belief Revisions and the Ramsey Test for Conditionals." *Philosophical Review* 95: 81–93.

Gärdenfors, P. (1988). *Knowledge in Flux*. Cambridge, Mass.: MIT Press.

Gärdenfors, P. (1990a). "Belief Revision and Nonmonotonic Logic: Two Sides of the Same Coin?" *Proceedings of the Ninth European Conference on AI*, edited by L. C. Aiello, pp. 768–73. London: Pitman Publishing.

Gärdenfors, P. (1990b). "The Dynamics of Belief Systems: Foundations vs. Coherence Theories." *Revue Internationale de Philosophie* 172: 24–46.

Gärdenfors, P., and Makinson, D. (1993), "Nonmonotonic Inference based on Expectations." *Artificial Intelligence* 65: 197–246.

Gibbard, A. (1981). "Two Recent Theories of Conditionals." In *Ifs: Conditionals, Belief, Decision, Chance and Time*, edited by W. L. Harper, R. Stalnaker, and G. Pearce, pp. 211–47. Dordrecht: Reidel.

Goodman, N. (1947). "The Problem of Counterfactual Conditionals." *Journal of Philosophy* 44: 113–28.

Hacking, I. (1990). *The Taming of Chance*. Cambridge: Cambridge University Press.

Hansson, S. O. (1991). "Belief Base Dynamics." Ph.D. dissertation, at Uppsala.

Hansson, S. O. (1992). "In Defence of the Ramsey Test." *Journal of Philosophy* 89: 522–40.

Hansson, S. O. (1994). "The Emperor's New Clothes." In *Conditionals and Artificial Intelligence*, edited by G. Crocco, L. Fariñas del Cerro, and H. Herzig. Oxford: Oxford University Press.

Harman, G., (1986). *Change in View: Principles of Reasoning*. Cambridge, Mass.: MIT Press.

Herzberger, H. (1973). "Ordinal Preference and Rational Choice." *Econometrica* 41: 187–237.

Hilpinen, R. (1968). *Rules of Acceptance and Inductive Logic*. Acta Philosophica Fennica 22. Amsterdam: North Holland.

Horwich, P. (1982). *Probability and Evidence*. Cambridge: Cambridge University Press.

Howison, C., and Urbach, P. (1989). *Scientific Reasoning: The Bayesian Approach*. Lasalle, Ill.: Open Court.

Jackson, F. (1990). "Classifying Conditionals." *Analysis* 50: 134–47.

Jeffrey, R. C. (1965). *The Logic of Decision*. New York: McGraw-Hill.

Jeffrey, R. C. (1984). "The Impossibility of Inductive Probability." *Nature* 310: 433.

Katsuno, H., and Mendelzon, A. (1992). "On the Difference between Updating a Knowledge Base and Revising It." In *Belief Revision*, edited by P. Gärdenfors, pp. 183–203. Cambridge: Cambridge University Press.

Keynes, J. M. (1921). *A Treatise on Probability*. London: Macmillan.

Kraus, S., Lehmann, D., and Magidor, M. (1990). "Nonmonotonic Reasoning, Preferential Models and Cumulative Logics." *Artificial Intelligence* 44: 167–201.

Kyburg, H. E. (1961). *Probability and the Logic of Rational Belief*. Middletown. Conn: Wesleyan University Press.

Kyburg, H. E. (1974). *The Logical Foundations of Statistical Inference.* Dordrecht: Reidel.

Kyburg, H. E. (1983). *Epistemology and Inference.* Minneapolis: University of Minnesota Press.

Kyburg, H. E. (1990a). *Science and Reason.* Oxford: Oxford University Press.

Kyburg, H. E. (1990b). "Probabilistic Inference and Nonmonotonic Inference." In *Uncertainty in Artificial Intelligence*, 4th ed., edited by R. D. Shachter, T. S. Levitt, L. N. Kanal, and J. F. Lemmer, pp. 319–26. Amsterdam: Elsevier Science Publishers.

Kyburg, H. E. (1992). "Normative and Descriptive Ideals." *Philosophy and AI*, edited by R. Cummins and J. Pollock, pp. 129–39. Cambridge, Mass.: MIT Press.

Lehmann, D., and Magidor, M. (1992). "What Does a Conditional Knowledge Base Entail?" *Artificial Intelligence* 55: 1–60.

Levi, I. (1965). "Deductive Cogency in Inductive Inference." *Journal of Philosophy* 63: 68–77.

Levi, I. (1966). "On Potential Surprise." *Ratio* 8: 107–29.

Levi, I. (1967). *Gambling with Truth.* New York: Knopf. Reprint, Cambridge, Mass.: MIT Press, 1973.

Levi, I. (1974). "On Indeterminate Probabilities." *Journal of Philosophy* 71: 391–418.

Levi, I. (1976). "Acceptance Revisited." In *Local Induction*, edited by R. Bogdan, pp. 1–71. Dordrecht: Reidel.

Levi, I. (1979). "Support and Surprise: L. J. Cohen's View of Inductive Probability." *British Journal for the Philosophy of Science* 30: 279–92.

Levi, I. (1980). *The Enterprise of Knowledge.* Cambridge, Mass.: MIT Press.

Levi, I. (1982). "A Note on Newcombmania." *Journal of Philosophy* 79: 337–42.

Levi. I. (1983). "The Wrong Box." *Journal of Philosophy* 80: 534–42.

Levi, I. (1984a). *Decisions and Revisions.* Cambridge: Cambridge University Press.

Levi, I. (1984b). "The Impossibility of Inductive Probability." *Nature* 310 433.

Levi, I. (1985a). "Epicycles." *Journal of Philosophy* 82: 104–6.

Levi, I. (1985b). "Common Causes, Smoking and Lung Cancer." In *Paradoxes of Rationality and Cooperation*, edited by R. Campbell and L. Sowden, pp. 234–47. Vancouver: University of British Columbia Press.

Levi, I. (1985c). "Illusions about in Uncertainty." *British Journal for the Philosophy of Science* 36: 331–40.

Levi, I. (1986a). *Hard Choices: Decision Making under Unresolved Conflict.* Cambridge: Cambridge University Press.

Levi, I. (1986b). "Probabilistic Pettifoggery." *Erkenntnis* 25: 133–40.

Levi, I. (1986c). "Estimation and Error Free Information." *Synthese* 67: 347–60.

Levi, I. (1988). "Iteration of Conditionals and the Ramsey Test." *Synthese* 76: 49–81.

Levi, I. (1989). "Possibility and Probability." *Erkenntnis* 31: 365–86.

Levi, I. (1990). "Chance." *Philosophical Topics* 18: 117–49.

Levi, I. (1991). *The Fixation of Belief and Its Undoing.* Cambridge: Cambridge University Press.

Levi, I. (1992). "Feasibility." In *Knowledge, Belief and Strategic Interaction,* edited by C. Bicchieri and M. L. Dalla Chiara, pp. 1–20. Cambridge: Cambridge University Press.

Levi, I., and Morgenbesser, S. (1964). "Belief and Disposition." *American Philosophical Quarterly* 1, 221–32.

Lewis, D. (1973). *Counterfactuals.* Cambridge, Mass.: Harvard University Press.

Lewis, D. (1976). "Probabilities of Conditionals and Conditional Probabilities." *Philosophical Review* 85: 297–315.

Lindström, S. (1990). "A Semantic Approach to Nonmonotonic Reasoning: Inference Operations and Choice." Unpublished manuscript.

Lindström, S., and Rabinowicz, W. (1994). "The Ramsey Test Revisited." In *Conditionals and Artificial Intelligence,* edited by G. Crocco, L. Fariñas del Cerro, and H. Herzig. Oxford: Oxford University Press.

Mackie, J. L. (1962). "Counterfactuals and Causal Laws." In *Analytic Philosophy,* edited by R. Butler, Oxford: Oxford University Press.

Mackie, J. L. (1973). *Truth Probability and Paradox.* Oxford: Oxford University Press.

Makinson, D. (1987). "On the Status of Recovery in the Logic of Theory Change." *Journal of Philosophical Logic* 16: 383–94.

Makinson, D. (1989), "General Theory of Cumulative Inference," In *Non-Monotonic Reasoning,* edited by M. Reinfrank, J. de Kleer, M. L. Ginsberg, and E. Sandewall, pp. 1–18. Lecture Notes on Artificial Intelligence no. 346. Berlin: Springer-Verlag.

Makinson, D. (1994). "General Patterns of Nonmonotonic Reasoning." In *Handbook of Logic in Artificial Intelligence and Logic Programming,* vol. 3: *Non-Monotonic Reasoning and Uncertain Reasoning,* edited by D. M. Gabbay, C. J. Hogger, and J. A. Robinson, pp. 35–110. Oxford: Oxford University Press.

Makinson, D., and Gärdenfors, P. (1991). "Relations between the Logic of Theory Change and Nonmonotonic Logic." In *The Logic of Theory Change,* edited by A. Fuhrmann and M. Morreau, pp. 185–205. Berlin: Springer-Verlag.

McCarthy, J. (1980). "Circumscription – A Form of Nonmonotonic Reasoning." *Artificial Intelligence* 13: 27–39.

McGee, V. (1985). "A Counterexample to Modus Ponens." *Journal of Philosophy* 82: 462–71.

McGee, V. (1989). "Conditional Probabilities and Compounds of Conditionals." *Philosophical Review* 98: 485–541.

Moore, R. C. (1983). "Semantical Considerations on Nonmonotonic Logic." *Proceedings of the 8th International Conference on AI (IJCAI),* pp. 270–76. San Mateo, Calif.: Morgan Kaufman.

Morreau, M., (1992). "Planning from First Principles." In *Belief Revision,* edited by P. Gärdenfors, pp. 204–19. Cambridge: Cambridge University Press.

Niiniluoto, I. (1986). "Truthlikeness and Bayesian Estimation." *Synthese* 67: 321–47.

Pearl, J. (1992). "Probabilistic Semantics for Nonmonotonic Reasoning." In *Philosophy and AI*, edited by R. Cummins and J. Pollock, pp. 157–87. Cambridge, Mass.: MIT Press.

Poole, D. (1988). "A Logical Framework for Default Reasoning." *Artificial Intelligence* 36: 27–47.

Popper, K. R. (1962). *Conjectures and Refutations.* New York: Basic Books.

Popper, K. R., and Miller, D. (1983). "A Proof of the Impossibility of Inductive Probability." *Nature* 302: 687–8.

Popper, K. R., and Miller, D. (1984). Replies to Levi, Jeffrey, and Good. *Nature* 310: 434.

Ramsey, F. P. (1990). *Philosophical Papers.* Edited by D. H. Mellor. Cambridge: Cambridge University Press.

Reiter, R. (1980). "A Logic for Default Reasoning." *Artificial Intelligence* 13: 81–132.

Reiter, R., and Criscuolo, G. (1981). *Proceedings of the 4th International Conference on AI (IJCAI)*, pp. 270–76. San Mateo, Calif.: Morgan Kaufman.

Rott, H. (1989). "Conditionals and Theory Change: Revisions, Expansions and Additions." *Synthese* 81: 91–113.

Rott, H. (1993). "Belief Contraction in the Context of the General Theory of Rational Choice." *Journal of Symbolic Logic* 58: 1426–50.

Savage, L. J. (1954). *The Foundations of Statistics.* New York: Wiley.

Schervish, M. J., Seidenfeld, T., and Kadane, J. B. (1984). "The Extent of Nonconglomerability of Finitely Additive Probabilities." *Zeitschrift für Wahrscheinlichkeitstheorie und verwandte Gebiete* 66: 205–26.

Sen, A. K. (1982). "Choice Functions and Revealed Preference." In *Choice, Welfare and Measurement*, pp. 41–53. Cambridge, Mass.: MIT Press. Originally published in *Review of Economic Studies* 38 (1971): 307–17.

Shackle, G. L. S. (1949). *Expectations in Economics.* Cambridge: Cambridge University Press. 2nd ed., 1952.

Shackle, G. L. S. (1961). *Decision, Order and Time in Human Affairs.* Cambridge: Cambridge University Press. 2nd ed., 1969.

Shafer, G. (1976). *A Mathematical Theory of Evidence.* Princeton: Princeton University Press.

Shimony, A. (1955). "Coherence and the Axioms of Confirmation." *Journal of Symbolic Logic.* 20: 8–20.

Shoham, Y. (1987). "Nonmonotonic Logics: Meaning and Utility." *Proceedings of the 10th International Conference on AI (IJCAI)*, pp. 388–93. San Mateo, Calif.: Morgan Kaufman.

Spohn, W. (1988). "A General Non-probabilistic Theory of Inductive Reasoning." In *Causation in Decision, Belief Change and Statistics*, edited by W. Harper and B. Skyrms, pp. 105–134. Dordrecht: Reidel.

Stalnaker, R. C. (1968). "A Theory of Conditionals." In *Studies in Logical Theory*, pp. 98–112. Oxford: Blackwell.

Stalnaker, R. C. (1984). *Inquiry*. Cambridge, Mass.: MIT Press.

Tversky, A., and Kahneman, D. (1983). "Extensional versus Intuitive Reasoning: The Conjunction Fallacy in Probability Judgment." *Psychological Review* 90: 293–315.

van Fraassen, B. (1980). *The Scientific Image*. Oxford: Oxford University Press.

Williams, M. (1994). "Transmutations of Knowledge Systems." In *Proceedings of the Fourth International Conference on Principles of Knowledge Representation and Reasoning*, pp. 619–629. San Mateo, Calif.: Morgan Kaufman.

Yosida, K., and Hewitt, E. (1952). "Finitely Additive Measures." *Transactions of the American Mathematical Society* 72: 46–66.

Zadeh, L. (1978). "Fuzzy Sets as a Basis for a Theory of Possibility." In *Fuzzy Sets and Systems* 1: 3–28.

Name Index

Adams, E., 107, 253–6
Ahtisaari, M., xi
Alchourrón, C., xi–xiii, 5–8, 14, 22, 30,
   37–8, 116–19, 205, 263, 297 n.6,
   n.8, 309 n.1, 316 n.6. 322 n.19
Arló Costa, H., xi, xii, 97, 100, 305 n.3,
   306 n.7, 323 n.2
Aviely, D., xiv

Borel, E., 244
Boutilier, C., 302 n.7

Carnap, R., 160, 235, 248, 249, 310 n.7,
   318 n.2, n.4. 320 n.10
Chisholm, R., 292 n.5
Cohen, L. J., xii, 180, 313 n.6, 314 n.11,
   318 n.5
Collins, J., xi, 50, 66, 76, 78, 79, 80, 100,
   293 n.5, 294 n.6, 295 n.8, 303 n.9,
   n.10, 305 n.13
Cramer, H., 244
Criscuolo, G., 207, 229–30

Danaher, J., xi
Danziger, S., xiv
Darwiche, A., 302 n.2
de Finetti, B., 32 n.11
Dempster, A. P., 187
Dewey, J., 285
Doyle, J., 224, 226
Dubois, D., xii, 259–61, 267–8, 313 n.6
Dudman, V. H., 13, 46–50, 69, 71, 74,
   294 n.7, 298 n.9

Earman, J., 320 n.10

Fisher, R. A., 304 n.13
Freund, M., 302 n.7, 323 n.2
Fuhrmann, A., xi, 14, 309 n.5

Gabbay, D., xii, 125, 204, 256
Gärdenfors, P., xi–xiii, 5–8, 14, 15, 17,
   28, 30, 31, 35, 37–41, 58–61,
   63–4, 66, 68–9, 88, 93–105,
   116–19, 122–3, 125–6, 148–50,
   173, 179, 188, 205, 210, 211, 222,
   256, 260–1, 263–4, 266, 271–3,
   277, 278–85, 291, 294 n.5, n.6,
   295 n.3, 297 n.6, n.8, 299 n.3, 303
   n.8, n.10, 304, n. 13, 305 n.14,
   306 n.7, 307 n.12, 308 n.14, n.15,
   313 n.6, 316 n.2, 317 n.6, 322
   n.19, 323 n.1, 325 n.6, n.7
Gentzen, 204
Gibbard, A., 49, 50
Goldszmidt, M., 302 n.7
Goodman, N., 292 n.5

Hacking, I., 229
Hansson, S. O., xii, 6, 39, 68, 69, 73,
   105, 117, 292–3 n.5, 299 n.3
Harman, G., 224–6
Heidegger, M., 254
Hilpinen, R., 315 n.14
Horwich, P., 320 n.10
Howison, C., 320 n.10

Jackson, F., 298 n.9
Jeffrey, R. C., 235, 318 n.2
Jeffreys, H., 320 n.10

Kahneman, D., 239–43, 319 n.8,
   318 n.9

333

Subject Index

abduction, 4, 161–2, 309 n.4
  logic of, 161
AGM as an account of belief change
  and suppositional reasoning,
  3–8, 114–19
ampliative inference, 121–4, 142–6,
  310 n.7, 318 n.2
  is nonmonotonic, 144–6
  only kind of nonmonotonic inference,
  125, 158–9
ampliativity and inductively extended
  inclusion (K*i7), (K*i3) and
  (K*i3m), 146–51
ampliativity and inductively extended
  restricted weakening (K*i8),
  (K8i4) and (K*i4m), 151–4
ampliativity condition, 144
ampliativity thesis, 144
attitudinal states and language, 43–6
avoidance of error, 51–3
  and truth value, 53

basic partition, 26, 305 n.4
belief
  full, 1
  Kyburgian degree of, 245, 256–7
  probabilitistic degree of, 234–7
  Shackle degree of, 234–7, 256–61
  and supposition, 1–8
belief revision system (brs), 61, 88, 90
  according to Gärdenfors (brsg and
    brsgw), 90
  according to Ramsey (brsr), 90
  inductively extended, 173–6
  weakly trivial, 61

boldness and caution, as a trade-off
  between risk of error and
  informational value, 171–2
bookkeeping, 151, 188–91, 314 n.12
  with maximum boldness, 190, 215,
  272–4, 315 n.13
  without maximum boldness, 190–1,
  215–16, 229–30, 274, 315 n.14
cautious monotony, 153, 209, 216, 272
  inductively extended, 154
certain and almost certain, 311 n.4
choice consistency and ordering of
  maxichoice contractions, 297 n.7
closure under expansion, 32, 95, 306 n.6
closure under strict expansion, 306 n.7
cognitive values, as aims of inquiry, 51
commitment and demands of
  rationality, 85
conditional judgment of credal probability, 35, 296 n.5
conditional judgment of serious possibility, 12–13, 34, 45–6
  not truth value bearing, 57
conditional logic, 15
  for iterated Ramsey test conditionals
    and modus ponens in particular,
    101–12
  and keeping the M-function and
    basic partition fixed in iteration,
    302 n.7
  for noniterated inductively extended
    Ramsey test conditionals,
    146–59, 191–3
  for noniterated Ramsey test condi
    tionals, 101–5

335